JOURNEYS

INTO DEEP SPACE

THE INNER SOLAR SYSTEM

GEOFFREY BRIGGS

ABOUT THE AUTHOR

The author, who has degrees in physics from Durham University in the UK, has, until his retirement from NASA in 2011, participated in the exploration of the planets since the 1960s (part of a 'brain drain' that led a number of Brits to join the US space program) – first as a member of the technical staff of a Bell Laboratories company (Bellcomm) tasked to support NASA and, later, as a member of Caltech's Jet Propulsion Laboratory. There he was a science team member of the first Mars orbiter mission, Mariner 9, of the Viking Mars orbiter & lander missions and, briefly, of the Voyager mission to the outermost planets. At NASA HQ he served as Deputy Director and then Director of NASA's Solar System Exploration Program in the late 1970s & 1980s and from the 1990s was Director of the Center for Mars Exploration that he established at NASA Ames Research Center. There, in the early 1990s he was co-lead of the NASA-wide study team tasked to create the first Design Reference Mission for the exploration of Mars by astronauts (still kept up to date by NASA). He is the recipient of NASA's medals for Exceptional Scientific Achievement, Exceptional Service and Outstanding Leadership. Asteroid 4209 Briggs is a carbonaceous asteroid discovered by Eleanor Helin. He lives in Bath back in the UK where he is a member of the William Herschel Society.

CONTENTS

ACKNOWLEDGEMENTS

This history is intended as acknowledgement of the huge number of participants from many nations who made possible this great adventure in exploration and science. For each project the managers and science principal investigators are recalled – each supported by their own teams of colleagues and, in the case of academics, by graduate students. At the program level at NASA HQ the author was ably supported by deputies Al Diaz and Frank Carr, and by the managers of each flight program, of advanced studies and of each planetary science discipline: Bill Brunk, Bill Quaide, Rod Mills, Dudley McConnell, Steve Dwornik, Joe Boyce, Bevan French, Bill Piotrowski, Bob Murray, Henry Brinton, Jurgen Rahe, Tom Morgan, Ann Merwarth, and Don Ketterer. Don Pinkler from the NASA Controller's Office provided essential help in managing budgets and communicating them within the Agency. The secretarial and support staff, who generally receive little recognition but make an essential contribution included Audrey Murray, Geraldine Paige and Monica Mayden (who, together, daily kept me out of trouble), Sandra Bembry, Bernadette Bannister, Linda Shields, and at SAIC: Chris Butera and Bill Harmon. Also at SAIC Dan Spadoni, and John Niehoff provided critical cost analyses of potential program options. Julius and Pat Dasch provided the Program with valuable educational outreach support.

The author is much indebted to colleagues who read this history at various stages noting omissions and correcting mistakes: David Morrison, Steve Clifford, Fred Taylor and Michael Carr. Any remaining errors of fact or judgement are strictly those of the author.

As recounted at some length in one chapter, recovery of the Program from the Reagan plan to terminate it, led to the formation of a committee, the Solar System Exploration Committee, whose members, John Naugle, Noel Hinners, David Morrison, Arden Albee, Kinsey Anderson, James Arnold, Charles Barth, Thomas Donahue, Michael Duke, Lennard Fisk, Lawrence Haskin, Donald Hunten, Harold Klein, Eugene Levy, James Martin, Harold Masursky, John Niehoff, Toby Owen, Donald Rea, Larry Soderblom, Edward Stone, Joseph Veverka, and Laurel Wilkening deserve particular acknowledgement and thanks. In addition, JPL's Gene Giberson and consultant Jim Martin (Viking Project Manager) were key advisors throughout.

Diane Rausch of NASA's International Affairs Office was a much-appreciated central player in developing and assuring continuing excellent relations with our Soviet Program equivalents.

Tony Symes of the William Herschel Society and David Baker of the British Interplanetary Society each provided helpful comments.

PREFACE

Since the Apollo landings half a century ago, humans have not ventured again into deep space – beyond low Earth orbit. Meanwhile, robotic spacecraft have completed the reconnaissance of our Solar System from Mercury to Pluto and beyond. Two spacecraft, the Voyagers, have departed the Solar System and are now retreating from us in Interstellar Space. There have been orbital encounters with the minor planets Ceres and Vesta as well as visits to numerous asteroids and comets (on one of which a European Space Agency spacecraft has landed). Most recently, reported in 2020, observations of microwave emissions from Venus were interpreted as evidence of phosphine ($PH3$) and of microbial life forms in the clouds. Alas, subsequent analysis of the data has cast doubt on this discovery, one that would have been history-making.

During these last decades Planet Earth has been studied intensively by scientists, including from orbiting monitors, so that its extreme environmental danger, exemplified by melting ice caps and glaciers, ever more powerful hurricanes, heat waves and ever more frequent forest fires, are now better understood and accepted – but, alas, only by those with a science background or an open mind and by the young whose future is at stake. Consequently, our planet has started on a path that could turn it into a second Venus.

The reconnaissance of all the planets, of all of their moons, of representative asteroids and comets is now complete. This effort has been motivated in large part by an effort to discover if life may have evolved elsewhere in our Solar System and, by inference, may be common throughout our galaxy. But, after half a century, the search for evidence of life elsewhere has still just begun. Future and ongoing scientific exploration will be driven by astrobiology priorities. Mars with much evidence of surface water in the past (and subsurface water today) has been the focus of most astrobiology research. Following the observations of the Voyager, Galileo and Cassin-Huygens spacecraft, the moons Europa and Enceladus have been found to have oceans deep beneath their frozen surfaces; they will be the principal destinations for future missions – discussed in a second volume. Saturn's moon Titan also stands out as a priority target for astrobiology exploration.

There is in the continuing exploration at most a limited role for astronauts. The US National Academy of Sciences has published a recent report on this complex subject: *An Astrobiology Strategy for the Search for Life in the Universe (2019)*. The new Venus observations were made too recently to be included.

For two decades astronauts on space stations in low Earth orbit have mainly been accumulating an improved understanding of the effects of long-duration spaceflight on crew health while conducting micro-gravity experiments that include life-support systems. It has long-since been taken as a given that the next big step in the destiny of mankind will be a crewed journey to Mars. Now, a half-century since Apollo, this continues to receive re-evaluation but, for a number of fundamental reasons – purpose/justification in a new era of artificial intelligence, risk to the lives and health of crew, forward and backward bio-contamination issues and, of course, great cost – may never happen. By contrast, the future direction of the robotic exploration of the Solar System will be science-driven and increasingly efficient because of artificial intelligence (AI) advances. The nature and direction of future crewed missions are, however, slowly emerging and, most likely, will be limited to further exploring and exploiting the resources of our Moon. If a crewed mission to Mars is undertaken it will likely be limited to controlling landed robotic assets from a base on one of the Martian moons. Meanwhile, we can all experience the immersive virtual reality experience of slowly travelling on the surface of Mars courtesy of NASA's Curiosity rover.

This history describes the evolution of deep space exploration from its beginnings in the 1960s Cold War stand-off between the US and the USSR to its present state where planning for a return by astronauts to the Moon is well underway and where robotic missions to both the inner planets and to the outer Solar System have become largely international in their implementation. Moreover, we now have an increasingly large catalogue of planets – *exoplanets* – in orbit about other stars in our Milky Way galaxy whose study will also have an astrobiology focus.

All of this has come about in the span of half a century. Prior to the 1960s, deep space exploration was the domain of science fiction writers – notably Jules Verne, HG Wells, Arthur C Clarke and Ray Bradbury – and of just a few visionary scientists and engineers. Since then, planetary science has enjoyed a Golden Age. Geologists, geoscientists, atmospheric scientists, and astrobiologists have enthusiastically participated – with a huge increase in our knowledge of the present state and evolution of our Solar System. The US taxpayers who have funded NASA's space exploration enterprise continue to be generally satisfied and supportive.

Throughout human history, exploration has been an imperative that has seen – for better or worse – our species expand into every corner of the globe – driven by motivations other than curiosity – mainly territorial conquest, exploitation of resources including land, minerals and slave labor, and intense competition with rivals. The exploration of our Solar System began in the 1960s for the last-mentioned reason and, thankfully, has evolved to include a significant degree of international cooperation and, mostly, fundamental scientific curiosity.

This history is divided into two parts beginning, in this volume, with the Inner Solar System of terrestrial planets and moons (only 3 in number including our own) whose exploration has been undertaken by robotic spacecraft of ever-increasing capability and, in the case of the Moon, by human explorers. A second volume will record the exploration by robotic spacecraft of the Outer Solar System of Giant Planets & Pluto, of Minor Planets, Asteroids and Comets and lastly, with the discovery and characterization of planets orbiting other stars in our Milky Way Galaxy. The intention is to provide the reader with an account of how, why and by whom the exploration has been carried out thus far – and, of course, what we have learned.

There are a large number of national space agencies – in the US, Russia, Japan, China, India, Israel, & the United Arab Republics – together with the multi-nation European Space Agency that are all part of the ongoing exploration of deep space. Of these, the US National Aeronautics and Space Administration, NASA, has made by far the largest investment and effort both in terms of crewed and robotics missions. It might be expected that the US would have a well-defined strategy for carrying out robotic deep space missions in a logical and orderly manner. In fact, as this history records, there have been a number of disorderly bumps in the road as a consequence of the fact that, in the US, Administrations change every 4 or 8 years and each new president arrives in the White House with an agenda in which space exploration is generally little more than an afterthought – though one that could, as in the case of John Kennedy, bring lasting fame. Typically therefore, at intervals of several years a blue ribbon panel of space experts is brought together to advise the new president on how best to proceed. Then plans are made with cost estimates and proposed missions are brought to the US Congress for the essential next step – approval (authorization and appropriation of funds). Since space exploration, whether by astronauts or by robotic spacecraft, is not a high priority for most in Congress, NASA's appropriated budget – greatly reduced since Apollo – usually shows little change (other than inflation) from year to year and, now as in the past, is insufficient to carry out any bold presidential initiatives like a crewed mission to Mars.

The latter vision has its roots in both science fiction and in the expectation that astronaut exploration of Mars would naturally follow on from the Apollo success. Support for, and implementation of, deep space exploration by *robotic* spacecraft has fared better because the US National Academy of Sciences provides NASA with balanced, expert advice about science priorities that NASA's Solar System Exploration Office and its other science programs rely upon. Of course, the science programs are much less expensive and risky than the crewed spaceflight program. Nevertheless, the exploration *science* program has been subject to some crises including even a proposal by President Ronald Reagan in 1980 to terminate the exploration program and apply the skills of NASA's Jet Propulsion Laboratory to missile defense ("Star Wars"). Chapter 8 describes this fraught episode and the recovery of a revitalized program. Lesser bumps have been mainly driven by the 1986 *Challenger* disaster (which impacted all of NASA) and, later, by the over-enthusiasm of a NASA Administrator.

Regarding manned exploration missions of Mars, perhaps the most likely time for its implementation would have been in a decade or so after the Apollo landings when NASA still had the Saturn V, plus a nuclear propulsion system under development and importantly, while the risk imposed by cosmic radiation in space was not a big issue. The decision to proceed in this direction would have been that of President Richard Nixon who chose otherwise. With present (2020) propulsion technology the crew of a mission to land on Mars would embark on a return trip of almost a thousand days' duration. We still do not fully understand the health risk involved – though the increasing time accumulated in low Earth orbit (many hundreds of days by Russian and US astronauts) is improving that understanding. Earth is the only one of the terrestrial planets to possess a substantial magnetosphere – one that, combined with our deep atmosphere, is powerful enough to provide a shield for us Earthlings against the solar and cosmic radiation that fills deep space. As a result, apart from all the other unavoidable risks to health and to life itself (including, in the case of Mars, global dust storms that last many months) that are intrinsic to human space exploration, an accumulating radiation dose over many hundred days may well prove to be a Mars mission show-stopper. We may soon have to accept that our Moon and Mars orbit are as far as we can responsibly go. This limitation may be particularly hard to accept by those who believe that mankind must become a multi-planet species quite soon in order to avoid extinction by (choose): nuclear Armageddon, malign artificial intelligence, asteroid impact, population bomb, pandemic or, most urgently, global environmental devastation. In this thinking we need to find a Planet B to save us from ourselves. Our exploration of the Solar System, mainly by robotic spacecraft, has clearly demonstrated that there is no Planet B. There is international scientific consensus that there may be just a few decades before Earth's climate passes the point of no return.

Remarkably, two US billionaire space enthusiasts, Jeff Bezos and Elon Musk, have developed the technology of landing and re-using otherwise conventional rockets (as opposed to space plane launchers like the Shuttle) and are now pursuing their own independent plans for colonizing space; these dreams are briefly described in Chapter 7 on *Astronaut Spaceflight After Apollo*. Private funding by Mr. Musk may well be sufficient to enable the journey from Earth to Mars for a manned *flyby* or even to enter *orbit* about Mars. A manned Mars lander mission would not only be a much more expensive/risky proposition but would also require development and landing one or more nuclear reactors to supply power; this would require White House approval (just like the launch of small RTGs (radioactive thermal generators) for robotic spacecraft such as Voyager because a launch pad accident could irradiate part of Florida. Indeed, even a NASA mission would encounter the same approval challenge.

Other privately funded space ventures are aiming at mining minerals on certain asteroids – ones that have been captured into Earth-approaching orbits. The balance between state and commercial activities in deep space may soon change.

In the first two decades of competitive space exploration that culminated with the Apollo landings on the Moon and the Viking landings on Mars, the astronaut missions and the robotic missions were planned and executed as part of a single strategy by both the US and Soviets. Afterwards, the astronaut and robotic missions have followed largely separate paths with new science knowledge as the primary motivation of the robotic missions – particularly relating to the question of whether life originated elsewhere in our Solar System. The transition occurred in the late 1970s when planetary science took a major step forward with NASA's Pioneer and Voyager launches to the outer planets. The giant Outer Solar System planets, the asteroids and comets have now all been explored in increasing detail and this makes up the second volume of this history. Spectacular progress has also been made by astronomers in discovering thousands of planets around other stars in our galaxy – also in that second volume. The search for evidence of life elsewhere in the universe has, thereby, moved far out from just within our Solar System to the rest of the Milky Way Galaxy. Since this history (and a projection of the likely future) deals with both manned and robotic deep space exploration the present volume focuses on the Inner Solar System of our Moon, Mercury, Venus and Mars.

After starting out as a US-USSR Cold War competition, both astronaut Earth orbital activities and robotic deep space missions are now largely carried out through international cooperation. After half a century since Apollo, international crewed missions are advanced in their plans for a return to the Moon in the near future with a mixture of practical applications (e.g. exploitation of near surface water ice in deep, shadowed polar craters) plus both lunar science and radio astronomy in mind. A LOX-Hydrogen 'filling station' on the crater rim at one of the lunar poles is by no means a wild idea. This in turn may lead to commercial applications on both the Moon and in exploiting the near-Earth asteroids.

The possible return of international competition to space exploration in the near future – this time between the US and China – is a regrettable political failure that is presently hard to evaluate. The rapid progress of China in the exploration of the Moon by robotic spacecraft, including a mobile lander is described in Chapter 11.

In seeking to record the overlapping histories of both astronaut and robotic Solar System exploration with some coherence, the first chapters of this history describe the early exploration phase (Apollo and, mostly, the first Mars missions) in which the astronaut and robotic missions were linked. That ended with the Pioneer and Voyager launches to the Outer Solar System. Subsequent chapters in this volume describe both ongoing science-driven robotic missions and NASA's decades of planning to initiate the astronaut exploration of Mars. While further detailed understanding of the evolution of all the planets and their moons (through geology, geophysics and atmospheric research) remains central to planetary science, it will be the astrobiology implications that will set the priorities in deciding which new robotic missions to undertake both in the Inner and the Outer Solar System.

The ongoing rivalry between the US and China has been taking form, mainly, as a trade war rather than in military confrontation – although the potential for an accidental hot confrontation somewhere in the South China Sea cannot be ignored. The Mutually Assured Destruction that maintained the peace during the US-USSR Cold War has been largely replaced in the US-China confrontation by a Mutually Assured Trade Wreck. Might both sides at some time in the future see the benefit of collaborating in space just as the US and Russia did so? It seems more likely that an effective Moon and Mars space competition between the US and China will emerge in the coming decade.

Chapter 7, Astronaut Spaceflight After Apollo, summarizes NASA's planning over many decades for crewed spaceflight beyond low Earth orbit. Blue ribbon committees have made regular recommendations to NASA regarding astronaut exploration – always with landings on Mars as the long-range goal. Meanwhile, as many decades have passed since Apollo, the *robotic* exploration of the surface of Mars has made enormous progress so that it is no longer clear what the contribution of astronauts (or Chinese taikonauts) to that exploration might be. As the lifetime of robotic Mars missions is now measured in decades, as long-range surface mobility has been amply demonstrated and as artificial intelligence capabilities grow in leaps and bounds, the contribution that can be made by astronauts is problematic. In fact, rather to the contrary, human explorers on Mars would inevitably contaminate Mars after decades of effort (sterilizing landers and parking dying orbiters in long-lived orbits) to avoid that outcome – thereby potentially confusing the ongoing search for evidence of past or present life on Mars.

A decade ago the British Astronomer Royal Sir Martin Rees summarized his view of human space exploration in the following way: "The moon landings were an important impetus to technology but you have to ask the question, what is the case for sending people back into space? I think that the practical case gets weaker and weaker with every advance in robotics and miniaturization. It's hard to see any particular reason or purpose in going back to the moon or indeed sending people into space at all".[1] His point grows in relevance as time passes.

The author's expectation that astronaut landings on Mars may, in fact, never happen will likely *not* be reflected in NASA's declared plans. However, an astronaut mission to Phobos, the inner moon of Mars, to provide real-time control of landed robots is more plausible and could take place in the 2030s. Only an improbable worldwide international collaboration *including China* would provide geopolitical justification and shared risk for launching a crew to explore the surface of Mars.

One day there will be commercial exploitation of resources found in space – processing lunar water ice and mining metal-rich asteroids – but the role of astronauts will surely be minimal given that, for commercial applications, minimization of costs will be fundamental to maximizing profit. Should we therefore regard astronauts as belonging to one more profession threatened by advances in artificial intelligence and robotics? While in the author's opinion the case for sending astronauts to explore the surface Mars has now passed its sell date, a much better case can be made for the Moon. As already noted, space exploration has proven to be one area where badly needed international cooperation takes place affordably toward shared knowledge, capability and prestige. This cooperation has flourished at the same time that

---

[1] https://www.theguardian.com/science/2010/jul/26/martin-rees-space

traditional national rivalries have otherwise grown apace. Our ability to send astronauts to the Moon and return them safely is well established and, importantly, is affordable within the budgets that can reasonably be expected to be appropriated by the US Congress for NASA after support for the International Space Station ends (perhaps sold to a private entity to turn into the first orbital hotel).

The International Space Station and most robotic science missions including, notably, the 1986 Comet Halley Armada (an ESA, Russia, Japan and US collaboration) have all benefited greatly from such international collaboration. In collaborations the cost, risk, prestige and scientific progress have all been shared and this is true for astronaut space activities too as evidenced by decades of support for the International Space Station. The program managers, astronauts, and science & engineering communities involved have all learned to respect and, most importantly, *trust* one another. Admittedly, this positive progress in international relations may seem minor compared to the prevailing antagonisms – but it has the potential to grow. There is one particular obstacle, however, and that is the ban imposed by the US Congress on any collaboration with China, the world's new major space power. It is to be hoped that this position will change. In the 21st Century such collaboration is, in the author's view, the principal justification for continued spaceflight by humans.

As noted, the account here includes the history of lunar exploration by both astronauts and robots, the history of Mars exploration by scientists controlling robots of ever-increasing capability, together with all the other robotic missions launched into deep space. It is an account for the general reader who will certainly have access to the vast amount of online information of which this history provides a digest. The reader should note that, whereas the images and measurements returned by spacecraft will remain reliably factual, the interpretation of those observations by the science community is subject to change as time goes by. As one example, the interpretation of results of the 1976 Viking Mars lander biology experiments is controversial.

The world's several space agencies have made sure that, in addition to their own comprehensive sources of information about missions, Wikipedia has reliable summaries of their missions. The US National Space Science Data Center, NSSDC, provides online information bases of NASA and non-NASA data as well as of the spacecraft and experiments that generate NASA space science data. NSSDC also provides information and support relative to data management standards and technologies. The NASA History Office is another source. To keep up to date with much that is going on in this field, the JPL News jplnewsroom@jpl.nasa.gov is the place to go. NASAWatch.com is another source.

Online references are therefore provided only for direct quotes like that of Professor Rees above and for information in print and not online. This review of the evolution of crewed space flight includes orbiting space stations as well as planning for a lunar base and, planning over half a century, for a first crew mission to Mars. Half a century of Mars science progress is reviewed to provide the context for future Mars science goals.

The ilustrations in this history consist mainly of photographs acquired by the many national space agencies. JPL and the Lunar and Planetary Institute in Houston are sources. Maps are the product of the US Geolgical Survey in Flagstaff. Wikipedia has been the most convenient source of many of the illustrations. One or two figures are from academic publications.

INTRODUCTION:

BEFORE THE SPACE AGE

The Space Age began on 4 October 1957, when the Soviet Union successfully launched Sputnik 1 into low Earth orbit. The exploration of our Solar System by spacecraft began just two years later when the Soviet Union launched its Luna 1 mission achieving escape velocity and becoming the first spacecraft to enter deep space. The capability for mankind to leave Earth behind in this way was a direct consequence of technologies that had been developed during WW2 and the subsequent US-USSR Cold War.

By the end of 2015, a little more than half a century later, the several national space agencies, sometimes in collaboration, had completed the first phase of Solar System exploration: we have visited and mapped all the planets from Mercury to Pluto as well as all of the outer planets' larger moons and their ring systems. Part 1 focuses on the exploration of the Moon and inner planets. In the case of our own Moon, six manned surface explorations have been carried out along with three Russian robot sample return missions that together have returned hundreds of pounds of rocks. Multiple robotic orbital missions and an impacting probe have discovered that there is water ice trapped in craters near the lunar poles. Meteorites that originated from the asteroid belt, from the Moon, and even from Mars have been analyzed in exquisite detail. Robotic landings, some mobile, have taken place on our Moon, Venus and Mars. Balloons have been tracked as they circumnavigate the upper atmosphere of sister planet Venus whose diabolical evolution holds special lessons for us!

All in all, this is an extraordinary list of achievements – international in origin – brought about by many thousands of imaginative engineers, navigators, information technology wizards, highly motivated scientists and project managers. Many of their names are recorded here – with great admiration!

As noted, besides the satisfaction of knowing much more about all of the other planets in our Solar System, we also now have a new perspective on Planet Earth and a better appreciation of what L. Frank Baum's Dorothy meant when she said 'There's No Place Like Home'. Much more consequential: there is no Planet B to retreat to if we fail to avert the worst consequences of Earth's ongoing climate catastrophe.

The First Solar System Astronomers

Mankind has been making progress in understanding the nature of our own planet, Earth, and of the other planets since the time of the Greek philosophers some hundreds of years BC. It began with the realization that the Moon is a spherical body and that Earth is too. In the pre-science world the planets (attached to crystalline spheres) and the Sun moved in circular orbits about the Earth just as the Moon does. And it remained so until Nicolaus Copernicus (1473-1543), influenced by the great Arab astronomer and mathematician Al-Battani, deduced the true nature of the Solar System with the Sun at its center; this was a heretical position in the eyes of the Catholic Church, dominant in Europe at that time. Galileo Galilei (1564-1642) at great personal cost endorsed this heresy. Following publication of his *Dialogue Concerning the Two Chief World Systems* in 1632 (a plague year) he was summoned to Rome to testify before the Roman Inquisition where he admitted his errors. Returning home outside Florence he spent the rest of his life under house arrest. Even greater had been the cost of such thinking by Giordano Bruno (1548-1600), a Dominican friar celebrated for his cosmological theories. He proposed that the stars were just distant suns with their own planets; he even raised the possibility that these planets could support life and, as such, he may be regarded as the father of Astrobiology. Bruno argued that the universe is in fact infinite. Tried for heresy by the Inquisition on charges that included denial of several core Catholic doctrines, he was found guilty, and was burned at the stake in Rome. No doubt, Galileo took notice.

Over the centuries that followed the Copernican revolution, other great scientists provided essential insights into the nature of our system of planets. The big break-through in understanding was achieved by Johannes Kepler (1571-1630) who, while he studied mathematics in Tubingen, learned of the Copernicus model of the universe. Later, as a Protestant exiled in Prague, he worked with Tycho Brahe and, after the astronomer's death, gained access to the records of Brahe who had spent his life making detailed observations of the motions of the planets. Mars was a particular challenge because it appears to move backwards during part of its orbit – something known since the time of Ptolemy but now needing to be better understood with the Sun as the central body. Kepler was unable to reconcile Brahe's measurements with his assumption that the Martian orbit is a perfect circle – a philosophical requirement at that time. Eventually Kepler determined and was forced to accept that the orbit is in fact an ellipse with the Sun at one focus and from this he formed his three laws of planetary motion applicable to all the planets, not just Mars:

1. Planets move in an ellipse with the Sun as one of the foci.
2. A planet sweeps out equal areas in equal times.
3. The squares of the periods of the planets are proportional to the cube of their average distances from the Sun.

By observing a relationship between ocean tides and the lunar cycle, natural philosophers since ancient times had developed sound intuitions about the force of gravity. It fell to the genius of Isaac Newton (1642-1727) – who had spent 1666 avoiding the plague in the English countryside – to provide a full understanding of this relationship in his theory of gravitation. Almost two hundred years later Albert Einstein published his general theory of relativity that provides a deeper, fundamental understanding of gravity as a geometric property of space and time. Among many other things the theory predicts the existence of black holes and gravitational waves – the latter confirmed as recently as 2016.

Today we can see that our Solar System, its collapse from a molecular cloud triggered by a nearby supernova, is the product of time, gravity and various sources of heat – radioactive decay, insolation, tides and impacts – acting on gas and dust within our region of the Milky Way Galaxy to form the Sun and retinue of planets. The march of time continues: tidal forces are still reshaping many of the moons in the outer solar system and, locally, tidal forces cause our own Moon to gradually move further away from us. And we have learned that greenhouse gases – carbon dioxide, methane and water vapor – in the atmospheres of Venus, Earth and ancient Mars contribute greatly to the evolution, past and present, of these inner planets. Venus serves as a red light warning.

The telescope, marvelous invention of Dutch spectacle makers and Galileo Galilei in the early seventeenth century, transformed planetary astronomy and increased enormously in power over time. In 1610 Galileo became the first to observe that the Moon was not a perfect sphere but had mountains, valleys, plains and large bowl-shaped depressions. Large dark terrain he identified, in analogy to terrestrial seas, as *maria. mons, lacus, oceanus, sinus, vallis* etc. found similar application.

The first maps of the Moon began to be made – with those of Polish astronomer Johannes Helvelius (1611-1687) and Giovanni Cassini (1625-1712) the most accurate. The lunar craters – which at the time had few terrestrial analogues – were generally presumed to be volcanic in origin. This supposition was first challenged by Grove Karl Gilbert (1843-1918) of the US Geological Survey. He concluded they were caused by impact events rather than volcanoes. He did, however, wonder why, given that most would be the result of oblique strikes, the craters were round and not oval. As is sometimes the way, this problem had in fact been dealt with by a brilliant young Estonian astronomer, Ernst Julius Öpik (who would go on to postulate the origin of comets in the Öpik-Oort cloud), in an 8-page note published in Russian and, probably because of the language barrier, barely known in the West. Öpik made it clear that his physical considerations were based on a model of explosion lunar craters proposed earlier by the Russian political revolutionary, physicist, and astronomer, Nikolai Alexandrovich Morozov (1854-1946), who appears also to have been unknown in the English-language literature. Öpik's main aim was to capture the cratering process postulated by Morozov numerically. Ever since having been cited by Baldwin, Öpik's 1916 paper has been referred to and fairly widely recognized as the first to document that a hypervelocity impact was similar in fact to an explosion, and, therefore, the resultant structures would be circular irrespective of the angle of incidence. In 1949 planetary scientist Ralph Belknap Baldwin (1912-2010) published his book *The Face of the Moon* in which he discussed the evidence for an impact origin of the large craters – and the probable basaltic nature of the maria surfaces.

With the advent of the telescope the ability to study not just the Moon but all of the planets became a possibility. Christiaan Huygens (1629-95) was able to sketch the first maps of Mars – including prominent Syrtis Major – and was led to speculate about the possibility of life on the other planets. By the late 17th Century all of the planets visible to the unaided eye had been observed and were entering the realm of early scientific understanding. It took a significant improvement in telescope performance, however, to extend the boundaries of the Solar System beyond Saturn. Science was making great strides now during an era that author Richard Holmes has described in his fascinating history as *The Age of Wonder*. A young German musician and amateur astronomer, William Herschel (1738-1822), had migrated to Britain in the wake of military upheaval in his native Hanover. His sister Carolyn (1750-1848) – who developed into a notable comet astronomer – joined him later. William was inspired to build his own telescope – a reflector with a larger metal mirror than those available at the time, one whose grinding and polishing he undertook himself. With this telescope, sited in his back garden in the spa town of Bath, Herschel carried out a systematic survey of the night sky, focused mainly on cataloguing double stars and nebulae. Uranus, first observed and noted by Herschel in March 1781, was thought perhaps to be a comet and, only later with more observations to establish its orbit, did its true planetary nature come to be agreed. Herschel's other great contribution to astronomy was the discovery of the infrared wavelengths that extend beyond the visible. No planetary spacecraft would be launched without an IR radiometer and spectrometer in its payload. The European Space Agency's Herschel Space Telescope with its 3.5meter mirror, operational between 2009 and 2013, is the largest infrared telescope launched to date.

The Lagrange Points

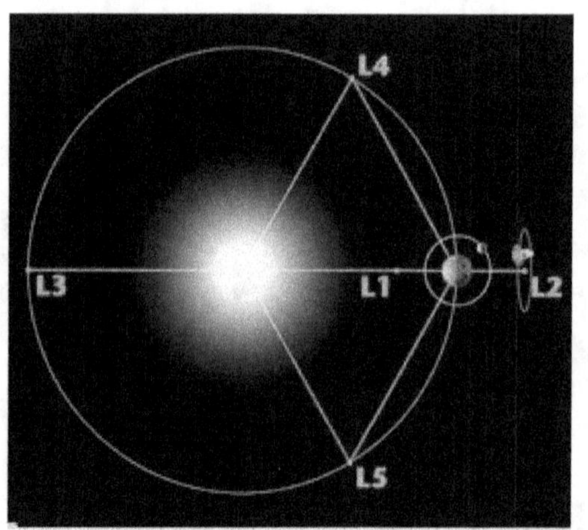

In 1772 the mathematician/astronomer Jean-Louis Lagrange (who succeeded Leonhard Euler as the director of the Berlin Academy) solved the general 3-body problem locating the positions where the gravitational pull of two large masses equals the centripetal force required for a small object to move with them. There are 5 of these libration points, two of which are stable (L4 and L5). The significance of Earth's L1 and L2 'parking places' would only become important much later. Solar space observatories are conveniently located at L1 and many space telescopes orbit L2 where they have a clear view of deep space along with consistent solar power and easy communication. Because Earth's L1 and L2 points are unstable on a time scale of approximately 23 days, spacecraft orbiting L1 and L2 must undergo regular course and attitude corrections.

Jupiter's L4 and L5 Lagrange points are among the most important elsewhere in the Solar System because some of the Main Belt asteroids that orbit the Sun between Mars and Jupiter have found a home around them. These are the Jupiter *Trojans*.

Carl Linnaeus, Charles Darwin, Alfred Russell Wallace and Gregor Mendel

In the middle of the 19th Century Charles Darwin (1809-1882) – who had been encouraged to pursue his scientific voyage on the Beagle by reading the *Personal Narrative* of Alexander von Humboldt's travels in Central and South America – and Alfred Russell Wallace (1823-1913) conceived the theory of natural selection as the basis for the evolution of life on Earth. Earlier the Swedish scientist Carl Linnaeus had developed the binomial nomenclature of taxonomy – genus and species. A man with a wry sense of humor, he proposed *Homo sapiens* for us and this has stuck even though *Homo ferox* would be much more accurate. Darwin's *On the Origin of the Species* was published in 1859. He evidently gave a good deal of thought to the question of the origin of life but was reluctant to publish his views. He did, however, write to his friend Joseph Hooker in 1871 and, in his letter, speaks of a 'warm little pond with all sorts of ammonia and phosphoric salts, light, heat, electricity etc. present that a protein compound was chemically formed, ready to undergo still more complex changes ...' in which the first molecules of life could have formed. One wonders if he gave much thought to the possibility that life might have come about elsewhere in the Solar System or the galaxy. Wallace did give the matter considerable thought and in 1904 published *Man's Place in the Universe* in which he attempted to evaluate the probability that life might originate on other planets elsewhere in the Milky Way concluding that Earth is, most likely, the only home to life. Today, given the vast number of stars in our galaxy and the exoplanets that circle them, it would be amazing if that were the case.

Along with the work of Gregor Mendel (1822-1884) in explaining how hybridization might be related to the emergence of new species, the modern era of biology had begun. In time astrobiology, the study of the origin, evolution and distribution of life in the universe would become a recognized science and underpin much of the motivation for exploring our Solar System.

Canals on Mars

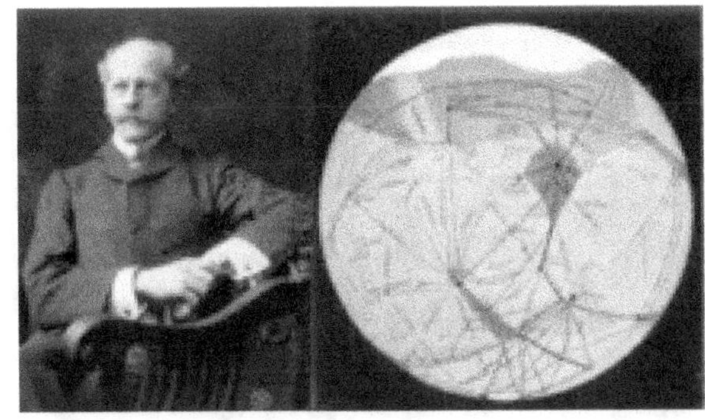

During the opposition of 1877 Italian astronomer Giovanni Schiaparelli (1835-1910) took advantage of this closest approach to map Mars as best he was able and, somewhat fancifully, he sketched a series of linear features that he termed *canali* meaning grooves or channels. In the late 1800s Percival Lowell (1855-1916), a wealthy Bostonian, developed a passionate interest in Mars whose study was the founding purpose and focus of the Lowell Observatory in Flagstaff, Arizona. With a 61cm refracting telescope the observatory was established in 1894 – telescopes for planetary astronomy were getting to be more powerful. Lowell was inspired by occasional brief flashes of excellent seeing conditions and recorded his observations of Mars as drawings of a planet-wide network of the same linear features that Schiaparelli had sketched. Wishful thinking then led him to conclude these must be canals constructed by intelligent beings to transport water equator-ward from the polar caps – something that even toward the end of the 19th Century was decidedly fanciful; the changing seasonal appearance of Mars might reasonably have been ascribed to causes other than seasonally changing vegetation.

The story of those telescopic observations has a certain irony because there really is on Mars a vast equatorial canyon that runs into chaotic terrain and huge flood channels. In fact, this canyon, Valles Marineris, did not figure in any of the telescopic mapping. The Lowell observatory observations of the bright region called Nix Olympica (the Snows of Olympus) were, however, subsequently found by Mariner 9 to have a basis in reality: clouds formed at the summit of Olympus Mons, the largest volcano in the Solar System. Lowell Observatory's Earl Slipher (1883-1964) documented the photographic observations that recorded large-scale, long-lasting dust storms and the waxing and waning of the Martian polar caps.

Mars: Its Moons and Its Dynamics: *The Photographic Story*

An important telescopic discovery about Mars was made in the late 19th Century: in 1877 Asaph Hall (1829-1907) at the US Naval Observatory in Washington DC observed the two tiny moons that circle Mars. They were given the names of the sons of Ares from Greek mythology: the inner moon became Phobos (panic/fear) and the outer moon became Deimos (terror/dread). In the 20th Century the scientific study of Mars through the 1950s had remained a cottage industry: most of the telescopic observations were still being made at the Lowell Observatory by Earl Slipher who, alas, was not to live to see the first rather foggy close-up images returned by Mariner 4 in 1965, the year after he died. His great contribution was the compilation of all of his photographic observations, made over several decades of the early 20th Century, in a 1962 book entitled *Mars: The Photographic Story*. Here, his best photographs (taken when the seeing through the Earth's dynamic atmosphere was optimal) provided the basis for a classic map of Mars (with regional features identified in great detail, all with Latin names, some of which survive even today) together with an extended record of the polar caps waxing and waning, of condensate clouds, of regional and global dust storms, and of changes in the prominent dark albedo (from Latin, meaning whiteness, is a measure of surface reflectivity) features that characterize the telescopic view. The obliquity of Mars (the tilt of its rotation axis relative to the plane of its orbit about the Sun) was determined to be quite similar to Earth's (25.2° vs our 23.5°) while the Martian year is almost twice as long (1.88x) as ours. Slipher's records provided invaluable support for the planning of the first Mars orbiter missions, Mariners 8 and 9 by the author and his colleagues.

Because the Martian orbit is more elliptical than ours the length of the seasons in the northern and southern hemispheres differ: northern spring lasts 7 months, summer 6 months, autumn 5.3 months and winter just over 4 months. Astronomers recorded the formation of a "polar hood" of clouds in both the north and south during the autumn and by the time the cloud cover had lifted in late winter the seasonal surface polar cap had formed. This polar cap then gradually retreated during the spring and summer.

Age Dating

The age of our planet cannot be determined directly by radiometric dating because the oldest rocks have been recycled by plate tectonics. Nevertheless, we can make an estimate by dating the most primitive meteorites and by assuming that the Earth and the rest of the solid bodies in the Solar System formed at close to the same time. Radiometric dating became an increasingly well-understood technique after World War 2 and was applied to the problem of estimating the age of the Earth and other bodies of the Solar System. After a rock is formed, tiny constituent amounts of unstable parent atoms (such a uranium-238, thorium-232) found in zircon crystals decay into daughter atoms – by steps, eventually into lead-206 in the case of uranium. These isotopes have half-lives of billions of years making them suitable for assessing the age of meteorites. The decay rates can be measured and provide the means to unravel the history of the rock. In collaboration with George Tilton (1923-2010), Clair Patterson (1922-1995) of Caltech (the California Institute of Technology in Pasadena) developed the uranium-lead dating method and, in 1956, by using lead isotope data from a Canyon Diablo meteorite, Patterson accurately estimated the age of the Earth to be 4.55 billion years.

Analysis of samples from identified locations on the Moon and Mars and from below the surface of a comet is one of the ultimate goals of Solar System science. Improvements in laboratory instrumentation were needed to take full advantage of the meteoritic samples that were in museum collections and samples of the Moon that were in prospect. Not long after his arrival at Caltech in 1955 Gerry Wasserburg (1927-2016) began the design of what became the *Lunatic I* – a mass spectrometer for making high-precision measurements. These were now an order of magnitude greater in precision than before and the instrument "revolutionized the field of geochemistry," according to Donald Burnett, one of Wasserburg's long-time colleagues.

The Discovery of Galactic Cosmic Rays

Until the beginning of the 20th Century it was supposed that the space above the Earth's atmosphere and between the planets was empty of everything besides light of various wavelengths – a perfect vacuum. That changed in 1912 when Austrian physicist Victor Hess (1883-1964) in a balloon flight to 5,300 meters discovered that the electrometers with which was equipped showed increasing levels of ionization as he ascended. He made his ascent during a near total eclipse thereby ruling out the Sun as the source of the radiation. He concluded, "The results of my observation are best explained by the assumption that a radiation of very great penetrating power enters our atmosphere from above." [2]

He had discovered Galactic Cosmic Rays – confusingly called *rays* since we now know them to be highly energetic *particles* (not *electromagnetic radiation*). These GCR originate outside the Solar System and are likely formed by explosive events such as supernova. Much later, the Sun was also discovered to be a source of charged particles – mostly protons. This 'solar wind' has a (variable) flux many times greater than that of the galactic cosmic rays but the particles are much less energetic. Space is not a simple vacuum and we have come to recognize that it is a hostile environment for humans and machines. Especially humans.

The Biochemical Nature of Life

Beginning in the 1930s the disciplines of molecular biology began to make progress toward understanding the fundamental biochemical nature of life in terms of the structure of nucleic acids that are found in all living things and, we now know, they function in encoding, transmitting and expressing genetic information. Deoxyribonucleic acid DNA had been isolated by biochemists as early as the mid to late 19th Century. Famously the helical structure of DNA had been determined by 1953 by Francis Crick (1916-2004), James Watson, Maurice Wilkins (1916-2004) and Rosalind Franklin (1920-1958). At the same time two scientists in California, Harold Urey (1893-1981) and his graduate student Stanley Miller (1930-2007), were carrying out an experiment to demonstrate that organic compounds vital to biology could be synthesized from inorganic substances by simple chemical processes. In this they were seeking to confirm the ideas of Alexander Operin (1894-1980) and JBS Haldane (1892-1964) that conditions on primitive Earth would have been favorable to such processes and would have led to the origin of life. This hypothesis focuses primarily on how life emerged and specifically postulates that life began as a set of chemical reactions starting from relatively simple molecules produced, for example, in the atmosphere by lightning or photochemistry, in the crust by water-rock reactions, or by aqueous alteration on the parent bodies of carbonaceous meteorites.

The Miller-Urey experiment sealed in a container a mixture of gases – water, methane, ammonia and hydrogen – through which electrical sparks served to simulate lightning. Analysis of the product of the experiments revealed that a number of amino acids had been synthesized including glycine, alpha-alanine and beta-alanine. After Miller's death, the product of the experiment was further analyzed and found to contain over 20 amino acids. Whether the experiment credibly simulated the conditions on primitive Earth isn't clear but it did describe conditions in the proto-planetary disc from which the planets, comets and asteroids were formed. The amino acid glycine has in fact been determined to be included in the composition of meteorites and two comets, perhaps the closest material from the proto-planetary nebula to which we have access.

As the recent National Academy report on *Search for Life in the Universe* points out, "A century following the Oparin-Haldane hypothesis the potential chemical and environmental pathways that gave rise to life have been greatly expanded and hotly debated." Meanwhile, astronomers had long since come to appreciate the insignificance of our Solar System in the Milky Way galaxy that contains more than 100 billion stars and in the wider universe that contains hundreds of billions of other galaxies. It seemed (and seems) vanishingly unlikely that Earth supports the only life forms in the universe – but there was (and is) no direct evidence to the contrary. Curiosity being one of the main, and generally positive, characteristics

---

[2] https://physicstoday.scitation.org/do/10.1063/PT.5.030993/full/

of mankind, once the opportunity arose to explore the other worlds in our Solar System this was impossible to pass up. Unfortunately, the opportunity only arose as a consequence of World War 2 and the following Cold War, prior to which the possibility of visiting other planets and their moons would have been (and was) just a fantasy.

Stellar Evolution

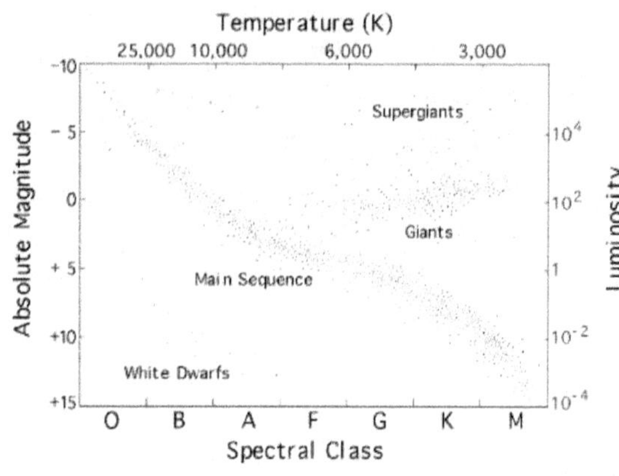

Our Sun, about which orbits the planetary debris left over from its formation, is a G type Main Sequence star whose own early history is key to understanding the origin and evolution of primitive Earth and the other planets. Basic insights into stellar evolution go back to the early part of the last century when Ejnar Hertzsprung (1873-1967) and Henry Norris Russell (1877-1957) created their famous HR Diagram plotting the luminosity of stars against their effective surface temperature (based on spectral classification). The resulting scatter of star types is remarkably orderly with about 90% of the stars lying in a diagonal band – with hot bright stars in one corner (top left) and cool dim stars in another (bottom right).

These 'Main Sequence' stars (which vary greatly in mass) are in the process of fusing hydrogen in their cores. Once all the hydrogen in the core has been consumed in this way the luminosity-temperature relationship changes and the star moves off the Main Sequence and assumes a different character. The Sun formed at the center of a collapsing cloud of gas and dust and initially its energy was the result of the gravitational contraction. Only after about 50 million years did the Sun have enough mass to begin the fusion of hydrogen in its core and join the Main Sequence. Today the Sun is middle-aged after about 4.6 billion years of life. When it has exhausted its hydrogen-burning, the Sun will grow substantially in size and will cool to become a Red Giant. Eventually the outer envelope of the Red Giant will blow off into space, exposing a hot, remnant core. Much heavier stars than the Sun – the kind found at top left of the HR diagram – end their short lives as supernovae and no longer populate the HR Diagram.

A physical understanding of the nature of stellar evolution that the HR Diagram represents emerged in the 1920s when the nature of stellar energy – nuclear fusion – came to be understood based on the insights of Arthur Eddington (1882-1944).

Origin of the Universe and Age of the Solar System

Radiometric dating of meteorites – a critically important technique that goes back to the work of Ernest Rutherford (1871-1937) and Bertram Boltwood (1870-1927) in about the same timeframe had established that the age of the Solar System is about 4.6 billion years. The research of Edwin Hubble (1889-1953) in the 1920s had established the expanding nature of the Universe and that our Solar System is relatively young. The now-widely-held understanding that our universe came into being from an infinitely dense singularity – the Big Bang – at some distant time in the past was proposed in 1927 in the *Annals of the Scientific Society of Brussels* by a Belgian scientist (who was also a Catholic priest but, given progress since 1600, was not in danger of being burned alive) by the name of Georges Lemaitre (1894-1966). He hypothesized a primeval atom or "Cosmic Egg of constant mass and growing radius accounting for the radial velocity of extragalactic nebulae". Albert Einstein, incidentally, was initially dismissive of this extraordinary idea.

Not surprisingly, the nature of the origin of the Universe remained controversial for decades. Indeed, singularities of this kind must be among the six impossible things that Lewis Carroll's White Queen in her youth was able to believe before breakfast. Dark matter, dark energy, black holes, colliding neutron stars, the Higgs field, and General Relativity are probably among the other impossible things Carroll (1832-1898) would have had in mind if he had lived a century later. In the 1940s Fred Hoyle (1915-2001) developed his Steady State model of the Universe, and, ironically, was the one to coin the disparaging term 'Big Bang' to describe Lemaitre's idea of a singularity at the start of time. Ralph Alpherin (1921-2007), Robert Herman (1914-1947) and George Gamow (1904-1968) in the late 1940s had thought through the physical consequences of Lemaitre's singularity; they correctly predicted the relative abundance of primordial light elements in the universe and, notably, that, now cool, relic radiation from the Big Bang would fill the expanding universe. It would not be until 1965, a year before Lemaitre's death that, following up on the research of Robert Dicke (1916-1997) and James Peebles, the Bell Labs radio astronomers Arno Penzias and Robert Wilson made the discovery of the Cosmic Microwave Background – a prediction of the Big Bang theory – that the issue was resolved. The CMB radiation is now very cold, less than 3 degrees above absolute zero. It fills the universe and is uniform in every direction (to better than one part in a thousand).

As Steven Hawking has commented, "The only reasonable interpretation of the background is that it is radiation left over from an early very hot and dense state. As the universe expanded, the radiation would have cooled until it is just the faint remnant we observe today." [3]

The Big Bang initiation of space and time has been deduced by astrophysicists to have taken place about 13.7 billion years ago. Based on the ages of the most primitive meteorites, our Solar System came into being about 4.6 billion years ago. A lot happened in the 9 billion years between these events including the creation of the elements of matter heavier than hydrogen and helium that make possible the formation of the terrestrial planets – and life on at least one of them. In the beginning, 13+ billion years ago and within instants of the Big Bang, the Standard Model of physics tells us that matter comprised only positively charged protons, neutrons lacking charge and negatively charged electrons. Carrying a positive charge, protons would become the nucleus of hydrogen atoms. Two protons and two neutrons would combine to form the nucleus of helium atoms. The 'strong' nuclear force of attraction that exists between protons and neutrons serves to overcome the repulsive electromagnetic force between positively charged protons so that stable nuclei made up of both protons and neutrons can be formed. These make up the periodic table of the elements from hydrogen to the heaviest, uranium, found in nature. The binding energy of a nucleus represents the difference in mass between that of the nucleus and its constituent protons and neutrons. This formation of hydrogen and helium nuclei had to happen quickly because free neutrons decay (by emitting an electron and neutrino to form a proton) with a half-life of about ten minutes.

Later these nuclei captured electrons to create *atoms* of hydrogen and helium. All the other elements in mass increasing up to that of iron (atomic mass 56) were created deep inside the first generations of stars (that were very much more massive and short-lived than our Sun) through fusion of lighter elements – a process that releases energy and powers the stars as Arthur Eddington (1882-1944) had argued in 1920. These early massive stars reached the end of their relatively short lives in the form of spectacular supernovae – as Fred Hoyle proposed in 1954 – and gradually seeded the galaxies with these heavier elements. Oxygen (fusion of 4 helium nuclei – atomic mass 16) and carbon (fusion of three helium nuclei – atomic mass 12) are the next-most-abundant elements after the hydrogen and helium that, still today, make up 98% of the baryonic matter in the Universe (the mysterious dark matter that seemingly makes up most of the Universe's mass differs in some fundamental way from the ordinary baryonic/atomic matter with which we are familiar). Next in abundance after oxygen (0.6%) and carbon (0.4%) come neon (0.15%), iron (0.1%), nitrogen (0.1%), silicon (0.06%), magnesium (0.05%) and sulphur (0.04%).

In the part of the nebula of gas and dust from which the Solar System formed, oxygen atoms joined with, and shared their electrons with, two atoms of hydrogen to form water H_2O) molecules while carbon joined with hydrogen to form methane (CH_4). Nitrogen, the seventh most abundant element, was 'reduced' to form ammonia (NH_3). Hydrogen, helium plus small quantities of water, methane and ammonia gases and even smaller quantities of heavy elements – cosmic dust – provided the ingredients from which the four giant planets formed. The ices that condensed from the gases make up the bulk of the moons (there are some puzzling exceptions) that orbit the giants plus Pluto-Charon and the yet-more-distant Kuiper Belt objects.

---

[3] https://www.berkeley.edu/news/media/releases/2007/03/16_hawking_text.shtml

The Compositions of the Planets

Mercury, Venus, Earth, Moon, Mars are mostly composed of the heavier elements and their oxides – mainly silicates of aluminum and magnesium – plus they have sizeable cores of iron and nickel. A comparison of the bulk elemental composition of the Sun and Earth demonstrates that they share these same refractory elemental abundances – something that is likely true for the other terrestrial planets. Earth, of course, accreted a significant amount of water as, probably, so did Venus, Mars, Ceres and the largest asteroids. The earliest Solar System history is largely hidden from us by the present evolved states of the planets so the analysis of meteorites continues to play its fundamental role along with the basic context provided by stellar astronomy and by exoplanet astronomy. We are making progress but with lots to go.

What about the elements that are heavier than iron? These are also created by fusion. But now the fusion requires energy. Most appear to be created when stars much heavier than the Sun collapse as supernovas at the end of their short lives; heavy element creation involves a series of neutron captures by Fe_{56} or other nuclei. Our understanding of the process by which the heaviest elements are created took a major step forward on 17 August 2017 when the collision of two neutron stars produced gravity waves detected for about 100 seconds by the US-based Laser Interferometer Gravitational-Wave Observatory and the Europe-based Virgo detector. Two seconds later a gamma ray burst was observed and, subsequently, X-ray, UV, optical, IR and radio wave emissions were observed – an historic event. "The new light-based observations show that heavy elements, such as gold and lead, are created in these collisions and subsequently distributed throughout the universe".[4]

A remarkably insightful picture of how the Solar System came into being goes all the way back to the 18th Century and the independent thinking of Emanuel Swedenborg (1688-1772), Immanuel Kant (1724-1804) and Pierre-Simon Laplace (1749-1827): The Sun and planets were understood to have been created as the result of the gravitational collapse of matter in part of an enormous molecular cloud. Swedenborg first proposed this model in 1734; Kant elaborated the model in 1755 and Laplace formulated a similar theory in 1796. In their model, the Sun captured most of the matter and, while this was happening, the bulk of what remained flattened into an orbiting disc from which, in stages, the planets and their moons formed.

Planet Earth

Long before missions into deep space became possible, scientists could study and try to understand only one planet – our own. An appreciation that Earth has a magnetic field that can be used for navigation using a compass goes back 2,000 years to the Han dynasty in China while the magnetic compass arrived in Europe only during the 12th Century. An English physician, William Gilbert, published *De Magnete* in 1600, concluding that the Earth behaves as a giant magnet. Joseph Larmor in 1919 proposed that a self-exciting dynamo could explain Earth's magnetic field. Walter Elsasser and Edward Bullard in the 1940s showed how motion in Earth's liquid core – a dynamo – would produce a self-sustaining magnetic field. By this time seismology and other studies had given a clearer picture of Earth's internal structure, as having a solid inner core, a liquid outer core, both with a composition more of metal (mainly iron) than rock, and a rocky mantle, all below a thin crust. Earth's dynamo is unstable and leads to magnetic reversals, when the polarity of the whole magnetic field changes over. Only after the start of the Space Age was it realized that a planet's magnetic field provides essential protection of both surface and atmosphere from solar and cosmic radiation.

Systematic stratigraphic geologic mapping of our continents as pioneered by William Smith (1769-1839) in England was widely carried out and a major breakthrough was the acceptance in the 1950s that continental drift, theorized in particular by Alfred Wegener (1880-1930), is indeed more than just a theory: plate tectonics has shaped our world. Earth's crust and upper mantle are broken into a number of continent-sized 'plates' and, driven convectively by internal heat, these are in constant relative motion, colliding and diving under one another – leading to earthquakes, volcanoes and mountain

---

[4] https://www.ligo.caltech.edu/page/press-release-gw170817

building. Frank Wigglesworth Clarke (1847-1931) is credited with determining the composition of the Earth's crust: he calculated that oxygen makes up ~47% of Earth's crust, the common rock constituents nearly all being oxides – silica, alumina, iron oxides, lime, magnesia, potash and soda. He concluded that the silica functions as an acid, forming silicates, the most common minerals of igneous rocks.

The early Earth would have been hot as a result of the short-lived radioactive isotopes plus gravitational heating from both the collisions in the late stages of its accretion (including the humungous impact that created the Moon) and the differentiation that then took place with iron and nickel sinking toward the planet's center. Subsequently, yet more heat was generated (and continues to be) by the decay of long-lived radioactive elements in its metal-rich core. In 1926, Harold Jeffries (1891-1989) based on his seismic studies, determined that the core of the Earth is liquid. Later it was evident that the core is divided into a solid inner core, radius of 1,220 km, and a liquid outer core, radius 3,400 km. Convection currents in the outer liquid core create the Earth's magnetic field, a dipolar field that is nearly aligned with the Earth's rotation axis. Only with the 1958 launch of the first US satellite, Explorer 1, did it become clear that our magnetic field creates a magnetosphere and serves to hold off the particle radiation streaming out from the Sun (the solar wind) preventing it from reaching the Earth's surface and atmosphere.

In the last decades the ability to study Planet Earth, its surface, atmosphere and oceans, from the vantage of orbiting spacecraft has provided us with a stunning appreciation of the beauty, complexity and fragility of our home planet. Television documentaries that have aired on PBS in the US and BBC in the UK should have made clear to everyone just how close we are to the precipices of environmental catastrophes.

The Atmospheres of the Planets

Serious thought about the atmospheres of Earth and the other planets had begun as early as 1940 when astronomer Rupert Wildt (1905-1976), then at Princeton University, realized that the large amount of carbon dioxide identified in the Venus atmosphere could lead to a greenhouse effect and might even raise the surface temperature to that of boiling water. In 1948 Harry Wexler (1911-1962), head of the US Weather Bureau and a man ahead of his time, began an interdisciplinary 'Project on Planetary Atmospheres'. There was, however, too little information to build on and the project ended after only four years. Telescopic study of the planets, including radio astronomy, took on more speed and the Martian case was clarified in finding that Mars has a thin cold CO_2 atmosphere while Venus is the opposite – having a thick atmosphere and a temperature of 600K or more. A man of action besides a visionary, Wexler was the first scientist to fly into a hurricane to study its nature.

The evolution of the atmospheres of the planets and their moons (most of which have entirely lost any atmosphere they may once have had) is a field of continuing importance; there are a range of different loss processes involved – Jeans escape, hydrodynamic loss, charge exchange, polar wind, photochemical processes and sputtering – as well as massive loss caused by asteroid impacts.

The Founding Fathers of Rocketry and Space Travel

In Russia, a true visionary and genius, Konstantin Tsiolkovsky (1857-1935), working in poverty in provincial Kaluga, dreamed and wrote about the possibilities of rocketry and space travel – many papers on space travel, designs for rockets and their steering, multistage boosters, space stations and closed-cycle biological systems to support space colonies. He established the fundamental scientific basis for rocketry – the Tsiolkovsky Rocket Equation – and in 1929 he published his theory of multistage rockets. Tsiolkovsky inevitably had a lot of influence on Germany's Herman Oberth (1894-1939) and Robert Goddard (1882-1945).

Goddard shown here holding the launching frame of his most notable invention – the first liquid-fueled rocket is with Tsiolkovsky and Oberth, a founding father of rocketry and astronautics with offspring including Wernher von Braun (1912-1977) and Sergei Korolev (1906-1966).

Korolev had been born and educated in the Soviet Union (specifically, in Ukraine). He had even endured some years of imprisonment including some months in a Gulag in the Far East of Siberia having been charged with being a member of an anti-Soviet counter-revolutionary organization. His is a remarkable story and he has been honored by the naming an important crater on Mars: Korolev Crater is a water-ice filled crater 81 km in diameter located at 73°N, 165°E. Korolev's enthusiastic interest in the development of rockets dated to the 1930s and had led to his appointment as head of a state-sponsored rocket development team. After WW2 ended, Korolev inherited a large number of German scientists and engineers; however, they were ones who had mainly been involved in the production of the V2s rather than their design. They were repatriated to Germany beginning in 1951 after his team had built and tested replicas of the V2 based on captured hardware. Improved versions of the V2 became the Scud missiles and now are to be found in the arsenals of a dozen countries, notably in North Africa, the Middle East and North Korea.

It was Nazi Germany's Wernher von Braun (1912-1977) who, perhaps inspired by thirteenth-Century Chinese war rockets, saw the potential of much larger liquid-fueled rockets as offensive weapons and made critical engineering advances that subsequently enabled travel beyond our atmosphere into orbit and into interplanetary space. In Germany, von Braun had been a student of visionary pioneer Oberth while in Russia Korolev had much the same inspiration. Von Braun's team of engineers at Peenemunde made spectacular advances in liquid-fuel propulsion (alcohol & liquid oxygen, LOX) and their single stage V2 vengeance rocket became a fully operational short range (300 km) ballistic missile in 1944 with over 3,000 launched against British and Belgian civilian targets. The V2 followed a parabolic trajectory and reached a maximum height of 95 km (not quite where 'space' is conventionally considered to begin). Given the relatively modest supersonic speeds reached, the missile's nose was not made of special material to deal with the heat of descent; however, some warheads evidently did explode on re-entry into the thicker atmosphere. The accuracy of targeting was very poor so the V2 had limited military value. Its arrival was unannounced by any sonic warning so the V2 served primarily to terrorize the civilian population. The attacks resulted in the deaths of an estimated 9,000 civilians (this, incidentally, could have included the author, who at that time was an infant resident in the suburbs of London) and military personnel, while 12,000 forced laborers and concentration camp prisoners died producing the weapons. Von Braun was clearly lucky not to have been tried at Nurnberg as a war criminal.

Operation Paperclip

The V2 technology – personnel and trainloads of hardware – was transferred into both the US (Operation Paperclip) and the Soviet Union as the war in Europe entered its final months in 1945. Now the communist Soviets came to be regarded as the principal adversary rather than as an ally. The lucky German technologists and engineers became key members of the US and Soviet intercontinental ballistic missile (ICBM) teams. The shooting war in Europe ended with the defeat of Germany by Soviet and Allied forces in May 1945.

Radio & Radar Astronomy

The development of radar during WW2 has served to create an essential technology – giant radio communications dishes – that makes deep space exploration possible. Prior to the war Bernard Lovell, who was studying cosmic rays at the University of Manchester and who was also working on the development of radar systems, became aware of sporadic echoes that were being detected by Britain's early warning radars. Wondering if they might be caused by cosmic ray particles, he acquired some army radar equipment and in December 1945 began making observations that soon revealed that the echoes were caused by the plasma trails of meteors as they burn up in the atmosphere. With engineer Charles Husband, Lovell planned and built at Jodrell Bank in Cheshire a 250ft (76m) radio telescope, a giant dish that could be pointed to any part of the sky – a fully steerable radio telescope. In October of 1957 Lovell's team were able to track the third stage of the Sputnik 1 carrier rocket. The means for commanding and receiving telemetry from US and USSR satellites had been created – "particularly those venturing far out into the Solar System".[5]

Intercontinental Ballistic Missile Development

An important technical challenge for the US and the Soviets was how to make thermonuclear weapons small enough to be launched on the intercontinental ballistic missiles (ICBMs) that were under development. Re-entry heating of the missile also became an important technical problem to be solved. The deployment of ICBMs in silos throughout the US and the USSR together with nuclear-armed long-range bombers and missile-carrying nuclear-powered submarines created the Cold War strategy of Mutually Assured Destruction (MAD). Each side built up a sufficiently large, and sufficiently dispersed, arsenal such that a first strike by either side would be met with a comparably devastating counter strike – a 'Mexican Standoff' that still keeps the peace.

In addition to the competitive military application of powerful rockets, the rivalry between the Soviet Union and the West inevitably extended into every other arena, both military and cultural. Non-military demonstrations of advances in rocket technology preoccupied both sides and plans were generously funded to carry out both scientific space missions and – much more challenging – missions with astronauts.

Post WW2 Cold and Hot Wars: The Doomsday Clock

With the death of President Franklin Roosevelt in April 1945 Harry Truman had become President. Two years later the Truman Doctrine was announced with the goal of containing communism wherever it threatened (as was the case in Greece and Turkey at that time). The war in the Pacific, which had begun for the US with the 1941 bombing of Pearl Harbor, included the second Sino-Japanese War that had been underway since 1937. The Pacific war ended in August 1945 with the destruction of Hiroshima and Nagasaki by atom (fission) bombs dropped from US B-29 aircraft. A civil war in China followed between the Nationalist Kuomintang (aided ineffectually by the US) and the Communist Party of China, and ended with the

---

[5] http://www.jb.man.ac.uk/history/mk1.html

retreat of the Kuomintang to Formosa (now Taiwan). In October 1949 Mao Zedong announced the founding of the People's Republic of China.

The United Nations was founded in 1945. The 1947 announcement of the Truman doctrine committing the US to the support of nations threatened by the Soviet Union generally counts as the year that the Cold War began. This was also the year in which the Bulletin of Atomic Scientists (founded in 1945 by Manhattan Project scientists who "could not remain aloof to the consequences of their work") set their Doomsday Clock at 7 minutes to midnight. The North Atlantic Treaty Organization (NATO), a system of collective defense by the Western powers in response to anticipated Soviet aggression came into being in April 1949. By this time the minute hand of the doomsday clock had moved 4 minutes closer – an estimated 3 minutes to the end of days. The Truman Doctrine (and associated European Recovery Program, the Marshall Plan 1948-1952) worked as intended in Europe but not in South East Asia.

Soon thereafter another hot war began when communist North Korea invaded South Korea in 1950 (the country having been partitioned through a de facto agreement between the US and the Soviet Union). China supported North Korea and the Soviet Union provided assistance while the US accounted for 90% of the United Nation's forces that defended South Korea. The war – a conventional conflict of armies and air forces that almost involved nuclear weapons – lasted for three years till July 1953 and ended in a stalemate rather than a peace treaty; the Korean peninsula remains divided today with tensions still very high. Some 4 to 5 million soldiers and civilians lost their lives during this relatively short war. A second, disastrous war of containment took place in Vietnam (precipitated by the French attempt to reassert control over their former colony), Laos and Cambodia (1955-1975) ending with defeat for the US.

The intensive development of very much more powerful nuclear weapons was carried out during this fraught time – the first test by the US of a thermonuclear (fusion) weapon took place in 1952. The first Soviet tests of a hydrogen bomb followed a couple of years later. It took about a decade more for China, with the help of the USSR, to begin testing nuclear weapons in late 1964. By September 1953 the Doomsday Clock had moved to 2 minutes to midnight – almost the closest we have approached to Armageddon thus far. This provides a measure of just how frigid the Cold War between the US and the Soviet Union Cold War was. As of early 2020, three years after the most recent election and subsequent impeachment of the 45th US President, the clock is now only 100 seconds to midnight!

Planetary Exploration by Spacecraft

When in the late 1950s exploration of the Moon, planets and small bodies became the subject for serious scientific planning, their systematic characterization would require an inventory of their composition/density, internal structure, geologic processes and relative ages of surface units, atmospheric composition/structure/circulation, magnetic field and magnetospheres. Flyby spacecraft can determine some of these properties, some require orbiters, some require landers; absolute age determination and definitive unravelling of history require the return of samples for analysis in specialized laboratories. And the search for evidence of extant or past extra-terrestrial life is an exceptional challenge met beginning with a search for characteristic gases (e.g. methane, oxygen) in a planetary atmosphere – next through *in situ* surface compositional analysis and eventually through the return of samples for analysis using the most powerful laboratory instrumentation. Some samples will have to be acquired from deep under the surface of Mars – a capability we do not yet have. Such exobiology issues led to concern by life scientists regarding both forward-contamination of the target planet in question and back-contamination of planet Earth (recall the 'Andromeda Strain' movie made in 1971). The Covid-19 experience has greatly added to everyone's sensitivity to matters of biologic contamination. The concern had been recognized early on and by 1959 planetary protection policy had been transferred to a Committee on Space Research (COSPAR) established by the International Council for Science in 1958; the Committee soon had codified the way in which this protection must be assured. In 1967, the US, USSR, and the UK ratified the United Nations Outer Space Treaty that provides the legal basis for planetary protection in Article IX of the treaty.

The Systematic Solar System Exploration Process

To be able to unravel the complex history of the Solar System as a whole and of each of its orbiting constituents – the planets, moons, asteroids and comets – NASA and the other space agencies have undertaken to conduct a thorough mass and compositional inventory of all these bodies along with an assessment of their geological, atmospheric and magnetospheric evolution. The ability to work backwards to the origin of our Solar System builds on the observations of astronomers who have recorded collapsing discs of dust and gas around very young stars – for it is from such a disc that our Solar System came into being. Primitive meteorites that had their origin close to that time provide the most direct insights into conditions at the earliest era. Our present era in Solar System history is roughly that of middle age, 4.6 billion years after formation. The present state of our Solar System has now been established in some detail by means of deep space missions all the way out to Pluto and beyond. Planetary scientists have modeled the composition of each body based on a calculation of its density – this by determining its size and its mass (a gravitational measurement made by passing and/or orbiting spacecraft). By measuring the magnetic fields (if any) of the planets and their moons the convective state of any metallic core can be assessed. In the case of our Moon (and even Mars in the form of a few special meteorites) we have uniquely informative samples to analyze. We also have large collections of other meteorites of mostly primitive composition. Some of these are made of iron presumably from the cores of broken-up asteroids. Dust samples of comets have been analyzed in terrestrial laboratories and one *in situ* compositional analysis has been made recently on a comet surface. The time-history of the sources of heat that were, and are, available to drive the evolution of the planets and their moons is of basic importance. Experiment and theory come into play. These heat sources include the decay of radioactive isotopes (of widely different half-lives) and, of course, the Sun's insolation. At the center of our forming Solar System the primitive Sun grew in mass until fusion began in its core – at which point its luminosity greatly increased. Other key heating processes include gravity (asteroid impacts, internal planetary differentiation, tides), and phase change (gas/liquid/solid). With ever-more powerful computers, scientists are developing plausible models of the rapid, complex dynamic evolution of the primitive solar nebula.

In the case of planets (and the rare moon) with atmospheres there are a number of loss mechanisms to take into account in understanding their evolution. Much depends on whether they have a magnetic field to ward off the solar wind. Thermal and non-thermal escape mechanisms come into play as do the consequences of asteroid impacts. We can now make an increasingly credible assessment of how and where each Solar System body formed and evolved. In particular we begin to understand why Earth and its sister planet Venus differ so profoundly as well as why most of the planetary moons are airless (albeit, how Saturn's giant moon Titan acquired and retained a significant atmosphere of nitrogen is still a puzzle). Putting all these considerations together to establish a grand history of our Solar System is the task for theoreticians; our exploration to date has already given them plenty of data and of work to do.

Initially lunar exploration was the inevitable focus of the US and Soviet manned spaceflight efforts – ending abruptly with the last of the Apollo missions, Apollo 17, in 1972. Thereafter only robotic missions, dozens, to deep space followed; it is these therefore that make up most of the history of mankind's exploration of the Solar System to date. The future of human flights into deep space beyond the Moon is, at best, highly problematic because the role of the astronaut is unclear and because particulate galactic cosmic rays are so energetic as to defy shielding. Journey time becomes a big problem for such flights.

A very positive development has been the fact that the exploration of the planets has become a worldwide enterprise with a considerable amount of international collaboration. The International Space Station that orbits in near-Earth space demonstrates what might be possible if/when manned missions to the Moon resume. Until recently, with the 2011 retirement of NASA's Space Shuttle, the required use of Russian launch vehicles to ferry international crews to the orbiting Space Station is the prime example of how nations can work together if they choose to. Another possible international effort – Back to the Moon – this time with China as a participant (unfortunately very unlikely), would introduce a little light into a world that is increasingly looking like a return to Cold War politics.

International space law has long-since been in place; the United Nations Committee on the Peaceful Uses of Outer Space oversees the treaty. The Treaty on Principles Governing the Activities of States in the Exploration and Use of Outer Space, including the Moon and Celestial Bodies was signed by the US, UK and Soviet Union in January 1967 while the Space Race to the Moon was in full swing. The treaty bars states from placing weapons of mass destruction in orbit or on the Moon (or any other celestial body). It limits the use of the Moon and other celestial bodies to peaceful purposes. Of late, commercial companies are paying attention to this international treaty because they see an opportunity opening in the relatively near future to mine water ice on the Moon. Asteroids are another potential resource.

Mars in Science Fiction

Astronomers and scientists were not the only ones to take an interest in Mars. Long before rocket technology had made possible deep space exploration, science fiction writers like HG Wells (*War of the Worlds*, 1898) and Edgar Rice Burroughs (ten Mars-based novels published from 1912 to 1948), Ray Bradbury (*The Martian Chronicles* in 1950) had entertained readers and (in the case of HG Wells, with the help of Orson Welles, terrified some radio listeners in the USA) with their stories about warrior inhabitants of Mars – a planet that surely had the potential to support intelligent life. Welles' famous radio broadcast took place in late 1938. In 1956, as the journeys into deep space were just becoming a possibility, an excellent MGM movie was released: *Forbidden Planet* loosely based on Shakespeare's *The Tempest*. Not a space cowboy production, it remains one of the best science fiction movies set in outer space thus far.

Returning to the fascination exerted by Mars on so many, by the early 1950s the German rocket engineer Wernher von Braun (1912-1977) who, because of his unique skills had escaped war crimes prosecution and likely execution, was now working in the USA, turned his thoughts again to the possibilities of human space exploration. He didn't pay much attention to robot spacecraft missions – likely because computational power was then primitive. Launch vehicle technology was, and remains, the enabling capability upon which all else depends. Von Braun had no peer in this arena and so his concepts were plausible projections of existing liquid-fueled rocket development. He laid out a systematic plan that would take crew to Mars by way of a space station, lunar precursor missions, and the use of multi-stage, winged launch vehicles.

Not one to think small, von Braun pictured a Mars mission that would involve the assembly of ten huge spacecraft in Earth orbit and the departure of a crew of no fewer than seventy astronauts. The analysis of the required launches, the trajectories to and from Mars, and other details were carried out as rigorously as was possible at the time. Von Braun's vision was shared by a number of notable space enthusiasts including Harvard astronomer Fred Whipple (1906-2004) and the science writer Willy Ley (1906-1969). Together with two other colleagues the team published a series of eight articles in Collier's Magazine beginning in 1952 and culminating in a 1954 article entitled Can We Get to Mars? Is There Life on Mars? The articles were excitingly illustrated by Chesley Bonestell (1888-1986). The illustrations show, among other anticipated technology advances, that Von Braun had recognized the need to be able to land launch vehicles on Mars in a vertical manner in order for them to take off with the crew when that moment came. (Colleague Fred Taylor reminds me that originally Von Braun had pictured the landings being made on skids – guaranteed disasters.) Vertical landings of launch vehicles and their re-use have only recently become a regular feature of SpaceX launches. In some respects such vertical landings of an almost empty vehicle are easier to control than are the immense forces of take-off.

Indeed, NASA's vision for space exploration half a century later still reflects this foundational thinking. The extremely hostile nature of the space environment – supposedly just a vacuum – to humans was not given much consideration so space exploration was regarded by most as the natural domain of human explorers.

CHAPTER 1

SPACE TECHNOLOGIES & NATIONAL SPACE AGENCIES

Powerful liquid-fueled rockets are the most obvious, but not the only, critical technology developments from WW2 that have made deep space missions possible. The development of the electronic digital computer by Alan Turing (1912-1954) and others provided another critical technology, one enhanced enormously by the development of semiconductors and integrated circuits in the 1950s & 60s following the first demonstration of the transistor by William Shockley, John Bardeen and Walter Brattain of Bell Labs in December 1947. Before electronic computers became commonplace, squadrons of young women with mechanical calculators served NASA as human computers (in time, let it be said, women would become senior scientists, engineers and project managers at JPL and NASA). "They were responsible for the number-crunching of launch windows, trajectories, fuel consumption and other details that helped make the U.S. space program a success."[6]

Human computers were also a key to the success of the early manned space program that led to Apollo. At NASA's Langley Research Center in Virginia the computers were young black women (racial segregation being the norm at that time). This episode in space exploration history is portrayed in the 2017 movie *Hidden Figures*. One of these exceptionally talented mathematicians, Elizabeth Johnson, died in February 2020 at the age of 101. She had received the Presidential Medal of Freedom (the nation's highest civilian honor). NASA Langley in Virginia was the original lead Center for Project Mercury and it was here that the Mercury Seven astronauts trained. The tracking and ground instrumentation were also designed and monitored here. A decade later NASA Langley would lead the successful Viking mission to Mars. Then on September 19, 1961, NASA Administrator James Webb announced that the new Manned Spacecraft Center (MSC) would be established on a 1,000-acre tract to be transferred to the Government by Rice University, near Houston.

Giant radio antennas and associated instrumentation are also vitally important; in the US these were an outgrowth of the radio astronomy experiments carried out by Karl Jansky (1905-1950) at Bell Labs and others – notably Grote Reber (1911-2002) – in the 1930s, initially to investigate the background noise hampering long distant radio communications at 15m wavelengths.

Further development of rockets by the Soviet team led first to ballistic missiles with increasing range; by 1953 Korolev had conceived of a rocket capable of launching a payload into Earth orbit. Similar interest was emerging in the US as plans for the 1957 International Geophysical Year (IGY) took shape. Amidst all the sabre-rattling, international cooperation among scientists still flourished and had led the International Council of Scientific Unions to propose an International Geophysical Year to be carried out in 1957-58 coincident with the solar cycle's approaching period of maximum activity. In part because of the establishment of the International Geophysical Year (IGY), the Doomsday Clock had been reset at 7 minutes to midnight by 1960. The time has moved back and forth since then with 1991, at 17 minutes-to, being the furthest from midnight – this was the result of the signing of the Strategic Arms Reduction Treaty and the dissolution of the Soviet Union.

---

[6] https://www.jpl.nasa.gov/edu/news/2016/10/31/when-computers-were-human/

In 2016, however, the clock was again set at an alarming 3 minutes to midnight and, following the 2016 US Presidential election reset in 2017 to 2½ minutes, now, 2020, just 100 seconds!

Inevitably, given their ongoing rocket developments, the US and Soviet participation in the IGY proved to be competitive rather than cooperative. The initial advantage in this game of one-upmanship was held by the Soviet Union as demonstrated by the launch of Sputnik 1 on 4 October 1957 followed by Sputnik 2 carrying the unfortunate dog Laika on 3 November of that year.

By comparison with the Soviets, the US got off to a rocky start in its attempt to launch a satellite as part of the IGY on 6 December 1957 – Vanguard TV3 managed by the US Naval Research Laboratory (NRL) – exploded in a spectacular and embarrassing fashion on the launch pad. As the NASA History Office notes: "This had a 'Pearl Harbor' effect on American public opinion, creating an illusion of a technological gap and provided the impetus for increased spending for aerospace endeavors, technical and scientific educational programs, and the chartering of new federal agencies to manage air and space research and development."[7]

US National Aeronautics and Space Administration

In some haste, President Eisenhower signed the National Aeronautics and Space Act into law on July 29, 1958 to bring the US civilian space agency – the National Aeronautics and Space Administration (NASA) – into being. The circumstances of NASA's formation are described in detail in a 1982 Special Publication NASA SP-4102 – *Managing NASA in the Apollo Era* – by Arnold Levine (with a preface by James Webb) who describes the reasoning behind the establishment of a new Agency that took over elements of Army and Air Force facilities as well as inheriting the National Advisory Committee for Aeronautics (NACA). The first objective of the Act creating NASA was the expansion of human knowledge of phenomena in the atmosphere and space. "The decision to put the space program in civilian hands was made by a President whose strongest feeling about the space program was that it must be kept from military control, both to keep space activity peaceful and to avoid creating a new, large, and expensive program located in the Pentagon."

NASA is an independent agency that reports to the President and, as such, the direction and pace of the agency's programs responds to the White House where the occupant changes regularly so enlightenment and continuity are not ensured. Rather the opposite.

In November 1960 John Kennedy was elected to succeed Dwight Eisenhower who had completed two terms in office (Kennedy's Republican opponent in the election was Vice-President Richard Nixon). Kennedy thereby found himself in the middle of the truly scary Cold War in which each side, the US and the Soviet Union, had the potential to trigger, by accident

or design, the end of civilization. The Kennedy years – and those of his Vice President Lyndon Johnson – were exceptionally full of incident: the Cuban Missile Crisis, the Bay of Pigs invasion of Cuba, the building of the Berlin Wall, assassinations of both John and Robert Kennedy and of Martin Luther King with associated riots across the country, the Civil Rights Movement, the escalation of the war in Vietnam and the protest movement that followed well into the 1970s. Kennedy had inherited the Space Race with the Soviet Union, the Apollo Program having already been conceived as a follow-up to the manned spaceflight Mercury and Gemini Projects that had been initiated in 1958.

Neither Eisenhower nor Kennedy were enthusiastic supporters of astronautic spaceflight – rather the contrary – so it took the historic flight of Yuri Gagarin in April 1961 to change Kennedy's mind, this based on the recommendations of Vice President Johnson. The new US space agency, NASA, established by Dwight Eisenhower and led by Administrator James Webb had developed extraordinarily ambitious plans for an early landing of astronauts on the Moon. With Kennedy's forceful commitment on

---

[7] https://www.space.com/38700-nasa-history.html

May 25, 1961, that "this nation should commit itself to achieving the goal, before this decade is out, of landing a man on the Moon and returning him safely to the Earth" the die was effectively cast – although there was mixed support from Congress and scientists. The competitive exploration of deep space would get underway with astronauts leading the effort and with robotic spacecraft in a supporting role. The race to land astronauts on the Moon had become a soft power element of the Truman Doctrine of communist containment.

Lyndon Johnson took over as President of the US following Kennedy's assassination in November 1963 and stood for election a year later – winning against his opponent, Republican Senator Barry Goldwater. Johnson chose not to run for re-election in 1968 in spite of his progressive domestic agenda – civil rights, war on poverty, voting rights, immigration reform. Growing crime rates and his escalation of the Vietnam War led to his withdrawal. The winner of that election was Richard Nixon who was to be US President during the period of the Apollo explorations that had been initiated by Nixon's great rival John Kennedy.

Soviet Space Program

In the Soviet Union there had been no equivalent to NASA; the organization was distributed, complex and not without rivalries. As Elizabeth Newton of JPL's International Affairs Office reported as late as 1990:

"Despite the number of political changes occurring, the Soviet civil space program remains a collection of disparate elements, not unified by any national, centralized space agency. The many space components fall into three categories: space industry, space sciences, and space services. In space industry, applied space research and the production of aerospace equipment occur in the industrial ministries, the largest of which is the Ministry of General Machine Building. Space sciences reside mostly in the Academy of Sciences, or, in the case of space medicine, the Academy of Medical Sciences. Operational services, such as tracking, launches, and flight control, are provided by the military's Strategic Rocket Forces. The USSR Academy of Sciences oversees many research institutes involved in space sciences, such as the Space Research institute (IKI), the principal research facility for space sciences; the Vernadsky Institute of Geochemistry and Analytical Chemistry; and the Keldysh Institute of Applied Mathematics, which has a Ballistics Control Center. The Academy formulates long term plans for fundamental research and development, which are combined with the applied research plans of the State Committee on Science and Technology (GKNT) for incorporation into the State Plan. Under the Academy's jurisdiction, an Inter-departmental Scientific Technical Council for Space Research formulates the future direction of space sciences research in a number of areas, including space physics, planetary exploration, and small body observations. The Ministry of Defense's Strategic Rocket Force operates the launch facilities, tracking stations, and mission control center for most missions. The Air Force directs cosmonaut training and spacecraft recovery." [8]

Academician Roald Sagdeev, a plasma physicist, headed the Space Science Institute, IKI, during the period before the Soviet Union broke up while geologist Valeri Barsukov (1928-1992) was Director of the Vernadesky Institute. In 1989 Sagdeev became Distinguished Professor of Physics at the University of Maryland.

Russian Space Agency: Roscosmos

The Russian Space Agency, Roscosmos, was formed in February 1992 by a decree of President Yeltsin and has subsequently experienced various ups and downs. Space exploration missions and other space sciences have been a minor priority. Launch and operation of the Mir space station and contribution to the creation of the International Space Station have

---

[8] https://ntrs.nasa.gov/archive/nasa/casi.ntrs.nasa.gov/19920015935.pdf

been the highlights of the Russian program. Roscosmos has continued to fly Soyuz and Progress missions including – a key contribution – providing NASA with crew launches to the International Space Station.

European Space Agency ESA

In Europe, the space agency, ESA, was established in 1975. The Agency is not an element of the European Union – ESA is an intergovernmental organization, whereas the EU, established in 1993, is supranational. The EU provides about 20% of ESA's budget. Rather like NASA, ESA has numerous Centers spread around the continent along with its Guiana Space Centre in Kourou, French Guiana, conveniently close to the Equator for launches. The Headquarters of ESA is in Paris. Astronomy and Planetary missions are controlled from the European Space Astronomy Centre, ESAC, near Madrid. The European Astronaut Centre, EAC, is located in Cologne. There are several other Centers with different responsibilities including ECSAT, the European Centre for Space Applications and Telecommunications, in Harwell in the UK that supports activities related to telecommunications, integrated applications, climate change, technology and science.

China's National Space Administration (CNSA)

In China, today the other rapidly emerging space power, the space program was initiated in 1956 based on the ballistic missile technology development underway. Formally, the China National Space Administration (CNSA) was created only in 1993. For a decade, starting in 1966 and continuing until Mao's death in 1976, China had been enmeshed in the violent 'Cultural Revolution' that had been launched by the Chairman. Unlike NASA and ESA, the CNSA is an agency subordinate to the State Administration for Science, Technology and Industry for National Defence and, that organization is, in turn, subordinate to the Ministry of Industry and Information Technology. Launches to deep space (presently the Moon and Mars) are carried out from the Wenchang Spacecraft Launch Site, 19 degrees north of the Equator. Taikonaut (astronaut) missions to the Chinese space station, Tiangong, are launched from the Jiuquan Satellite Launch Center in the Gobi Desert. The Chinese deep space network includes 50-meter dishes in Beijing and 40-meter dishes in Kunming, Shanghai and Urumqi.

As might be expected, the goals of China's space program are ambitious. The 2019 Report to Congress of the US-China Economic and Security Review Commission reads as follows:

> China's goal to establish a leading position in the economic and military use of outer space, or what Beijing calls its "space dream," is a core component of its aim to realize the "great rejuvenation of the Chinese nation." In pursuit of this goal, China has dedicated high-level attention and ample funding to catch up to and eventually surpass other space-faring countries in terms of space-related industry, technology, diplomacy, and military power. If plans hold to launch its first long-term space station module in 2020, it will have matched the United States' nearly 40-year progression from first human spaceflight to first space station module in less than 20 years. [9]

Keith Cowing of his online blog NASAWatch observes:

> China views space as critical to its future security and economic interests due to its vast strategic and economic potential. Moreover, Beijing has specific plans not merely to explore space, but to industrially dominate the space within the moon's orbit of Earth. China has invested significant resources in exploring the national security and economic value of this area, including its potential for space-based manufacturing, resource extraction, and power generation, although experts differ on the feasibility of some of these activities. Beijing uses its space program to advance its terrestrial geopolitical objectives, including cultivating customers for the Belt and Road Initiative (BRI),

---

[9] https://www.uscc.gov/sites/default/files/2019-11/Chapter%204%20Section%203%20-

while also using diplomatic ties to advance its goals in space, such as by establishing an expanding network of overseas space ground stations. China's promotion of launch services, satellites, and the Beidou global navigation system under its "Space Silk Road" is deepening participants' reliance on China for space-based services. [10]

At present collaboration between the US and China is prohibited by the US Congress. The Chinese space program came into being for much the same reason as the US and Russian space programs – as an offshoot of a ballistic missile program "in response to perceived American (and later Soviet) threats". The Chinese program began in late 1950s with the development of rockets initially based on a Soviet design. Qian Xuesen (1911-2009) became known as the "Father of Chinese Rocketry. He had graduated from Shanghai Jiao Tong University in 1934, studied at the Massachusetts Institute of Technology and later at the California Institute of Technology. During the 1940s Qian was one of the founders of the Jet Propulsion Laboratory. In 1950 the US government accused him of having communist sympathies and stripped him of his security clearance. He returned to lead the Chinese rocket program.

By the mid-1960s China had tested its first indigenous intermediate range ballistic missile and during the US-Soviet Space Race Chairman Mao initiated a program of taikonaut space exploration. The first successful launch of a space satellite took place in 1970 and by the 1980s China was using its Long March rocket as the basis of a commercial launch program. The emphasis of the Chinese space program has been on crewed spaceflights to their Mir-size space station (named Tiangong 1 that re-entered the Earth's atmosphere in 2018) with the goal of one day landing their taikonauts on the Moon; the first launch of a crew to Earth orbit took place on 15 October 2003. The Chinese first launch of a robot spacecraft to orbit the Moon took place on 24 October 2007.

US Beginnings in Space

Back in the 1950s America, in a somewhat ad hoc arrangement, the US Army Ballistic Missile Agency under Von Braun, together with the California Institute of Technology (Caltech)'s Jet Propulsion Laboratory under William Pickering (1910-2004), successfully developed the 14kg Explorer 1 spacecraft that was launched on 31 January 1958. (NRL's second attempt – Vanguard 1 – did, also, make it successfully to orbit.)

By its detection of the Van Allen radiation belt (named for James Van Allen (here with Pickering and Von Braun) who was the Principal Investigator of the University of Iowa's cosmic ray instrument). Explorer 1 made a major scientific discovery that went well beyond what would otherwise have been a compelling demonstration of technical capability like the Soviets'. The discovery of this trapped radiation established a new understanding of the Sun's influence in near-Earth

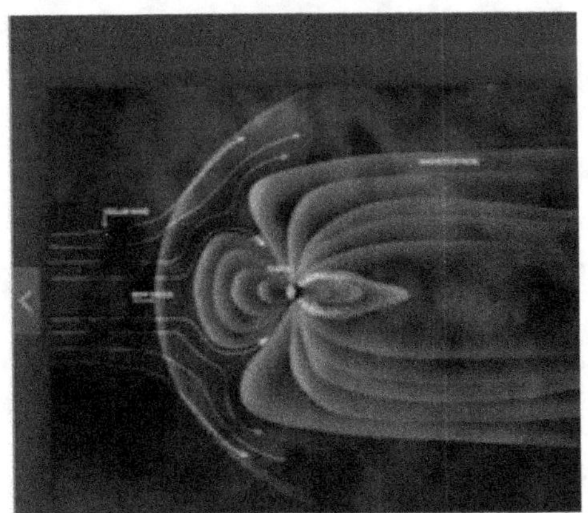

space and beyond. The Sun is, we now know, the source of a solar 'wind' (a term proposed in the 1950s by Eugene Parker of the University of Chicago) comprised mainly of energetic protons and electrons; on encountering a planetary magnetic field, their trajectories are controlled such that a region of space around the planet is formed that is termed a magnetosphere, a name suggested by Thomas Gold (1920-2004).

The solar wind varies in intensity and is subject to occasional violent outbursts known as solar storms. These we now know can be deadly both to unprotected humans and to inadequately shielded spacecraft. Particle energies range up to 10 MeV – low enough that shielding is possible. The flux of galactic cosmic rays (created by supernovae) is very much less intense but

---

[10] http://nasawatch.com/archives/2019/11/a-detailed-look.html

more deadly for long-duration flight in deep space because shielding is not possible given the enormous particle energies in question (100 MeV to 10 GeV). The Apollo astronauts, traveling beyond the protective magnetosphere, found themselves in a different radiation environment, one still incompletely understood in terms of risk to health (discussed more in Chapter 8). They experienced occasional flashes of light that are evidently created by cosmic rays passing through their eyes. Space Station astronauts have also reported this phenomenon.

In addition to protons and alpha particles, cosmic rays include heavy nuclei that greatly amplify the damage caused by the trails of destructive ionization as the charged particles pass through tissue. A leader in this field, Eugene Parker, in 2006 concluded that "Unlike most of the other challenges of venturing into deep space, which engineers should be able to solve given enough time and money, cosmic rays pose irreducible risks, and dealing with them involves fundamental trade-offs. They could be the show-stopper for visiting Mars". [11]

Particulate radiation experiments at NASA's Space Radiation Laboratory using mice conducted by neuroscientist and radiation biologist Charles Limoli of UC Irvine focused on damage to the brain associated with learning and memory. He comments: "Scientists have seen no sign that damaged dendritic complexity and spine density can repair themselves after cosmic ray exposure, and whereas it is premature to refer to such changes as permanent, we have no evidence that neurons recover from this type of injury".[12] Taken together, solar storms and galactic cosmic rays make the deep space environment extremely hazardous for humans.

In another paper Charles Limoli and colleagues conclude that,

Maintaining neurologic function on long-term voyages through space exposure is vital for both the health of the astronaut and the security of the mission. Studies and previous observations from returning astronauts and cosmonauts from long missions aboard space stations have demonstrated significant unanticipated concerns. Countermeasures to protect the astronauts from space exposure require further exploration and are vital components in ensuring safe and reliable journey to Mars.[13]

NASA Administrators and NASA Centers

As noted, NASA was established in something of a hurry in 1958 and has operated under eleven Presidents (Dwight Eisenhower, John Kennedy, Lyndon Johnson, Richard Nixon, Jimmy Carter, Ronald Reagan, George HW Bush, Bill Clinton, George W Bush, Barack Obama, and Trump) who, in turn, have appointed a dozen NASA Administrators (Keith Glennan, James Webb, Thomas Paine, James Fletcher, Robert Frosch, James Beggs, James Fletcher, Richard Truly, Daniel Goldin, Sean O'Keefe, Michael Griffin, Charles Bolden, James Bridenstine). The most famous was James Webb (1906-1992) who ran the Agency with enormous success during the Apollo Program. The giant space telescope that is due to replace the Hubble is named for him: the James Webb Space Telescope (JWST).

NASA was initially made up of three major research laboratories (Langley Aeronautical Laboratory, Ames Aeronautical Laboratory, and Lewis Flight Propulsion Laboratory) that together formed the existing National Advisory Committee for Aeronautics (NACA). NASA Headquarters is in Washington DC (where the Office of Management and Budget (OMB) and Congress keep the money). Soon NASA added the space science group of the Naval Research Laboratory in Maryland, the Jet Propulsion Laboratory managed by the Caltech, and the Army Ballistic Missile Agency in Huntsville, Alabama, where von Braun's team of engineers were developing increasingly large rockets (this is now the Marshall Space Flight Center). Subsequently NASA created several other Centers including the Kennedy Space Center in Florida, the Johnson Space Center in Houston and the Goddard Space Flight Center in Maryland.

---

[11] Eugene Parker, Shielding Space Travelers, *Scientific American*, March 1, 2006

[12] https://www.ncbi.nlm.nih.gov/pmc/articles/PMC5791508

[13] https://news.uci.edu/research/mars-bound-astronauts-face-chronic-dementia-risk-from-galactic-cosmic-ray

US National Academy of Science's Space Science Board (SSB)

The US National Academy of Science's Space Science Board (SSB) was appointed in the spring of 1958 at the request of the Executive Committee of the US National Committee for the International Geophysical Year with a charter to survey the scientific aspects of the exploration of space. The SSB provided advice on the continuation and expansion of the IGY rocket and satellite program, and later advised NASA on its plans for interplanetary probes and space stations including potential problems of manned spaceflight, the exploration of Venus and Mars, and other related matters. The SSB in turn organized a number of specialized committees to advise NASA on priorities for its several Space Science Programs: Astrophysics, Planetary Exploration, Heliophysics, Earth Sciences and Life Sciences. The broadest of the National Academy's studies takes the form of decadal surveys — looking out ten or more years into the future to prioritize research areas and potential missions. The Committee for Planetary and Lunar Exploration (COMPLEX) provides advice on a continuing basis consistent with the decadal survey.

NASA's Associate Administrator for Space Sciences (often also the Agency's Chief Scientist) has the responsibility of maintaining a healthy balance between the several Space Science Programs (Astrophysics, Solar System Exploration, Earth Sciences, Life Sciences, Heliophysics) — all of which are effectively competing for the budget that Congress appropriates each year. Generally speaking, the NASA Administrator appointed by each President, while taking a significant interest in the Space Science Programs, defers to the Associate Administrator's budget decisions. These, in turn, aim to be consistent with the advice provided by the National Academy of Science.

The Solar System Exploration Program Office has relied very closely on the advice of COMPLEX throughout the years. Regular meetings between NASA officials and COMPLEX take place each year in Washington DC and at other Academy facilities. Given the limitations of funds appropriated annually to NASA by Congress the challenge to NASA and COMPLEX has always been to maintain a healthy balance in the exploration program between space missions to inner planets, outer planets and small bodies together with ground-based research (planetary astronomy, Antarctic meteorite collection, laboratory geophysics and geochemistry, atmospheric modelling, instrument development, and mission data analysis). In the author's experience, the relationship between the Solar System Exploration Program and the SSB has worked well. Inevitably, there has always been far more to accomplish in all of NASA's Space Science programs than have resources been available so finding the right overall balance and health among the several science programs has been a continuing challenge for the NASA Associate Administrators for Space Science.

In the US Congress there have occasionally been Senators and Representatives who have taken a serious interest (some positive, some not) in NASA and its Space Science Programs. Richard Mallow, who for many years was the chief congressional staffer on the House appropriations committee (chaired by Edward Boland (1911-2001)) in charge of NASA's budget, was one of the most important individuals responsible for supporting the science direction of the Agency over the years.

In the beginning, the robotic exploration missions were managed by three Centers: JPL, NASA Langley in Virginia (also having an early responsibility for the manned program) and NASA Ames in California. NASA Johnson has responsibility for the manned program and continues to have responsibility for planetary materials curation. Then for a long time the missions were led by JPL until, more recently, other players have become involved: the Advanced Physics Laboratory of Johns Hopkins University and the Southwest Research Institute in Texas. Still, the history of robotic Solar System Exploration has been in many ways the history of the Jet Propulsion Laboratory, located in Pasadena, California in the shadow of the San Gabriel Mountains.

Jet Propulsion Laboratory JPL

Just before WW2 an aeronautical laboratory at the nearby California Institute of Technology evolved to create JPL in time to play a useful role in the war; the nearby Arroyo Seco ("dry stream" in Spanish — a 40km-long canyon) was a convenient place to test small rocket engines. During the war JPL worked under a US Army contract with Caltech and developed Jet-

Assisted-Take-Off (JATO) technology for heavily-burdened, under-powered military aircraft (hence the Laboratory's name) and, subsequently, the Corporal and Sergeant intermediate-range ballistic missiles. Later JPL joined forces with Von Braun's US Army team at the Redstone Arsenal in Huntsville, Alabama to develop ablative materials for payload protection during atmospheric re-entry. They used a 3-stage Jupiter missile as the launch vehicle for suborbital re-entry experiments. As described above, in collaboration with James Van Allen of the University of Iowa, JPL and the Huntsville team launched the first US satellite, Explorer 1, in late January 1958 – the start of the US effort to catch up with the Soviet Union's spectacular Sputnik successes.

When President Eisenhower established NASA in 1958, JPL moved from an Army contract to one with the new Agency. In comparison with the six other civil service NASA Centers there were, and are, mutual advantages for JPL, for NASA and for Caltech in this contractual arrangement: first, the many rigid civil service rules and constraints (including, e.g., citizenship as in the case of the author and many others) do not apply to JPL employees – making it easier for JPL to hire its staff who, in practice, have almost as much job security as civil servants. JPL also has many talented contractors. JPL can maintain in-house skills – such as building and testing spacecraft and operating the Deep Space Network – that other Centers contract out. Then too, Caltech, a world famous science university, is a source of exceptional talent to lead projects (e.g. Voyager's Project Scientist Edward Stone and Mars Global Surveyor's Arden Albee) and to lead the Laboratory (Directors Bruce Murray and Edward Stone). Further, JPL, unlike civil service Centers, can exert a degree of independence from NASA HQ because Caltech's President can lobby the Administration and Congress. In summary, JPL in several ways enjoys the benefits of being a federally funded institution with fewer of the restrictions that apply to the civil service – which probably doesn't make it especially popular with the other NASA Centers in competing for new projects at times of increasingly tight budgets.

The JPL Center Directors have been Bill Pickering 1958-76, Bruce Murray 1976-82, Lew Allen 1982-1990, Ed Stone 1991-2001, Charles Elachi 2001-16 and Michael Watkins since July 2016.

NASA Ames Research Center

NASA Ames is located at Moffett Federal Airfield in California's Silicon Valley. The Center was founded in 1939 as the second National Advisory Committee for Aeronautics (NACA) to conduct wind-tunnel research on the aerodynamics of propeller-driven aircraft. When that agency was dissolved in 1958 Ames was transferred to NASA where its role now includes spaceflight and information technology. It provides leadership in astrobiology; small satellites; robotic lunar exploration; the search for habitable planets; supercomputing; intelligent/adaptive systems; advanced thermal protection; and airborne astronomy. Ames managed the Pioneer missions to Jupiter & Saturn and to Venus, the Galileo Entry Probe and the Kepler Space Telescope. It is home to NASA's Astrobiology Institute and its Center for Mars Exploration.

The Ames Center Directors have been Smith J. DeFrance, 1940-1965, H. Julian Allen, 1965-1968, Hans M. Mark, 1969-1977, Clarence A. Syvertson, 1977-1984, William F. Ballhaus Jr., 1984-1989, Dale L. Compton, 1989-1994, Ken Munechika, 1994-1996, Henry McDonald, 1996-2002, Scott Hubbard, 2002-2006, Simon P. Worden, 2006-2015 and Eugene L. Tu, 2015-present.

Johns Hopkins University Applied Physics Laboratory APL

APL is located near Laurel, Maryland, and is a not-for-profit, university-affiliated research center that serves as a technical resource for the US Department of Defense, NASA, and other government agencies. Since the adoption of the *Discovery* and *New Horizons* programs by NASA's Solar System Exploration Program the APL has become an important player in that exploration having built and operated missions to an asteroid, a comet, Mercury and, famously, the *New Horizons* mission to Pluto. In 2019, the APL-proposed *Dragonfly* mission was selected as the fourth NASA *New Frontiers* mission. Dragonfly is a lander for Saturn's moon Titan that will fly using rotors in a dual-quadcopter configuration to move around the Moon's surface.

NASA's Solar System Exploration Program

NASA's Space Science program began with the main purpose of directly supporting the Apollo program and what was anticipated to follow: astronaut missions to Mars. After the Space Race had run its course, NASA's Space Science Program assumed a vigorous life of its own pursuing the recommendations of the National Academy's Space Studies Board (SSB). A dozen NASA Associate Administrator/Chief Scientists have been responsible for the Science Programs and they, in turn, have appointed engineers and scientists over the years to direct the Solar System Exploration Program. Some of the US Presidents and some of the NASA Administrators have demonstrated a serious interest in space exploration but, since the Apollo years and the absence of a serious challenge, US space endeavors have been a minor consideration in the life of most Americans. A handful of public enthusiasts along with the political muscle of a few key states (Texas, Florida, California, Alabama, Maryland, Virginia & Ohio) and of the aerospace industry have kept the US Space Program alive and, generally, flourishing.

* Associate Administrators: John Naugle, Noel Hinners, Burt Edelson, Len Fisk, Wes Huntress, Ed Weiler, Al Diaz, Mary Cleave, Alan Stern, John Grunsfeld, Geoffrey Yoder, Thomas Zurbuchen
* Chief Scientists: John Naugle, Frank McDonald, Noel Hinners, France Cordova, Kathie Olsen, Shannon Lucid, John Grunsfeld, James Garvin, Waleed Abdalati, Ellen Stofan, James Green
* Solar System Exploration Directors: Oran Nicks, Donald Hearth, Robert Kraemer, Thomas Young, Angelo Guastaferro, Jesse Moore, Geoffrey Briggs, Wesley Huntress, Jurgen Rahe, Colleen Hartman, James Green, Jim Watzin

When NASA was created early in the Cold War, the space program was a priority that was recognized throughout the nation so ample funding during the 1960s reflected that. A few years later, when the perceived Soviet challenge had diminished, funding had dropped to about 40% of the peak. Recovery from the Challenger disaster of 1986 led to a temporary bump in NASA's appropriation that then levelled out at about half of the peak. Funding for the Space Sciences has been sustained at about 20% of the total NASA budget; it was, however, actually larger than this because the cost of launch vehicles and the Deep Space Network were book-kept elsewhere. A major antagonist of spending on the US space program was Senator William Proxmire of Wisconsin, famous for his *Golden Fleece* awards presented each month from 1975 to 1988 and intended to embarrass government-funded projects that he considered self-serving and wasteful. This led to NASA ceasing to fund the radio telescope search for extra-terrestrial intelligence – ever since a privately funded effort led by the SETI Institute. It also led to an attempt to eliminate NASA's funding of meteorite geoscience – this was the author's baptism of fire when he first took up a position at NASA HQ in 1977. During the Administration of President Ronald Reagan (with James Beggs as Administrator and Hans Mark as his Deputy) the Solar System Exploration Program was placed under extreme – indeed existential – pressure (Chapter 8) but then grew modestly in the 1990s when Daniel Goldin was appointed Administrator. Most recently that budget has levelled off at $2.7B (now, unlike in the past, including the cost of launches and communications, a healthy amount by most past standards).

NASA Some General Fiscal Background

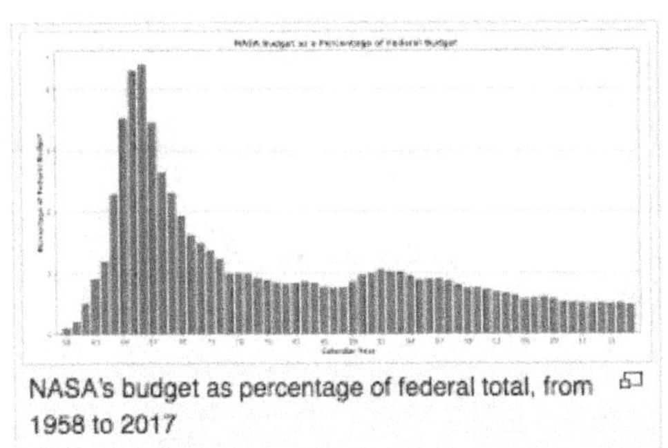
NASA's budget as percentage of federal total, from 1958 to 2017

No doubt, most are aware that NASA's budget at the time of Apollo was considerably larger than today. Perhaps few are aware of the contrast illustrated here. Total federal spending on space activity did not greatly diminish, however, as the military and intelligence agencies picked up the slack. There was in this later, some benefit to space astronomers as the great Hubble Space Telescope is derived from the KH-11 Kennen reconnaissance satellite.

NASA's budget peaked in 1964-66 when it consumed roughly 4% of all federal spending. Apollo was a top national priority and consumed more than half of NASA's budget. The workforce comprised more than 34,000 employees and 375,000 contractors from industry and academia. Today the number of employees at HQ and the 7 Centers is half as large: over 17,000. The budget of $22+B is about one half of a percent of the Federal Budget.

The budget for deep space crewed exploration – return to the Moon – in 2020 totals just over $6B and includes funding for the Orion crew vehicle, the Space Launch System, the Lunar Gateway orbital station and the Lunar Lander (see Chapter 17). In 2020 at the time of writing it is hard to anticipate how the Covid-19 catastrophe will impact future federal budgets including that of NASA.

There has been a major advance in launch vehicle technology – the ability to return launchers for refueling and re-use with major cost savings. This advance has come from the private sector – Elon Musk's SpaceX Company – to the unacknowledged embarrassment of NASA. How it will impact NASA's science programs is unclear. The giant rocket SLS that NASA is developing at great expense to return astronauts to the Moon is an expendable vehicle.

For comparison, the budget of ESA totals 6.68 billion euros in 2020. Of this, 6.645 million euros are for human and robotic exploration. The budget for Science Programs is 538 million euros.

The Planetary Sciences Community in the US

When the exploration of the Solar System began in the 1960s the size of the planetary sciences community in the US (and world in general) was quite small, consisting of some astronomers, meteorite geochemists, atmospheric scientists and a tiny number of astrobiologists (a term not yet invented). Mature since the 1970s, the planetary sciences are highly interdisciplinary. The overwhelming emphasis of NASA's initial effort was directed at the Moon but some visionary scientists could see a much bigger picture – to learn the histories of all the planets and the nature of the comets and asteroids toward establishing how the Solar System itself came into being and, of course, how life emerged in our Solar System (and perhaps others). It didn't take long for the planetary science community to grow. In anticipation of the Apollo landings there was a demand for geologists to plan the orbital mapping of the Moon, help select the best landing sites, train the astronauts, and analyze the mountain of data that would result. Seismologists and geophysicists were needed to develop instrumentation to deploy at the landing sites. Geochemists were needed with the right expertise to analyze the samples that would be returned by the astronauts. There were a number of centers of such expertise that were only too pleased to leap into action including the new Astrogeology Branch of the US Geological Survey in Flagstaff Arizona, meteorite geochemists and seismologists at Caltech in Pasadena and elsewhere. The NASA Johnson Space Center in Houston hired scientists to train the astronauts and to build a facility to receive, curate, distribute to the community and analyze the lunar samples.

While NASA's Solar System Exploration Program is mainly in the business of carrying out missions in deep space it also has built a major ground-based telescope for planetary science research – the Infrared Telescope Facility (IRTF) on the summit of Mauna Kea in Hawaii – located above most of the atmosphere's IR-absorbing water vapor. The telescope has a 3-meter mirror and was originally built to support the Voyager mission to the outer planets. The Voyagers launched in 1977 and the first light of the IRTF was two years later. "The IR spectral region, 1-1,000 microns, is uniquely suited for planetary research for two reasons. First, all planets absorb solar radiation and reemit most of this energy at infrared wavelengths. The earth, for example radiates thermally with a spectrum characteristic of a 300K blackbody, a spectrum that peaks in intensity at 10 microns… This reemission is precisely what makes infrared astronomy so difficult – the sky is bright in the infrared! Thus the planets are brightest at infrared wavelengths, making this wavelength range useful in the search for minor atmospheric constituents. [14]

Today, academic centers of expertise where most planetary scientists train include, in the US, a number of universities – Arizona, Arizona State, California Berkeley and California Los Angeles, Colorado, Washington, New Mexico State, Idaho, Washington U in St Louis, Brown, Hawaii, Caltech & Massachusetts Institute of Technology. Plus independent research

---

[14] Alan Tokunaga, *High Resolution Infrared Spectroscopy of Planets*, Astronomical Society of the Pacific, October 1983

institutes like the South West Research Institute and the Lunar & Planetary Institute both in Texas and the Planetary Science Institute in Arizona. Today planetary scientists are typically members (~1,500) of the Division of Planetary Science of the American Astronomical Society and/or of the American Geophysical Union (which includes Earth, atmospheric, oceanic, hydrologic, space and planetary scientists). In the US they work in academia, at JPL and NASA Centers, and at the private institutes.

Most of NASA's Space Science funding is spent in the aerospace industry and at NASA Centers; space scientists in academia competing to carry out space mission experiments and competing for NASA research grants don't find themselves especially privileged with respect to financial reward. Since the size of the community has the potential for exponential growth (as tenured faculty graduate new PhDs every few years) competition for peer-reviewed NASA grants is fierce. Funding requests for the planetary science disciplines – astronomy, atmospheres, geology, geophysics, geochemistry – mainly attempts to keep pace with inflation and to provide funds for the analysis of new data and the development of advances in instrumentation. However, it is worth remembering that the planetary science investigators (a few dozen for each mission) do get most of the public spotlight that results from the deep space missions made possible by thousands of largely anonymous project and industry personnel.

In the 1980s, NASA's Planetary Data System was created to archive and distribute the ever-growing data acquired by the deep space missions. That effort is divided among ten US centers by expertise (atmospheres, plasma, radio science, cartography and images etc.). In this way the long-term usability of the data is assured. Similarly, a program to develop new and improved instrumentation was initiated. Federal funding for the US National Science Foundation (NSF), the only federal agency that funds basic non-biomedical research and education across all fields of science and engineering, totals just $8.3B in FY2020. About 11,300 competitive awards were funded in 2019 with 366,000 individuals and students directly supported. Space science benefits from some of this as the NSF funds ground-based astronomy including optical telescopes in Arizona and Chile, the National Solar Observatory's telescopes, the National Radio Astronomy Observatory's telescopes and the giant (305 meter) Arecibo radio telescope in Puerto Rico (where NASA's Solar System Exploration Program funds planetary radar experiments). The Pentagon's Advanced Research Projects Agency had funded the Arecibo telescope's construction in the early 1960s as part of an effort to detect and intercept incoming Soviet missiles. NSF took over the facility in 1969. A catastrophic misfortune befell the Arecibo telescope in August 2020 when a support cable collapsed followed in November by another. In December the collapse was complete. Plans have been proposed for replacing the telescope but NSF's priorities and the funds required for a replacement are problems. [15]

Hurricanes are another concern: when Category 4 Hurricane *Maria* hit Puerto Rico in September 2017 it cut off electricity and communications island-wide and led to significant damage to some of the observatory's scientific instruments.

"So far, the only damage that's confirmed is that one of the line feeds on the antenna for one of the radar systems was lost," according to Administrator Nicolas White. "That part was suspended high above the telescope's main 1,000-foot dish, which lost some panels when it shook loose and fell down." [16]

The US Congress is notoriously indifferent to non-military, non-medical scientific research; perhaps *anti*-science (especially as it relates to the environment) would better describe the perspective of many Congressmen (and, alas, their constituents). So, in some ways it is surprising that Congress appropriates as much as it does to the space sciences. The larger part of the NASA budget supports its crewed element – launches to the International Space Station in low Earth orbit and plans for crewed missions beyond LEO; the latter requires massive technology investments including in the development of Saturn-equivalent launch vehicles while purchasing Soyuz launches to the ISS from Russia. Human spaceflight remains the part of NASA's program to which the public seems to most readily respond. In the UK as late as 2016 the participation of British astronaut Tim Peake aroused enormous public interest perhaps because a quarter century

---

[15] https://en.wikipedia.org/wiki/Arecibo_Observatory

[16] http://www.npr.org/sections/thetwo-way/2017/09/25/553594041/puerto-ricos-arecibo-radio-telescopesuffers-hurricane-damage

had passed since Helen Sharman had been the first Briton in space when she visited Russia's Mir space station in 1991 after 18 months of intensive training in Star City. Another British/US astronaut, Michael Foale, has also logged much time in space on both Shuttle flights and on Mir.

Science fiction movies set in space – recently *The Martian* and the latest episode of *Star Wars* – have been major hits with audiences worldwide whereas deep space exploration by robotic space science missions only occasionally finds much notice (the landing of ESA's Philae on Comet 67P and NASA's New Horizons' Pluto encounter in 2015 both did well in this respect). Science is therefore a lesser consideration for NASA than preparation for the eventual astronaut missions that it proposes to one day carry out in the inner Solar System – a return to the Moon and, it is repeatedly anticipated, exploring Mars.

Looking back, one can see that NASA's leaders remain caught up in the logic expounded by von Braun for how space exploration should proceed. In von Braun's visionary thinking – formulated in the 1930s, 40s, and 50s – the intelligence, versatility and oversight of the human explorer were central to the enterprise; capable robot spacecraft were entirely in the realm of science fiction. Logically, routine access to near-Earth space would be provided by a reusable launch vehicle. A space station in low Earth orbit would be the needed staging point for missions to the planets (recall Stanley Kubrick's 1968 epic *2001: A Space Odyssey*). Exploration of the Moon and Mars would follow after these capabilities had been put in place. What actually happened was the dramatic race to land humans on the Moon, after which, with a much reduced budget, there was an attempt by NASA to return to the von Braun scenario: scrap the expendable launch vehicles (*Saturn Vs, Atlases, Titans*), build a reusable launcher (*Shuttle*), build and operate a space station (*Freedom*) and then, maybe, return astronauts to the Moon or, instead, go on to Mars. In fact, by 2015 mankind's reconnaissance phase of the exploration of the Solar System had already been substantially realized – but by means of robot spacecraft of ever-greater capability, endurance and accomplishment. The reusable launch vehicle – the Space Shuttle – though brilliant in concept and an amazing technology, proved to be much more expensive to operate and more fragile than anticipated. The Russian space station *Mir* and the *International Space Station* have served to prove that long-duration zero-gravity human space flight can be carried out successfully in a relatively protected radiation environment (low Earth orbit) and, equally important, that large-scale international cooperation on space projects is possible and mutually beneficial.

As time has passed, in addition to the national space agencies and long-established aerospace giants, there are now, in the US, several new commercial players like SpaceX, Virgin Galactic, Blue Origin, Bigelow Aerospace, Vulcan Aerospace, and SpaceDev that have ambitions for participating in human spaceflight. SpaceX in particular has, in the last few years, demonstrated heavyweight capability including in 2020 the first US launch of astronauts to the ISS in many years; SpaceX and Blue Origin have demonstrated the capability of landing large rockets vertically for reuse. The capability of controlling smaller rocket landers had been demonstrated decades ago by the Soviet Union and by NASA in lunar applications including, of course, Apollo. The Viking landings on Mars date back to the 1970s. So, why did it take so long for the landing and reuse of launch vehicles? Perhaps launch vehicle manufacturers preferred the economic benefits of selling launchers for one use only.

CHAPTER 2

SPACE RACE TO THE MOON BEGINS

Luna Orbiters, Lunokhods, Rangers, Lunar Orbiters & Surveyors, Vostok, Voskhod, Mercury & Gemini

Apollo 1, 4, 5, 6, 7, 8, 9, 10

Following the early US-Soviet competition that took place during the International Geophysical Year in 1957, the stage was set for the next act – the race to land men on the Moon, our nearby planetary body that, surprisingly, had not received much in-depth scientific attention until this time. Given the moderately high-resolution images of the near side of the Moon that can be acquired by telescopes – something possible only using spacecraft elsewhere in the Solar System – the geologic history of the Moon was already being studied by some prescient scientists including at the US Geological Survey. But, of course, science was not the driver of the coming Space Race – that focused on launching humans to Earth orbit and beyond in order for the US and the Soviet Union to compete in demonstrating their global leadership. In 1961, President John F. Kennedy committed the US to the extraordinarily ambitious goal of landing a man on the Moon – and returning him safely – by the end of the decade. One wonders if the President really appreciated the enormous risk that the US was taking and the low probability that the goal could be achieved by the end of the decade. George Mueller, Associate Administrator for Manned Spaceflight, when he arrived at NASA in September 1963, commissioned an assessment of meeting the President's goal and was told the odds were one in ten. Consequently "Between 1963 and 1966 Mueller, drawing on his experience of US Airforce methods transformed NASA".[17]

James Webb, appointed by John Kennedy, led the management of NASA. Mr. Webb had an unusually broad background of experience as an attorney and businessman, also, he had served as Director of the Bureau of the Budget and as Undersecretary of State in the Truman administration. In addition, he also had served as president and vice president of several private firms and served on the board of directors of the McDonnell Aircraft Company. So, although he did not have a science or engineering background, he had a rare combination of experience and judgment that would see NASA succeed in the President's almost impossible assignment.

Administrator Webb was inspired to recruit George Mueller, a physics PhD and an electrical engineer who had worked at Bell Labs and later at TRW. While working on missile systems Mueller became convinced that all-up testing was essential as "you don't want to be testing piece-wise in space. You want to test the entire system because who knows which one's going to fail, and you'd better have it all together so that whatever fails, you have a reasonable chance of finding the real failure mode, not just the one you were looking for." [18]

---

[17] https://www.theguardian.com/science/2015/oct/21/george-mueller
[18] https://en.wikipedia.org/wiki/George_Mueller_(NASA)

It was this approach to the development and testing of the Saturn V rocket that saved enough time to allow the Apollo program to meet Kennedy's "end of the decade" pledge. George Mueller, in turn, recruited senior US Air Force Officer General Samuel Phillips along with a staff of officers under Phillips, and together took the necessary steps to make President Kennedy's challenge successful. Phillips became the Apollo Program Director, managing Apollo from January 1964, until it achieved the Apollo 11 landing in July 1969, after which he returned to Air Force duty. Another key senior manager at NASA was Robert Seamans who in 1960, joined NASA as Associate Administrator. In 1965, he became Deputy Administrator, retaining many of the general management-type responsibilities of the Associate Administrator and also serving as Acting Administrator. Under Webb's reorganization of NASA, the directors of the Manned Spacecraft Center (Robert Gilruth), Marshall Space Flight Center (Wernher von Braun), and the Launch Operations Center (Kurt Debus) reported to Mueller. Von Braun and Debus had come to the US via Operation Paperclip.

Bellcomm Inc.

A key Apollo management role was also played by Bellcomm, a small company that James Webb had brought in to "provide some technical capability that he perceived was not present in NASA."[19] Carved out of Bell Labs in 1962, the role of Bellcomm evolved under George Mueller to become his systems engineers with, quite unusually, "an in-line role" that "also created a little bit of difficulty in the Center understandings because they were not used to having a contractor working on the overall system design. But we finally got the concept understood, and it worked very well. ... One thing that Bellcomm did do, which they did superlatively, was to manage to cause system engineering to get done. Not necessarily to do it, although they did a good deal, but they caused the Centers to develop their own system engineering talent and to be sure that it was getting done and the proper set of trade-offs made and the proper understanding of what the design consequences were." [20]

Origin of the Moon

Richard Corfield in an August 2009 *Chemistry World* article describes the major controversy about the origin of the Moon that Apollo was to settle:

"In 1958, three years before President John F Kennedy would commit the US to landing a man on the Moon by 1970, Nobel laureate and Manhattan project veteran Harold Urey approached NASA associate administrator Homer Newell and suggested the scientific study of the Moon should be a major focus of the newly minted US space effort. Urey believed that the Moon's craters had been formed solely by meteorite bombardment (the so-called 'cold-Moon' theory). This was in contrast to the competing theory (spearheaded by Gerard Kuiper of Kuiper Belt fame) that the Moon, like its neighbor Earth, had once been intensely volcanically active. These 'hot-Mooners' believed that an ancient period of intense volcanism, now long ceased, had turned the Moon into an overcooked cinder, forming the Moon's craters in the process. In fact the two competing theories could be tested relatively easily. All that was necessary was to go to the Moon and perform some chemistry. William Hartmann was a young cosmochemist at the time of the first Moon landings. For much of his career he has been interested in the central questions that surround the Moon – what are the maria and the highlands, how old are they, how was the Moon formed, and perhaps most importantly, what can it still tell us. Hartmann is clear that all of the Apollo missions were scientifically important even if, as he says, the first two missions were really to establish the engineering and the safety of the program. They gave us access to the lava plains and the dating of the lava plains. In fact, most of the rocks that they picked up there were

---

[19] NASA History Office oral interview with Mueller
[20] NASA History Office oral interview Collins with Mueller

related to the emplacement of those lava plains.' Urey too knew that an early trip to the lava plains would prove his theory one way or the other. And so it proved. As Armstrong walked on the Moon he spotted something that looked like a vesicular rock – a rock likely to have been formed by volcanism. Armstrong was unsure however and Urey was relieved when Armstrong changed his mind. However, the first batch of samples returned from the Moon were delivered to the lunar receiving laboratory later in July 1969 and then investigations proceeded so quickly that by September 1969 NASA was able to announce the preliminary findings, and they held bad news for Urey. The rocks returned by Apollo 11 were unequivocally basalts, a rock that was well known to be formed volcanically. The cold Moon theory was disproved."

Corfield continues:

"A geologist at Bellcomm, Farouk El-Baz, was a key player in the geology of the Apollo missions and arguably the man most directly responsible for selecting the landing sites. He too is in no doubt as to the importance of the science of Apollo. 'It was Apollo 11 that answered that question [the hot Moon/cold Moon debate] early on. As soon as we got the samples from Apollo 11 and found out that it was basalt there was no question that it was from volcanism.' He recalls wryly that Urey was not at all happy with this turn of events. 'Harold still could not accept it though. He thought that perhaps the force [of impacting meteorites] had been so severe that they had penetrated the crust and released lava from within the Moon [thereby forming the maria]. So he did not give up right away. But to us, as geologists, it was confirmation that it was volcanic basalt, and we know how basalt forms.'"[21]

It was generally assumed that, given the huge investment involved, there would be a sizeable number of lunar landings rather than just one to satisfy President Kennedy's goal. As David Portree reports: "Many expected that the basic Apollo program would end with Apollo 20, the tenth Apollo landing mission, then advanced lunar missions would occur in the Apollo follow-on program President Lyndon Baines Johnson had set in motion in 1964. Dubbed the Apollo Applications Program (AAP) in 1966, many expected that it would include two-week lunar surface missions to increasingly complex and interesting landing sites. Scientist-astronauts would live in Apollo Lunar Module-derived "camper" landers, wear advanced hard-shell space suits, pilot one and two-man lunar flying vehicles, drive long-distance traverse rovers, and use sophisticated geological tools such as deep drills to get at the moon's many secrets."[22] In fact, of course, Apollo 17 in 1972 was the last of the manned lunar missions.

The US science community of geologists and meteoriticists quickly organized itself to support the Apollo program. As it happened, Gene Shoemaker (1928-1997) was a visionary who had been preparing himself to be the first to make a geologic map of the Moon and, thereby, unlock its history. He developed the methodology in the early 1960s.

USGS Astrogeology Branch

Shoemaker applied the basic superposition principle of geologic mapping (younger rocks lie on top of, or intrude into, older rocks) to determine the relative ages of the different lunar terrains. Later, when higher resolution images had been obtained from lunar orbit, much more detailed relative age dating became possible by measuring the areal density of impact craters, of all sizes, on each unit. A long-time resident of Flagstaff, Arizona, in 1961 Shoemaker brought into being the Branch of Astrogeology within the U.S. Geological Survey and he founded the USGS Field Center in Flagstaff in 1963. Using telescopic images the geologic mapping of the Moon began with great effect and provided the basis for the Lunar Orbiter and Surveyor analyses upon which the selection of the Apollo landing sites was carried out.

---

[21] https://www.rsc.org/images/Moon_tcm18-158679.pdf?dom=prime&src=syn
[22] https://www.wired.com/2013/09/ending-apollo-1968/

In 1960 a second branch of the USGS – the Branch of Military Geology – published the first lunar geology map. Arnold Mason and Robert Hackman (1923-1980) were the authors of this pioneering documentation of the 3 principal lunar terrain types: the relatively young post-mare craters, intermediate-age maria, and, the oldest terrain – the highlands.

The USGS Astrogeology Branch was created to support the planning of the Apollo program and, as now seems inevitable, the subsequent exploration of the terrestrial planets and the moons of the outer planets. Shoemaker and his colleagues at the Branch have been central figures in the planning and execution of NASA's Solar System Exploration program from its beginning: the names of Hal Masursky (1922-1990), Jack McCauley (1932-2012), Don Wilhelms, Danny Milton, Michael Carr (a Yorkshire-man by the way), Larry Soderblom, Ray Batson (1931-2013), Hugh Keiffer and Mike Duke (all much admired) immediately come to mind along with Jack Schmidt of Apollo 17 fame. And, of course, the Branch has been deeply involved in the analysis of the returned data and production of the maps that record the geologic history of each body.

Meteor Crater in Arizona

In the early 1960s, as noted earlier, there was still disagreement about the nature of the lunar craters that some argued were volcanic in nature. Certainly, the dark, smooth maria are indeed the solidified floods of basaltic lava. The astronomer Gerard Kuiper, Gene Shoemaker and scientist Robert Dietz (1914-1995) came together to support the lunar geologic mapping effort and argued for the asteroid impact hypothesis which is indeed, we now know, how the basins and craters on the Moon formed. The 1.25 km diameter Barringer Crater (also called Meteor Crater) near Flagstaff where the USGS Astrogeology Branch is located provided a perfect example of the type of feature that is created when a bolide (in this case 30-50 meters across) crashes into a terrestrial planet.

Shocked quartz identified in the Barringer Crater by Shoemaker demonstrated that craters like this, on Earth were formed by impacts because volcanoes could not generate the pressure required. The smaller lunar craters have just the same appearance as Barringer. To understand the details of the crater-forming process, including the ejecta, NASA built a Vertical Gun Range (still operational) at its Ames Research Center (Principal Investigator Pete Schultz of Brown University) that can fire marble-sized pebbles at 25,000 km/hour into targets and record the results with high-speed photography. The Earth may well have still been molten at the time that it suffered the cataclysmic collision that led to the formation of the Moon – this being the most popular theory regarding how the Moon came to be. Al Cameron (1925-2005) of Harvard University was a notable advocate of this theory. In such a collision with a Mars-sized body (that even has a name: *Theia*) a substantial part of the Earth's crust and mantle would have been torn away, subsequently to re-accrete in orbit about the Earth forming a body lacking a sizeable metallic core like Earth's.

First US Spacecraft to Escape Earth

Important developments in robotic exploration necessarily preceded the crewed lunar missions. The US Air Force was responsible for a series of Pioneer Able missions that were a first attempt to achieve Earth escape velocity. Beginning in 1958, they were launched on Thor-Able and Atlas-Able rockets with, with one exception, a notable lack of success: seven Pioneer launches were failures because of launch vehicle problems of one kind or another. Only Pioneer 5, launched on 11 March 1960 got properly underway and made useful measurements of the interplanetary medium.

Deep Space Communications

NASA-JPL had not yet built its Deep Space Network of parabolic-dish antennas so the UK's Jodrell Bank Observatory received most of Pioneer 5's data. After JPL was transferred from the US Army to NASA the Lab was given the responsibility for creating the DSN communications system for all US deep space missions. The first dishes built at Goldstone in California, near Woomera in Australia and near Johannesburg in South Africa were 26 meters in diameter. Later, stations were built near Madrid and near Canberra. Together these dishes could cover the whole sky. The first 64-meter dish was built at Goldstone in 1966. "About the size of a 20-story office building, the dish has been in service for 48 years. Some parts of the 70-meter antenna, including the transmitters that send commands to various spacecraft, are 40 years old and increasingly unreliable. The Deep Space Network (DSN) upgrades are planned to start now (in 2020) that Voyager 2 has returned to normal operations, after accidentally overdrawing its power supply and automatically turning off its science instruments in January." [23]

Pre-Apollo Missions

On the US side the pre-Apollo robotic missions that paved the way were a series of Ranger Impactors, Lunar Orbiter and Surveyor Lander missions. On the Soviet side the preparation missions were much more ambitious; they included Lunakhod rovers and Luna sample return as well as orbiters. There were several impressive Soviet "firsts": the Luna 3 flyby in October 1959 returned the first images of the far side of the Moon, Luna 9 achieved the first soft landing, Luna 10 in 1966 was the first lunar orbiter, Luna 17 and 21 carried the Lunakhod rovers to the Moon where Lunokhod 2 operated for about 4 months, traveled a remarkable 37 km and returned 80,000 images and, most impressive of all, Luna 16 in 1970, 20 in 1972 and 24 in 1976 collected and returned surface samples to Earth. They were a brilliant achievement and a model for the kind of robotic sample return missions – from Mars, the lunar far side and asteroids – that will provide uniquely valuable science return. The 101 gm, 50 gm and 170 gm samples were acquired from depths of up two meters at three locations – Mare Fecunditatis, a highland site between Mare Crisium and Mare Fecunditatis, and Mare Crisium.

The Luna-10 orbiter tracking data indicated that the lunar gravitational field was very 'rough'. Two years later, in 1968, Paul Muller and William Sjogren of JPL analyzed the tracking data from NASA's Lunar Orbiters and identified the consistent correlation between very large positive gravity anomalies and depressed circular basins on the Moon; they had discovered the Moon's mass concentrations: 'mascons'. The lunar mascons are of sufficient size that they greatly influence the stability of spacecraft in low-altitude orbits: the sub-satellite released by Apollo 16 had an expected lifetime of a 500 days but survived only 35 days before crashing; the definitive mapping of the Moon's gravity field would have to await decades for NASA's GRAIL mission. The work of Muller and Sjogren allowed mission planners to make allowance for this gravitational distortion in their analyses of spacecraft trajectories to chosen landing sites. The new understanding of the lunar gravitational field had major consequences for the Apollo program because it reduced the size of the landing error ellipse by a factor of ten from 2 km to 200 meters – obviously important for safety. The Apollo 12 crew was thereby able to land just 163 meters from their target, the Surveyor 3 spacecraft.

The Soviet effort to land cosmonauts on the Moon ahead of the US began with a series of six orbital flights carrying a single crew (similar to NASA's Mercury program) as part of the Vostok program between 1961 and 1963. This was followed over the next two years by the Voshkod flights with three- and two-man crews (similar to NASA's Gemini program). Development of a rocket capable of sending cosmonauts all the way to the surface of the Moon – the 3-stage N1 rocket – had begun in 1959.

---

[23] https://www.jpl.nasa.gov/news/news.php?feature=7611

Death of Sergei Korolev

The death of Sergei Korolev in 1966 was a significant setback and, given that the NASA effort remained on track, the failure of four N1 test launches (made known to the world only in 1989) effectively ended the Space Race to the Moon. Thereafter the Soviet manned spaceflight program was limited to increasingly advanced developments of Earth orbital space stations: first the Salyut program carried out between 1971 and 1986. Subsequently, the Mir space station was assembled in low Earth orbit and operated by crews of three from 1986 to 2001. Valeri Polyakov spent 438 days in Mir between 1994 and 1995 – a record.

NASA's Johnson Space Flight Center

NASA's Manned Space Flight Center in Houston, now named for President Lyndon Johnson had, and still has, the lead role in human space flight on the US side. While the robot precursor missions were being planned and executed by other NASA Centers, JSC was learning how to support astronauts in space and also survive their fiery re-entry (an area of NASA Ames Research Center's special expertise) and parachute descent to an ocean landing, beginning with the suborbital pathfinder flight of Alan Shepard (1923-1998) and then a series of orbital flights. The first Mercury flights, each with a single astronaut (one of the famous Mercury Seven), were carried out in 1961. The initial home of Project Mercury was NASA's Langley Research Center in Hampton, Virginia. The Mercury Seven achieved considerable fame: Scott Carpenter (1925-2013), Gordon Cooper (1927-2004), John Glenn (1921-2017), Gus Grissom (1926-1967), Wally Schirra (1923-2007), Alan Shepard (1923-1998) and Deke Slayton (1924-1993).

The Mercury spacecraft, built by McDonnell Aircraft and controlled from Cape Canaveral (from now on referred to as just 'the Cape'), were quite basic in design and had supplies to support the astronaut for only one day. Launched by Redstone and Atlas missiles they returned to Earth using retrorockets to de-orbit, an ablative heat shield to slow down and, finally, a parachute to splash them into the Atlantic for recovery by helicopter: a hair-raising ride!

Twelve two-man (hence the name Gemini) flights followed in 1965 and 1966. Michael Collins, who was one of them, describes the crew selection process in amusing detail in his book *Carrying the Fire* written after he returned from the hyper-historic Apollo 11 flight. They were launched on Titan II rockets, and were significantly more ambitious and capable. McDonnell Douglas again built the spacecraft. They were the first to be controlled from NASA JSC's new Mission Control Center in Houston. The goal was to demonstrate the ability to keep a crew alive and well in space for at least 8 days – the time it would take to carry a crew to the Moon and back. Gemini, critically, also served to demonstrate the orbital maneuvers involved in rendezvous and docking (this would be required by Apollo in lunar orbit) and the ability for one of the crew to carry out a 'space-walk' outside their craft (first carried out by Gemini 4's Edward White).

Gemini 3 was the first to carry astronauts. Gemini 4 included White's first spacewalk. Gemini 5 maintained the crew in orbit for more than a week. Gemini 6 and 7 missions demonstrated orbital rendezvous. Gemini 8 conducted the first docking of two spacecraft in orbit – with an Agena upper stage. This mission also served to demonstrate the ability to recover from the first critical in-space system failure of a U.S. spacecraft – one that threatened the lives of astronauts Neil Armstrong (1930-2012) and David Scott – both making their first space flights. After undocking from the Agena, the Gemini spacecraft starting tumbling end-over-end rapidly – the result of one of their maneuvering system thrusters being stuck on. This thruster system was shut down and the re-entry system thrusters used to regain control. The mission was aborted. Re-entry took place over China and landing in the Pacific to the east of Okinawa.

Although disaster on this flight was avoided, there would be one Gemini-related fatal disaster – but it was not directly related to the hazards of space flight. On 28 February 1966 two astronauts – Elliot See (1927-1966) and Charles Bassett (1931-1966) – who were scheduled to fly the Gemini 9 mission crashed their T-38 jet in bad weather at Lambert Field in St. Louis, Missouri. This tragic, chance event would turn out to be the reason that, later, Buzz Aldrin would be assigned a flight on Gemini and, subsequently, be one of the first two men to land and walk on the Moon.

(L-R) Young, Collins

The Gemini 10 mission that launched on a Titan II on July 18 1966 with John Young, making his second spaceflight, and Michael Collins making his first – his later flight would, of course, be as the Command Module Pilot for Apollo 11. After that historic flight Collins wrote a book about the whole experience of training as an astronaut and the two flights he made. The book – *Carrying the Fire* (Farrar, Straus, Giroux 1974, still in print) – is full of minute-by-minute colorful detail from launch to recovery and a must read.

The goal of the Gemini 10 mission was to rendezvous with an Agena target vehicle (launched from an adjacent pad just before the crew's) and, also, to carry out a "spacewalk' EVA. They were successful in all of this and also demonstrated a second successful rendezvous with the derelict Agena left over from the aborted Gemini 8 flight. Collins' tethered EVA used a nitrogen-gas-powered maneuvering gun that took him to the dormant Agena from which he retrieved a cosmic dust-collecting panel.

The crew of Gemini 11 – Pete Conrad (1930-1999) and Richard Gordon – has the distinction of carrying out the only artificial gravity experiment in space. Their Gemini capsule was tethered to a spent Agena rocket that had been launched separately. If they could have maintained tension in the tether, they would have been able to generate artificial gravity by rotating the combined vehicles. It had been expected that the Earth's gravity gradient would maintain the necessary tension but that proved not to be the case. Firing their side thrusters, the crew was able to slowly rotate the tethered pair and this way they did generate 0.00015 g!

The three-day Gemini 12 mission with the crew of James Lovell and Buzz Aldrin marked the successful conclusion of the manned Gemini flights. That launch took place on 11 November 1966. Aldrin undertook three EVAs, lasting a total of 5½ hours, that served to demonstrate that astronauts can work effectively outside of their spacecraft and thereby marked an important milestone toward the Apollo goal.

NASA JSC subsequently led both the science training of the Apollo astronauts and the science mission operations carried out on the lunar surface. As Michael Collins describes in his book, the training included jungle and desert survival training to guard against the possibility of a badly off-nominal recovery of the returning Gemini spacecraft. Starting in 1963, the USGS Astrogeology Research Program played an essential role in the astronaut training and in supporting the testing of equipment for both manned and unmanned missions. Harrison (Jack) Schmitt, USGS geologist and Apollo 17 astronaut, is still the only scientist to have set foot on the Moon.

Saturn Launch Vehicles

Simultaneously with the Mercury and Gemini flights, the Marshall Space Flight Center led the development of the Saturn I, IB and V launch vehicles. The Saturn launchers – ten Saturn Is, nine Saturn IBs and thirteen Saturn Vs – were the foundation of the Apollo Program and, even today, are the most powerful launch vehicles ever built. Indeed, the five huge first stage F-1 engines of the Saturn V need to be seen in person to be believed.

Boeing built the Saturn V first stage, North American Aviation the second stage and Douglas Aircraft Company the third stage. IBM led the creation of the Saturn's digital computer. The integrated, inline, three-stage design of the Saturn V has in recent decades evolved into launcher designs that are somewhat simpler: a liquid-fueled core stage together with a number of solid (or more recently liquid) fuel boosters that can be dropped a few minutes into the flight and subsequently recovered for refurbishment and reuse. These boosters are effectively the first stage of the rocket; everyone will remember how this worked out for the Space Shuttle Challenger on one freezing morning in January 1986.

The Saturn rockets were amazingly reliable – which is just as well since, according to *The Space Review* (3 April 2006), "If the Saturn 5 exploded, it could do so with the force of a small atomic bomb, the equivalent of half a kiloton, or about

1/26 the size of the bomb that destroyed Hiroshima. Naturally, this was a significant concern for Apollo program officials".[24] Those watching the launches were unaware.

Soviet Lunar Exploration Plans

The launch failures of the comparably large Soviet N1 must have been both amazing and depressing to witness. An elegant, tapering design whose development began under Sergei Korolev, the N1 had thirty Nk-15 engines for its first stage in contrast to the five F-1 engine Saturn V approach. The first test launch was carried out in February 1969 and the last in November 1972. None were successful. While the N1 failures effectively ended the astronaut space race for the Soviets, the competition did not entirely end there. The Soviets carried out a series of robotic lunar missions – orbiters and mobile landers – that were intended to climax with the robotic return of samples ahead of the Apollo landings. With their Proton launch of the Luna 15 spacecraft on 13 July 1969 they came close to succeeding as is described in the detailed online site of Andrew LePage and summarized in Chapter 4. [25]

The Soviet plans for their cosmonaut missions to the Moon were not made public but Wikipedia has an entry *Soviet crewed lunar programs* that is, no doubt, reliable. "The Soviet government publicly denied participating in such a competition, but secretly pursued two programs in the 1960s: crewed lunar flyby missions using Soyuz7K-L1 (Zond) spacecraft launched with the Proton-K rocket, and a crewed lunar landing Soyuz spacecraft launched with the N1 rocket." [26]

K (Lunniy korabl – "lunar craft" or lunar ship).

Soyuz 7K-LOK rendering.

---

[24] http://www.thespacereview.com/article/591/1
[25] https://www.drewexmachina.com/2019/07/19/luna-15-thesoviet-unions-last-lunar-gamble/
[26] https://en.wikipedia.org/wiki/Soviet_crewed_lunar_programs

"The crewed landing plan adopted a similar method to the single launch and lunar orbit rendezvous of the Apollo project. For mission safety, weeks before the crewed mission, an LK-R uncrewed L3 complex and two Lunokhod automated rovers would be sent to the Moon to work as radio beacons for crewed LK, with the LK-R used as a reserve escape craft. The Lunokhods were also equipped with manual controls for the cosmonauts, both for transfer to LK-R in necessity and for regular research. The N1 rocket would then carry the L3 Moon expedition complex, with two spacecraft (LOK and LK) and two (Block G and Block D) boosters. A variant of the Soyuz craft, the "Lunniy Orbitalny Korabl" (LOK) command module, would carry two men, with three modules like the regular Soyuz 7K-OK, but was heavier by a few tons. The 7K-OK was half the mass of the three-crew Apollo orbital command ship. The "Lunniy Korabl" (LK) accommodated only one cosmonaut, so in the Soviet plan, only one cosmonaut would land on the Moon. The mass of the LK was 40% of the mass of the Apollo lunar lander."

"During the L3 complex's journey to the Moon, there would be no need to undock and re-dock the orbital and landing craft as was done in Apollo, because the cosmonaut would transfer from the LOK to LK by a 'spacewalk'. On the Apollo missions, the transfer was done using an internal passage."

"On the Moon, the cosmonaut would take Moon walks, use Lunokhods, collect rocks, and plant the Soviet flag. After a few hours on the lunar surface, the LK's engine would fire again using its landing structure as a launch pad, as with Apollo. To save weight, the engine used for landing would blast the LK back to lunar orbit for an automated docking with the LOK. The cosmonaut then would spacewalk back to the LOK carrying rock samples. The LK would then be cast off, after which the LOK would fire its rocket for the return to Earth." [27]

The US-Soviet Space Race to the Moon could have had a quite different outcome, instead of the one that has in some ways shaped the world.

Evolution of the NASA Field Centers

In 1959 the US Department of Defense had transferred personnel and its Missile Firing Laboratory at Florida's Cape Canaveral to Alabama's Marshall Space Flight Center so that MSFC initially had responsibility for NASA's launch operations. In November 1963 Lyndon Johnson, who had succeeded to the Presidency after the assassination of John Kennedy, made this Florida facility a new NASA Center and named it the Kennedy Space Center with continuing responsibility for assembly of launch vehicles, the mating of payloads to launch vehicles and the associated launch operations.

While all this was going on, the Jet Propulsion Laboratory and NASA's Langley Research Center developed and carried out a series of Ranger, Lunar Orbiter and Surveyor missions to identify and certify landing sites that were both of scientific priority and safe. Even though USGS lunar geology maps of the near side were available, much higher resolution images were needed to assess safety. There were also uncertainties about the bearing strength of the lunar regolith and how far into the regolith the astronauts would sink. The first two Atlas-Agena launches of JPL's Ranger were tests of both the launch vehicle and of the spacecraft system. There were failures, the first serving to demonstrate the inadequacy of the spacecraft's stabilization system. Navigation, instrumentation and other problems bedeviled the next four attempts casting some doubt on JPL's capabilities.

Rangers 7, 8 and 9 finally achieved the intended purpose of these early lunar missions. Ranger 8 is notable in that it was targeted to Mare Tranquillitatis and in February 1965 returned 800 images of the region in which Apollo 11 landed in 1969. This baptism of fire for the JPL team is noted in the Center's records as "after a long trouble-plagued start that taught the system engineers a great deal and the scientists virtually nothing, Project Ranger finished with three flights that greatly advanced the lunar scientists' knowledge of the surface and whetted their appetites for a closer look." [28]

---

[27] https://en.wikipedia.org/wiki/Soviet_crewed_lunar_programs
[28] https://solarsystem.nasa.gov/resources/2287/nasa-facts-rangers-and-surveyors-to-the-moon/

Surveyor Lunar Landers

The Surveyor program of instrumented lunar soft-landers was undertaken at about the same time as Ranger. It was managed by JPL with the Hughes Aircraft Company as the spacecraft system contractor. The project required a more powerful upper stage (the liquid-fueled Centaur) on top of the Atlas launcher; the lander itself was a tripod design featuring

three steerable engines. They were launched between May 1966 and January 1968. The first mission in June 1966 was a major success, landing in Oceanus Procellarum and acquiring panoramic coverage of the site as well as very high-resolution images of a landing pad in the lunar soil. There were, however, still important lessons to be learned as two other Surveyors were failures. But four more performed just as planned. Surveyor 1, 3, 5, and 6, returned data from selected mare sites while Surveyor 7 provided data from a rugged highland region. Two years after it landed in the Ocean of Storms, Surveyor 3 was visited in November 1969 by the Apollo 12 crew: Charles Conrad (1930-1999) and Alan Bean (1932-2018). In addition to sending back thousands of images, the Surveyors also returned data on the surface bearing strength, temperatures, and radar reflectivity as well as digging trenches with a surface sampler. Several small parts of the Surveyor 3 lander were returned by the crew for study of the long-term exposure effects of the harsh lunar environment.

Surveyor 6 carried out a unique 'hop' that took it 2.5 meters away and thereby provided views of the surface disturbances that had been produced by the initial landing including the effects of firing rocket engines. Surveyor 7 had the luxury of being targeted to a site selected for purely science interest – the rim of Tycho Crater – rather than as a potential Apollo landing site. The landing took place at a rugged, rock-strewn ejecta blanket near the crater after which it operated for two lunar days. In addition to acquiring a wide variety of lunar surface data Surveyor 7 also obtained pictures of Earth, performed star surveys and observed the solar corona.

A third program of robotic lunar missions was underway in support of Apollo, namely the Lunar Orbiter program managed by the Langley Research Center in Norfolk, Virginia. The missions took place in 1966 & 1967 and by mapping the entire lunar surface (including the far side) with a resolution of 60 meters or better, provided NASA with the means to greatly improve their geologic maps, finalize selection of the Apollo landing sites and provide the science community with the data needed to support Apollo sample analysis in unravelling the history of our Moon. All were launched on Atlas Agenas. The first three missions were inserted into low inclination orbits with the goal of mapping potential landing sites.

The last two missions were focused on broader scientific objectives and were inserted into high-altitude polar orbits from which the entire Moon could be mapped. Unlike later robotic spacecraft that were equipped with vidicon or charge-coupled device (CCD) image sensors, the Lunar Orbiter cameras (supplied by Eastman Kodak) used film that was processed onboard, scanned and the images transmitted back to Earth. Image motion compensation was achieved by moving the film during each exposure! Remarkably, all of the 5 Lunar Orbiter missions were successful and represent a singular achievement by the management team at the NASA Langley Research Center (led by James Martin (1920-2002) and Tom Young – two of the most capable managers NASA was fortunate to employ) – and by the Boeing-Eastman Kodak team that built and operated the spacecraft.

This success no doubt accounts for the fact that NASA Langley was entrusted with the management of the first US Mars lander mission – Viking – some years later. Tom Young would later be appointed Director of NASA's Solar System Exploration Program and would invite the author to be his Deputy. Jim Martin in retirement would be a key advisor to the Program Directors (including the author).

Soviet Robotic Lunar Missions

Given the intense competitiveness of the Lunar Space Race, the Soviets carried out an intensive series of robotic lunar missions – more than 40 in all – that were intended to support their own cosmonaut ambitions. About one third were successful. The cosmonaut missions were, inevitably, dependent on the development of a Saturn V-class launcher – in which they were not successful. The robotic missions included impactors, flybys, orbiters, soft landers, rovers and sample return. Two of the 3 Lunokhod rover missions were successful: Luna 17 – the first – in November 1970 and then Luna 21 in January 1973. Lunokhod 1 traveled 10.5 km in 322 days and returned more than 20,000 images. Lunokhod 2 traveled 42 km in about four months.[29]

The three successful sample return missions (out of 11 attempts) took place beginning at the same time as the Apollo 11 mission and are described in the following chapter. If they had not been overshadowed by the Apollo landings, these sample return missions would have received the wider recognition they deserved.

Popular Science Fiction of the Era

At about this time, in 1966, the science fiction television series *Star Trek* was aired in the US on the NBC network. Movie versions followed and Star Trek has become a cult phenomenon. The valiant crew of the star-ship *Enterprise* journeys not among the planets of our Solar System but among the stars of the galaxy. The continuing role of a human crew was, of course, intrinsic to the show and to its audience – consistent with the plans being developed by NASA for Apollo. Indeed, the show made NASA's plans look very unambitious – though at least a beginning in the right direction. Advanced, aggressive extra-terrestrial life was another basic ingredient of the show. In 1968 Stanley Kubrick's movie *2001: A Space Odyssey* was released just a year before Apollo 11. Unlike the Star Trek shows, the 1968 book by Arthur C. Clarke on which the movie was based was a serious extrapolation of technical capabilities that were under development including even artificial intelligence. In the movie, hibernation of most of the crew was a feature of the long journey out to Jupiter; this has in fact been studied by the European Space Agency.[30] Many considered the movie to be forecasting where NASA was doubtless heading.

Even more popular than the *Star Trek* phenomenon were the *Star Wars* movies that were launched in 1977 five years after the Apollo program ended. They are still going strong four decades later. The movies again feature an intrepid crew and fanciful technology – and, doubtless, have shaped how the US public anticipates space exploration will eventually proceed. There will be great public disappointment (as well as among scientists) if we find no evidence of life elsewhere in our Solar System. More generally, the essential role of astronauts in the exploration of our Solar System is surely implanted in the minds of the US public.

Apollo Program Ambitions

In planning the Apollo program NASA had the continuing support of a small multidisciplinary company – Bellcomm Inc – that, as discussed above, had been carved out of Bell Labs in 1962 to provide systems engineering and analyses for NASA's Office of Manned Space Flight. In 1968 in advance of the first Apollo missions Bellcomm's technical staff led by Noel Hinners proposed an ambitious program of Apollo missions that would continue through 1976 – four more years than actually took place. The first five Phase 1 missions would have been much like those that were, in fact, carried out. Four Phase 2 missions would have been carried out in 1972 and 1973 with an increase in the landed science payload and with three days on the lunar surface with six moonwalks at each site. Intriguingly, the development of a one-man lunar flying unit with a range of 5 to 10 km was part of the plan. Phase 3 would consist of a single 28-day lunar-orbital survey mission in 1974 and would

---

[29] https://en.wikipedia.org/wiki/Luna_programme#Sample_return
[30] http://www.esa.int/Enabling_Support/Space_Engineering_Technology/Hibernating_astronauts_would_need_smaller_spacecraft

commence a series of Apollo Applications missions in 1975-76. Two very ambitious Apollo missions would follow to complete the program: these were 'dual launch' missions requiring two Saturn V rockets! A crew in lunar orbit would remotely pilot an unmanned cargo lander (carrying a Moon rover, a spare advanced space suit, a 100-foot core drill, a transportable 10-foot core drill, life support consumables, and a geophysical station with 10-year life) to a chosen landing site to which a second crew would descend and spend two weeks carrying out an in-depth exploration of the site.

Of course, the Apollo missions were in fact ended after 1972 but some of the thinking described may carry over into a future return to the Moon by astronauts. Bellcomm returned to its parent company Bell Labs in 1972 as funding for the Apollo program was phased out. The author was a part of a Bellcomm team (Principal Investigator Bill Thompson) that was supporting the first Mars orbiter and lander missions (Chapter 7). When the Bell Labs contract was ending he was invited to join the staff at JPL. From 1972 to 1974 Noel Hinners (1935-2014), a senior Bellcomm alumnus, served with distinction as NASA's Director of Lunar Programs (years later a part of the Solar System Exploration Office) and subsequently, from 1974 to 1979, he was NASA's Associate Administrator for Space Science. Yet another Bellcomm alumnus, Bill Piotrowski, would serve as manager of the Magellan Venus radar-mapping mission in the HQ Program Office.

Apollo Landing Site Selection

The selection of the Apollo landing sites was carried out in a rigorous fashion (albeit with some difficulty in reaching a consensus) by a team of scientists, engineers and managers. They identified a list of sites representative of the major provinces of the Moon – the dark maria and the older cratered highlands – and where the stay-time and mobility of the astronauts would suffice to make observations and collect samples that would serve to discriminate between a range of theories about the origin and evolution of the Moon. The choice of sites was also intended to allow characterization of typical lunar features such as fresh impact craters, their central peaks, and sinuous rilles (small volcanic features that somewhat resemble terrestrial river valleys).

Bellcomm geologist Farouk El-Baz played a leading role in narrowing down the selection of sites and in the science training of the astronauts. Indeed, Richard Corfield comments "Farouk El-Baz was a key player in the geology of the Apollo missions and arguably the man most directly responsible for selecting the landing sites". [31]

The Apollo Site Selection Board recommended ten different sites but, in the event, only six were reached before President Nixon's Administration terminated the Apollo program. The first H-class missions relied on the crew being able to reach specific locations near the touchdown point simply by hiking to them. The later J-class missions provided the crew with a battery-powered rover that greatly extended their reach.

The landing sites were all on the Moon's near side where real-time communications could be maintained with the crew – involving a balance between scientific exploration priorities and practical considerations. To provide absolute ages (rather than just the already-known relative ages) of the principal lunar strata, rock and soil samples were needed. To minimize the prospects of a crash-landing, the selected sites needed to be of acceptable roughness – avoiding hills, escarpments, craters, boulders, and steep slopes. Also, the selected sites had to be resilient to the implications of possible launch delays and navigation errors. Another consideration involved the preferably wide distribution of the seismometers contained in the Apollo Lunar Surface Experiments Packages (ALSEP) that would tell us about the internal structure of the Moon.

A Site Selection Board identified the major questions that the Apollo landings were to address – first: the structure of the lunar interior, second: the composition and structure of the surface of the Moon and the processes modifying the surface, and third: the history or evolutionary sequence of events by which the Moon has arrived at its present configuration. In turn, these led to a dozen of more detailed questions. The Board had the difficult but enviable task of recommending an optimal set of landing sites that would allow field observations, measurements of heat flow, the collection of rock samples and emplacement of long-lived geophysical instrumentation. The landed Apollo science effort

---

[31] https://www.rsc.org/images/Moon_tcm18-158679.pdf?dom=prime&src=syn

was carried out through pre-planning and by real-time communications between astronauts on the lunar surface and geologists at Mission Control in Houston. On three missions the Lunar Rover provided the crew with mobility to carry out long-range directed geologic analyses (imaging and sampling). The vehicle was battery-powered and had a range of 50 miles. The crews deployed long-lived geophysics stations (ALSEPs) to measure lunar seismicity. They used drills successfully on two missions to acquire samples and measure heat flow.

With six landings in three years the value of human participation in the scientific exploration of the Moon was established even though only one of the dozen astronauts was a professional scientist. All of the Apollo landing sites are on the near-side of the Moon so that real-time communication between the science team in Houston and the lunar astronauts was assured. Looking far ahead: any astronaut missions to any destination beyond the Moon in deep space would have to deal with communication time delays of tens of minutes so that a crew would surely need to include more than one trained scientist.

The alarm in the US that was caused by the early successes of the Soviet space program had created an opportunity for Von Braun to leapfrog over his original development plan and lead the Saturn launcher development for the Apollo program. This, however, short-circuited the systematic architectural approach that Von Braun had envisioned – with consequences that are now playing out. Nevertheless, NASA has, for the last four decades, attempted to implement a Von Braun-type development of space access infrastructure – beginning with a reusable space plane and a space station – but, in the case of the Space Shuttle, has now found this to be overly challenging and is faced with recovering Saturn V equivalent launch capability it had back decades ago.

Apollo Mission Architecture

Given the performance challenge of achieving escape velocity from Earth coupled with the need to enter and leave the Moon's gravity field, considerable thought went into the architecture of the Apollo missions. The most direct approach would be one in which a three-man crew would leave Earth in a spacecraft that would then land on the Moon and subsequently take off and return to Earth. The launch vehicle performance needed to carry this out would have required more than one Saturn V launch and was clearly not ideal. Tom Dolan, an engineer at Vought Astronautics proposed a single Saturn V mission approach. Earlier, two Canadians, Jim Chamberlin (1915-1981) and Owen Maynard (1924-2000), had studied this approach; they had been part of the cancelled (in 1959) Avro Arrow (a delta-winged jet interceptor) design team and were now working at NASA's Langley Research Center in Virginia. Adopted by Apollo the approach involved a rendezvous in lunar orbit.

A *command and service module* (CSM) supporting three astronauts would make the journey from Earth coupled with a *lunar excursion module* (LEM). The LEM would comprise both a descent stage and an ascent stage. The CSM would remain in lunar orbit piloted by one astronaut while the two-man lunar excursion module would separate and descend to the surface. Later the landed crew would ascend back to lunar orbit and rendezvous with the orbiting CSM (a capability demonstrated by the Gemini missions); the ascent module would then be abandoned and the CSM would return back to Earth. There the blunt cone crew capsule (command module) would separate and enter the Earth's atmosphere behind a very capable ablative heat shield composed of phenolic epoxy resin. The compelling logic (in terms of launch energy

requirement) of this somewhat complex approach was accepted; John Houbolt (1919-2014) at Langley led the team that fully developed it.

Given the extreme risks that manned spaceflight involves it is perhaps remarkable that, besides the death of astronauts Elliot See and Charles Bassett in their crashed jet, only three other astronauts – Gus Grissom (1926-1967), Ed White (1930-1967) and Roger Chaffee (1935-1967) – were killed during the Mercury, Gemini and Apollo programs. Their deaths came about on 27 January 1967 as a result of a fire in their Apollo capsule while being tested at the Cape. They were in the command module, mounted on an un-fueled Saturn IB rocket as if ready for launch. The Investigating Review Board was not able to determine conclusively the specific initiator of the fire but it did identify the conditions that led to the disaster so that major modifications in design, materials, and procedures were implemented.

This was a tragedy that shook the nation but, given the intensity of the competition with the Soviets during the Cold War, it did not put a brake on the effort. The test was labeled Apollo 1; manned flights were then suspended while the cause of the accident was investigated and changes made to the spacecraft and safety procedures. James Webb (1906-1992) reported the investigation findings to Congress and chose to step down as Administrator; he was presented with the Presidential Medal of Freedom. Astronauts James Lovell, Jack Swigert and Fred Haise also came close to losing their lives on their amazing Apollo 13 mission (Chapter 4). During the Soviet lunar planning stage Yuri Gagarin (1934-1968) and other cosmonauts also lost their lives. In April 1967 Vladimir Komarov (1927-1967) became the first man to die on a space mission when the parachute on his Soyuz 1 failed on its landing.

On 13 June 1971 three Russian cosmonauts were found dead in their space capsule after it landed in Kazakhstan: Georgi Dobrovolsky, Vladislav Volkov and Viktor Patsayev were dead in their seats on the Soyuz 11, apparently the result of cabin depressurization during preparations for re-entry. Of course, many more lives were lost in the Challenger (January 28, 1986) and Columbia (February 1, 2003) disasters – the Challenger crew: Francis Scobee, Michael Smith, Ronald McNair, Ellison Onizuka, Judith Resnik, Gregory Jarvis, and Christa McAuliffe, and the Columbia crew: Rick Husband, William McCool, Michael Anderson, Kalpana Chawia, David Brown, Laurel Clark and Ilan Ramon.

Tests of an escape system for the Apollo crews were carried out and a series of unmanned test flights of the Saturn IB took place. The first flight of the Saturn V – designated Apollo 4 – took place on 9 November 1967 and successfully demonstrated the restart of the third stage as well as testing the re-entry heat shield at the velocity that would be encountered on returning from the Moon. A Saturn IB launched the Apollo 5 mission that tested the Lunar Module. Apollo 6 in April 1968 was a second test flight of the Saturn V and encountered some launch vehicle problems that were considered sufficiently understood and fixed so that the Saturn V was now declared 'man-rated'. Apollo 7 in October 1968 was the first manned Apollo flight and the first manned flight of the Saturn IB – in effect taking on the intended Apollo 1 mission: 11 days in Earth-orbit to check out the redesigned Command Service Module.

The first humans to visit the Moon were the crews of the Apollo 8 and 10 missions in preparation for the landing missions. The three-stage Saturn V took the crew to Earth orbit where they spent a few hours before a 12-minute burn of the third stage placed them on a trajectory to the Moon (after which the stage was jettisoned).

Apollo 8

The Apollo 8 crew was made up of Frank Borman, James Lovell and William Anders. They spent 3 days traveling to the Moon where they braked (involving a 4-minute burn of the Service Propulsion System (SPS) – "the longest four minutes of their lives") into a 12-degree inclination orbit and then spent the next 20 hours (famously, on Christmas Eve 1968) making ten orbits at an altitude of 110 km. This time in orbit was spent in taking photographs of the Moon – including the far side – and with special emphasis on Mare Tranquillitatis. Following an even more critical burn of the SPS they returned to Earth on 27 December, splashing down in the Pacific Ocean south of Hawaii.

Though Apollo 8 did not involve a landing, it was a series of historic – and heroic –*firsts*– the first crewed spacecraft to leave low Earth orbit and the first to reach the Moon, orbit it, and return. The crew was the first humans to fly to the Moon, to witness an Earthrise, and to escape the gravity of a celestial body. Amazing!

Apollo 9

Left to right: McDivitt, Scott, Schweickart

Apollo 9 had to carry out yet more testing of the system before a commitment was made to the first lunar landing. The crew (James McDivitt, David Scott and Rusty Schweickart) spent 10 days in low Earth orbit conducting a range of essential tests: of the Lunar Module engines, the life support and navigation systems, and the docking maneuvers required to un-dock and re-dock the lunar lander with the command vehicle. McDivitt and Schweickart also test-flew the Lunar Module (LM), practicing separation and docking maneuvers and using the descent stage engine to fly the LM 180 km from the Command Module before jettisoning the descent stage and using the Lunar Module's ascent stage to return.

Apollo 10

The Apollo 10 crew – Thomas Stafford, John Young and Eugene Cernan (1934-2017) – carried out the final Apollo test flight, "a full-dress-rehearsal", before the landing missions began. Every system was exercised as it would be on a landing mission including flying the Lunar Module down to 15 km from the lunar surface! After Apollo 10's trans-lunar injection by the third stage of the Saturn V the command/service module was released from its location at the top of the stacked units and turned around to dock at the top of the Lunar Module. At this point the third stage was jettisoned. Injection into lunar orbit took place using the engine of the Command Module. Stafford and Cernan were the crew of the Lunar Module while Young remained in the Command Module. The Lunar Module descended toward the surface but the crew was aware that the ascent module was less-than-fully fueled lest they be tempted to carry out a landing! Violent rolls that the crew had accidentally initiated marred the return of the ascent module; as a result, the mission came close to disaster. Happily, the remainder of the time in space passed without incident and the crew splashed down in the Pacific Ocean east of American Samoa. Another amazing adventure!

Apollo Quarantine

NASA has always taken very seriously and rigorously the protection of Planet Earth from possible hazards that might be the result of returning material from the Moon or other planets – Mars in particular. The following description of the Apollo quarantine effort is instructive in terms of what might be needed to support the return of samples from Mars. Indeed, because it will be necessary to protect Martian samples from terrestrial contamination as well as to protect the terrestrial environment, a Mars Sample Receiving Laboratory will be much more challenging to design and operate especially now that we live in the post Covid-19 era.

The construction of a Lunar Receiving Laboratory (LRL) at NASA JSC began in 1966. It was completed in 1968 in good time to support the first return of astronauts and samples from the Moon. In addition, a mobile quarantine facility was built to house and transport the Apollo crewmen from the recovery ship to the LRL. This facility was located on the hangar deck of the recovery ship and could house up to six people (the crew, a crew surgeon and recovery engineer) for ten days and, subsequently, support them at the Lunar Receiving Laboratory. Quarantine was assured by maintaining negative internal pressure and by filtration of effluent air. On the basis that most terrestrial disease agents lead to evident symptoms within 21 days after exposure of the host, plans were made to quarantine the crew for at least 21 days after their return. Shades of Covid-19!

In addition to the assessment of the crew's health, the returned lunar materials were also assessed to ensure that they could be safely released for scientific analysis. The *Lunar Receiving Laboratory* at NASA JSC was of impressive size ($7,700m^2$) and included a crew reception area, a vacuum laboratory and sample laboratories together with support functions. Airflow into the sample-handling areas was controlled, liquid waste was sterilized and 'contaminated' air incinerated. The primary biological barrier consisted of the vacuum complex and Class III biological cabinets. A secondary barrier was achieved by maintaining the areas at negative pressure with respect to the atmospheric pressure external to the building. Biological and physical/chemical testing of the lunar samples was performed within biological cabinets – gastight enclosures through which all manipulations were performed using neoprene gloves.

Following the experience of dealing with the crews and lunar materials returned by Apollo 11 through Apollo 14 the quarantine requirements were relaxed so that the only concern became the protection of the samples from terrestrial contamination. Meteoriticist Michael Duke (ex-Caltech & USGS) became the curator of the Lunar Receiving Lab after the aborted Apollo 13 mission when there was still much to do in terms of organizing the characterization, documentation and security of the lunar samples that were being made available to the science community for analysis.

"Only at Ames and JSC did NASA build lunar receiving facilities to analyze soil samples returned from the Moon. JSC would identify and isolate hazardous materials in the samples; Ames would explore the essential composition of the lunar materials. So Ames built a very clean laboratory and outfitted it with unique equipment. They observed the carbon chemistry of the samples, and concluded that they did not contain life. This led them to question what kind of carbon chemistry happens in the absence of life. They discovered that the Moon was being constantly bombarded with solar wind and micrometeorites, which left the Moon with a carbon chemistry dominated by the energetic interaction of the Sun, the Moon and cosmic debris." [32]

As discussed in the following chapter, the analyses of the lunar samples have, as expected, greatly illuminated our understanding of the evolutionary history of the Moon. The unique experience and capability of the NASA JSC sample curatorial facility continues to provide the science community with an essential service in support of the continuing collections of extra-terrestrial materials. These take the form of 1) meteorites harvested annually in the Antarctic (a collaborative effort organized by Case Western Reserve University and supported by NASA, The National Science Foundation and the Smithsonian) and 2) cosmic dust collected by NASA aircraft flying in the stratosphere. As of 2020 it still appears that there will be a long wait before samples are returned from Mars. Samples collected by astronauts from the region of the lunar South Pole and from the Moon's far side may well be the next major source of planetary material to benefit from the unique NASA JSC facility.

---

[32] https://www.nasa.gov/ames/apollo

CHAPTER 3

APOLLO LANDING MISSIONS & LUNA SAMPLE RETURNS

Apollo 11, 12, 13, 14, 15, 16, 17

Luna 15, 16, 20, 23, 24

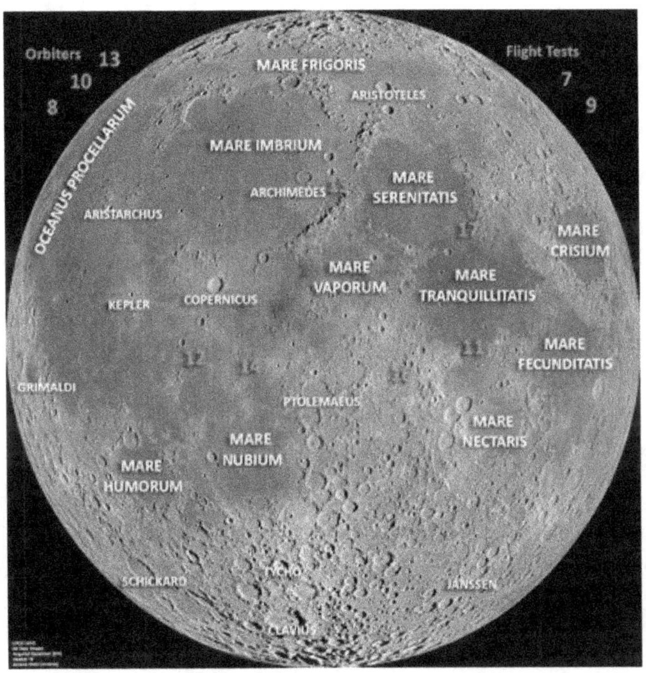

Apollo 11 Mission Commander Neil Armstrong (1930-2012), Command Module Pilot Michael Collins and Lunar Module Pilot Buzz Aldrin

Neil Armstrong and Buzz Aldrin, in a never-to-be-forgotten moment in history, carried out the first manned lunar landing on 20 July 1969; their journey to the Moon had taken three days. The Earth return vehicle – the command and service spacecraft – remained in orbit, crewed by Michael Collins. The landing took place on the south west side of Mare Tranquillitatis, providing a representative sample of the basaltic lava that fills most of the basin. One of the Ranger impactor spacecraft had acquired images of Mare Tranquillitatis in February 1965 and Surveyor 5 had landed in the basin in September 1967. Also, the crews of Apollo 8 and 10 had acquired useful photographs from an altitude of 110 km. This

first crewed landing on the Moon required the utmost coolness on the part of the crew because, nearing the surface, the target area was observed to be strewn with boulders. With Aldrin calling out altitude and velocity, Armstrong redirected the LEM to a safe landing (20:17:40 UTC on Sunday 20 July 1969) but with only 25 seconds of fuel in reserve.

The crew spent only a little over two hours outside the lunar lander collecting samples, amounting to 21 kg. Armstrong's description of the samples provided the first discrimination between different theories about the nature of the Moon. *New Yorker* correspondent Henry Cooper (1933-2016) provided insightful annual "Letters from the Space Center" and reported Harold Urey's reactions as Neil Armstrong described the samples he picked up. Professor Urey had been an advocate of the lunar maria being the remains of oceans that had long since been lost; he was therefore disappointed to hear Armstrong speak of vesicles in the samples.

In addition to collecting samples during their relatively brief time on the lunar surface (slightly less than a day) Armstrong and Aldrin carried out simple mechanics experiments on the soil – whose bearing capability had been somewhat uncertain in spite of the previous Surveyor landings. Sampling tubes could easily penetrate 10 to 20 cm but required hammering to reach 70 cm. Another simple experiment was carried out to collect a sample of the solar wind by exposing a sheet of aluminum foil on a pole facing the Sun. The astronaut pair deployed a passive seismometer that radioed back data for three weeks, not sufficient to get a handle on lunar seismicity but a useful learning experience.

The Apollo 11 Laser Ranging Retro-reflector was the first of three units that would also be deployed by Apollo 14 and 15. The corner cube mirrors were subsequently illuminated by laser beams from the McDonald Observatory in Texas (and elsewhere) providing a measure of distance to the mirror accurate to about 3 cm. Such precision has improved our knowledge of the Moon's orbit and the rate at which is receding from Earth. Measurements of variations in the rotation of the Moon have led to information about the distribution of mass inside the Moon and an estimation of the size of a small core.

Michael Collins has recorded his participation – and interactions with Armstrong, Aldrin and Mission Control – in great detail in his wonderful book *Carrying the Fire*. This is the book to read for anyone primarily interested in the processes, personalities, complexities & excitement of human spaceflight in the 1960s.

The Apollo 11 mission accomplished the goal that had been announced by John Kennedy. As it turned out, by the time that the first lunar landing took place Richard Nixon was President – a somewhat ironic situation as he was the antithesis of John Kennedy and not eager to burnish the memory of the candidate who defeated him in the November 1960 election. The lunar landings would end with Apollo 17 and the Apollo Applications Program that Lyndon Johnson had supported would be forgotten.

Luna 15

The Luna 15 sample return mission was launched on 13 July 1969 three days ahead of Apollo 11 and on 17 July entered a 100 km altitude orbit about the Moon. Navigation difficulties were encountered because of the lumpy nature of the Moon's gravitational field so that a series of adjustments were made to its orbit before it was able to make a successful landing in Mare Crisium. "During its 54th orbit of the Moon, Luna 15 jettisoned its nearly empty outboard propellant tanks and started its de-orbit burn at 15:46 GMT on July 21. Neil Armstrong and Buzz Aldrin had already completed their historic first moonwalk 10 hours earlier and were preparing Eagle for liftoff in two hours". "At 15:50 GMT, contact with Luna 15 was lost just four minutes into its expected six minute descent with a radar altimeter reading of three kilometers. According to the team at England's Jodrell Bank, which had been monitoring the Luna 15 mission, the spacecraft had crashed into the lunar surface at a speed of about 130 meters per second. The crash site was at 12° N, 60° E about 800 kilometers east of the Apollo 11 landing site".[33]

Apollo 12 Mission Commander Pete Conrad (1930-1999), Command Module Pilot Richard Gordon and Lunar Module Pilot Alan Bean

Apollo 12 was launched on November 14, 1969 toward a landing site in Oceanus Procellarum previously visited by Surveyor 3 – the largest of the lunar maria – more than 2,500 km across – on the western side of the Moon's near side. A thick flat layer of magma had flooded this giant impact basin in the distant past. The landing site was toward the south-eastern edge of Oceanus Procellarum and to the south of the prominent Copernicus crater.

The launch was particularly exciting as lightning struck the Saturn V 36.5 seconds after lift-off, "discharging down to the Earth through the ionized exhaust plume. Protective circuits on the fuel cells in the service module (SM) detected overloads and took all three fuel cells offline, along with much of the command and service module (CSM) instrumentation".[34]

There was a second strike at 52 seconds! The Saturn V nevertheless continued to fly normally "but the loss of all three fuel cells put the CSM entirely on batteries, which were unable to maintain normal 75-ampere launch loads on the 28-volt DC bus. These power supply problems lit nearly every warning light on the control panel and caused much of the instrumentation to malfunction"

Alan Bean, the Lunar Module Pilot "put the fuel cells back on line, and with telemetry restored, the launch continued successfully. Once in Earth parking orbit, the crew carefully checked out their spacecraft before re-igniting the S-IVB third stage for trans- lunar injection . The lightning strikes had caused no serious permanent damage."[35]

In spite of the trajectory-distorting effects of the lumpy lunar gravity field, touch-down was precise, only 200 meters from where the Surveyor had landed in April 1967 and provided the Project with confidence that subsequent missions could be targeted to scientifically interesting sites in relatively rough terrain.

The crew was out on the surface within 3 hours of landing and their first EVA (extravehicular activity) lasted 4 hours during which they visited the nearby Surveyor (where Alan Bean photographed Pete Conrad), collected samples from nearby including a core tube sample, deployed an aluminum foil solar wind collector, and also their ALSEP: magnetometer, passive seismometer, solar wind spectrometer, dust collector, and supra-thermal ion detector. The crew carried out a second ~4-hour EVA whose main focus was on carrying out a geology traverse to acquire documented samples of rock, soil, and cores from five nearby craters.

Apollo 13 Lunar Module Pilot Fred Haise, Commander James Lovell and Command Module Pilot Jack Swigert (1931-1982)

The aborted Apollo 13 mission is almost as famous as Apollo 11 owing to the nail-biting performance of the crew in surviving the crippling explosion of an oxygen tank in the Service Module – the spacecraft element that provided their primary

---

[33] https://www.drewexmachina.com/2019/07/19/luna-15-the-soviet-unions-last-lunar-gamble/
[34] https://en.wikipedia.org/wiki/Apollo_12
[35] ibid.

propulsion and maneuvering capability, together with most of the crew's consumables. The explosion was in the #1 oxygen tank. The service module bay #4 cover was blown off, all oxygen stores were lost along with water, electrical power, and use of the propulsion system. After the explosion, the crew was confined to the Lunar Module with limited power, loss of cabin heat, shortage of potable water and accumulating carbon dioxide in their cabin. So, there was a critical need to adapt the Command Module's cartridges for the carbon dioxide scrubber system to work in the LM. Happily, the crew and mission controllers were successful in improvising a solution.

After launch on 11 April 1970 the crew returned to Earth six days later having swung by the far side of the Moon at an altitude of 250 km. This remarkable human drama served to remind the public audience in the US – an audience that, tellingly, had already become slightly blasé on account of the earlier successes – of the almost superhuman nature of these missions. The Apollo 13 movie serves as a quite accurate documentary of what happened to Lovell and his crew. One can only wonder about the different history that might have followed if the Apollo 11 crew had had a similar near-catastrophic experience.

Little noticed but an important indication of the potential of space exploration to become an international endeavor, Premier Aleksei N. Kosygin sent a message to the US government saying: "I want to inform you the Soviet Government has given orders to all citizens and members of the armed forces to use all necessary means to render assistance in the rescue of the American (Apollo 13) astronauts".[36] During the Apollo 13 re-entry, Soviet ships in the Pacific had been alerted to aid in the recovery of the astronauts if needed. The Soviet government ordered all Soviet radio transmitters using frequencies close to those of Apollo 13 to maintain silence from the time of spacecraft's entry into the earth's atmosphere to its splashdown.

Indeed, offers of assistance with ships to aid in the recovery came from many nations. The Associated Press quoted the Russian news agency Tass as saying that four Soviet ships were steaming toward the splashdown area, one of them the Chumikan, a missile tracker equipped with a helicopter. Tass said the Chumikan and fishing trawler No. 8452 were ordered to join the cargo carriers Academician Rykachev and Novopolotsk converging on the Pacific target area.[37] The space age has produced an exemplary degree of cooperation between nations.

The intended landing site had been chosen to explore the Fra Mauro formation, a site only about 170 km from where Apollo 12 had landed five months earlier but a first highland site. The 80 km Fra Mauro crater is found there. It was expected that the crew would be able to bring back samples of material that had been ejected from the Imbrium basin to the south. The next Apollo mission subsequently achieved the objectives.

In spite of the abandonment of a landing on the Moon by the astronauts, Apollo 13 did literally make an impact: the spent third stage (S-IVB) of the Saturn launcher that was used to propel the docked Apollo Command Module and Lunar Module from Earth orbit into its lunar trajectory separated from the Command Module and impacted the Moon on 14 April 1970; the rayed crater located north of Mare Cognitum is about 30 meters in diameter. Only the seismometer at Apollo 12 was available to record the impact that occurred at 135 km from that seismic station.

---

[36] https://www.quora.com/Did-Russia-offer-help-to-America-to-rescue-the-Apollo-13-crew
[37] https://er.jsc.nasa.gov/seh/pg13.htm

Luna 16

Luna 16 was the second Soviet attempt to return samples from the Moon and was the first of three successes – a remarkable achievement that went largely unnoticed in the West. Launched on a Proton from the Baikonur on 12 September 1970, Luna 16 landed in Mare Fecunditatis and using a drill on the end of a sampling arm it collected a 38 cm core sample from a few tens of cm deep. The arm then served to place the 101 gm sample into a return capsule on top of the lander. The lander's upper stage then lifted off and returned to Earth on a direct trajectory.

The Soviet Luna 16 probe landed in a field in Kazakhstan after its voyage to the Moon, on September 26, 1970.

Apollo 14 Lunar Module Pilot Edgar Mitchell (1930-2016), Commander Alan Shepard1923-1998) and Command Module Pilot Stuart Roosa (1933-1994)

The Apollo 14 crew inherited their target landing site and mission goals from the earlier attempt by Apollo 13: to investigate the Fra Maura formation which, based on telescopic mapping, was evidently formed as ejecta from the Imbrium Basin event – the impact of a large asteroid or comet – way to the north. This ejecta was widely distributed across the lunar nearside and acts as a convenient stratigraphic marker. Also, it was expected that the ejecta might include material excavated by the Imbrium impact from kilometers depth. Specifically, the landing site was near a small (370 meters diameter) fresh crater, Cone Crater, the source of recent ejecta and thus a source of material from depth. Data from Apollo 14 would serve to date the Imbrium event to ~4.25 billion years ago.

Launch took place on 31 January 1971. The crew stayed on the Moon for over 30 hours and carried out 2 EVAs that together lasted for over 9 hours and traversed 3.5 km in reaching 13 designated locations. Various experiments were carried out including the deployment and later collection of a solar wind composition foil. Also: deployment of the ALSEP Seismic Experiment and a Laser Ranging Reflectometer. Apollo 14 was the last of the two-day "H" missions after which the crews would stay on the Moon for longer and would be equipped with a Lunar Rover.

Apollo 15 Mission Commander David Scott, Command Module Pilot Alfred Worden and Lunar Module Pilot James Irwin (1930-1991)

Apollo 15 was the first mission – a "J" mission – to carry a lunar roving vehicle that could carry the 2-person crew, their equipment and science payload for greatly extended geological exploration: 26 km in all during 3 seven-hour EVAs. Launch took place on July 26, 1971. The landing site in the Hadley Apennine region was chosen to lie on the eastern margin of the Imbrium Basin (centered ~650 km to the northwest) where it was expected that samples from the basin rim could be acquired that would have originated deeper in the crust than those collected by the preceding mission. The Apennine Mountains themselves are a consequence of the Imbrium Basin impact, now known to have taken place ~3.85 billion years ago. The basalts that fill Mare Imbrium formed 500 million years later.

Apollo 15 was successful in returning a rock sample that predates the Imbrium event – the crew collected a rock at Spur Crater on Mount Hadley Delta that proved to be an anorthosite (an igneous rock) and has come to be known as Genesis Rock because it originated at a very early stage in the Moon's history when the lunar surface was still largely molten; as it gradually cooled, plagioclase-rich anorthosites (mainly sodium-aluminum & calcium-aluminum silicates) would have floated on the surface rather like icebergs.

Hadley Rille is a prominent and unusual lunar feature: a deep sinuous channel that winds along the western margin of the Apennine Mountains. The Apollo 15 observations indicate that the 135 km long, 1.5 km wide channel probably originated as a lava tube whose roof later collapsed. Volcanic features of similar origin also occur on Earth but are not as large.

The completion of the Apollo Program was now in sight. Given the Apollo 13 experience there was a move to end the Apollo missions on the high note provided by Apollo 14 and 15. President Nixon needed to be persuaded otherwise by OMB Deputy Director Caspar "Cap" Weinberger but he did authorize only two more missions.

Luna 20

The second successful Soviet sample return mission, Luna 20 was launched on 14 February 1972. Using the same sampling approach as Luna 16 the spacecraft acquired its 55 gm core sample that had a mineral composition similar to those returned by the Apollo 16 crew.

Apollo 16 Command Module Pilot Thomas Mattingly, Mission Commander John Young (1930-2018) and Lunar Module Pilot Charles Duke

Apollo 16, the second J-class mission, was targeted to land in the ancient lunar highlands to sample two geologic units (the Descartes Formation and the Cayley Formation) both of which are widespread on the nearside of the Moon, covering about 11% of the surface. The Descartes landing site is in a highlands region of the Moon's southeast quadrant, characterized by hilly, grooved, furrowed terrain. It was selected as a location for sampling two volcanic constructional units of the highlands – the Cayley formation and the Kant Plateau.

Launched on 16 April 1972 the Apollo 16 landing site was on the Cayley Plains, a unit that fills in topographic lows throughout this region. The rugged uplands east and south of the landing site are the Descartes Mountains. The units were incorrectly expected on the basis of telescopic observations to be of volcanic origin – formed from magmas more viscous than those that form the basin-filling maria. Analysis of the Apollo 16 samples has, however, shown that the rocks are in fact composed of broken fragments of minerals – breccias – produced by asteroid impacts. Remote sensing has its limitations. These Apollo 16 samples provided the biggest scientific surprise of the entire Apollo program.

Young and Duke spent almost three days on the lunar surface, during which they conducted three EVAs totaling 20 hours. They drove the Lunar Roving Vehicle 26.7 km and collected 96 kg of samples. The first EVA activities lasted 7 hours and included deploying the ALSEP experiments package. The geologic traverse to the region west of the landing site covered 4.2 km with stops at two field stations. Back at the lander, they deployed the Solar Wind Composition Experiment. Also during this EVA, 19 small explosive charges were fired using a "thumper," which was part of the Active Seismic Experiment.

The second EVA comprised a geological traverse south of the landing site to the Cinco Crater area of Stone Mountain. The traverse covered 11 km, included stops at six field stations and lasted over 7 hours. The third EVA covered another 11 km but was reduced to 5 hours 40 minutes, concentrating the work at three field stations. In all the Apollo 16 crew collected 731 individual rock and soil samples, including a deep drill core that included material from 2.2 meters below the Moon's surface.

Between the Apollo 16 and 17 missions, beginning on 2 August 1972, the Sun was very active and led to a weeklong storm – "legendary at NASA" of high-energy protons that would have been potentially lethal to unprotected astronauts.[38] Critically, the eruptions that took place repeatedly for over a week and produced 'a record-setting fusillade of solar proton radiation' that happened in the gap between missions. This experience – quite separate from the hazard presented by galactic cosmic rays – serves as a serious warning to those making plans for future human missions into deep space.

Apollo 17 Lunar Module Pilot Harrison Schmitt, Mission Commander Eugene Cernan and Command Module Pilot Ronald Evans (1933-1990)

The Apollo 17 spacecraft was launched from the Cape just after midnight on 7 December 1972 – perhaps the most spectacular launch ever (the cover). The Serenitatis basin landing site was selected as a location where rocks both older and younger than those previously returned from other Apollo missions, might be found. Prior reconnaissance had observed cinder cones and steep-walled valleys with large boulders at their base so that the crew had the opportunity to sample both young volcanic rock excavated from depth as well as older highland material. Since this was to be the last Apollo mission some other attractive sites had to be passed over – e.g., near the central peaks of Copernicus crater or in the southern highlands near Tycho crater.

The Apollo 17 Lunar Module landed in a deep narrow valley called Taurus-Littrow that is located in the mountainous highlands that form the eastern rim of the Serenitatis basin. The site is in a dark deposit between massifs of the southwestern Taurus Mountains and south of the crater Littrow. Rocks obtained from walls of the valley were expected to provide samples of ancient highland material and allow the age of the Serenitatis impact event to be determined. Cernan and Schmitt were on the lunar surface for 72 hours during which time they collected the richest sample of rocks from any of the Apollo landing sites.

The crew conducted three EVAs that totaled 22 hours. They used the Lunar Rover to conduct traverses totaling 36 km and collected samples at 22 locations in the Taurus-Littrow Valley. They deployed and/or performed 10 science experiments. These included three explosive packages to carry out a seismic profiling experiment. The crew collected 741 individual rock and soil samples, including a deep drill core that included material from 3 meters below the lunar surface, with a total mass of 111 kg. The sample collection included beads of chemically distinctive orange glass (the "orange soil") and black devitrified glass. These silicate glasses are rich in titanium, magnesium and iron and appear to have been formed by volcanic processes deep in the Moon's interior.

The striking photograph below shows Schmitt stands next to a large boulder during EVA-3. E.W. Wolfe of the USGS reports:

Boulder 1 at Station 2 is one of three boulders sampled by Apollo 17 at the base of the South Massif, which rises 2.3 km above the floor of a linear valley interpreted as a graben formed by deformation related to the southern Serenitatis impact.

---

[38] https://www.nasa.gov/mission_pages/stereo/news/stereo_astronauts.html

The boulders probably rolled from the upper part of the massif after emplacement of the light mantle. Orbital gravity data and photo-geologic reinterpretation suggest that the Apollo 17 area is located approximately on the third ring of the southern Serenitatis basin, approximately 1.25 times larger than the analogous but fresher Orientale basin structure. The massif exposures are interpreted to represent the upper part of thick ejecta deposited by the southern Serenitatis impact near the rim of the transient cavity. Basin ring structure and the radial grabens that give the massifs definition were imposed on this ejecta at a slightly later stage in the basin-forming process. There is no clear-cut compositional, textural, or photo-geologic evidence that Imbrium ejecta was collected at the Apollo 17 site.[39]

Eugene Cernan, who had been a member of the Apollo 10 mission that had tested the Lunar Module, was the last person to set foot on the Moon. He and Schmitt blasted off the Moon on 14 December, rejoined Ronald Evans in orbit to transfer their samples and equipment to the Command Module and then perform the burn to return them to Earth. The ascent stage of the Lunar Module was deliberately crashed onto the Moon to create a signal for the array of seismometers emplaced by the several Apollo landings. Cernan, Schmitt and Evans splashed down in the Atlantic on 19 December 1972.

Apollo 17 also carried out an orbital radar mapping experiment to study the Moon's surface and subsurface, revealing structures beneath the surface in both Mare Crisium and Mare Serenetatis and helped understand the formation of long low ridges that are found in many of the maria by motions along faults. As described in later chapters, orbital radar experiments have provided a powerful means of mapping cloud-shrouded Venus and of probing the subsurface of Mars in search of water ice. In the case of Mars another form of EM sounding is being developed to achieve much deeper penetration in search of a possible hydrosphere.

A side note: Jack Schmitt later resigned from NASA to seek election as a Republican to the United States Senate representing New Mexico in the 1976 election. He served one term during which he was the chairman of the Science, Technology, and Space Subcommittee of the US Senate Committee on Commerce.

Luna 23 and 24

Both Luna 23 and 24 were sent to Mare Crisium. Luna 23 was damaged during its landing on November 6, 1974, and failed to collect any samples, though it did return data for three days. The last successful Soviet sample return missions was Luna 24 launched almost 4 years after the Apollo 17 mission on 9 August 1976. "According to Barsukov (1977) the drill penetrated into the regolith 225 cm on an angle with a vertical depth of ~ 200 cm. However, the actual length of the returned core was 160 cm, with total weight of 170 grams. Samples were exchanged with the US, Great Britain, India and other countries".[40]

Summary of Post-Apollo Analyses

After six highly successful manned missions the value of human participation in the scientific exploration of the Moon – mostly under real-time communication with geologists in Houston – had been established, albeit at great expense and risk to the crews. Only one professional scientist had been among the dozen astronauts who had walked on the Moon but many

---

[39] https://pubs.er.usgs.gov/publication/70001244
[40] https://curator.jsc.nasa.gov/lunar/lsc/luna24core.pdf

had been involved in the selection of landing sites and in planning the EVAs. The science return from the missions would, however, not be achieved until the returned samples had been subject to years of analyses in many specialized laboratories and, also, after the analysis of data returned from the ALSEP experiments. Scientists were glutted with data of all kinds from the many robotic orbital and lander missions as well as the Apollo missions. Analysis of the 2,200 returned samples (382 kg total) has preoccupied them for many years. Having established that the lunar samples are sterile and present no back-contamination threat, the lunar samples were (and remain) at the Johnson Space Center under rigorous isolation; they are kept in glove boxes that are maintained at slightly elevated pressure to avoid any possible terrestrial contamination and from there are released for analysis by the science community – ~400 samples are distributed annually.

The analyses of the samples and other data returned from the Apollo missions have been carried out over many years. Results were presented and debated at the annual Lunar and Planetary Science Conferences that are held at the Johnson Space Center. The results appeared in the traditional science journals and were also described in books and a series of excellent articles – *Annals of Space* – that Henry Cooper (1933-2016) wrote for the *New Yorker*. Cooper's excellent coverage extended to many other aspects of space exploration.

Among the many scientists involved in the lunar sample analyses were Caltech's Gerry Wasserburg (1927-2016), Bob Walker of the Washington University in St Louis, Jim Arnold (1923-2012) of the University of California San Diego and Paul Werner Gast (1930-1973) who was chief scientist of the Apollo Lunar Science Staff. Professor Wasserburg created his famous *Lunatic Asylum* at Caltech where he pioneered high-precision, high-sensitivity isotopic analyses of meteorites, lunar and terrestrial samples. He and his colleagues were major contributors in establishing a chronology for the Moon and proposed the hypothesis of the *Late Heavy Bombardment* of the whole inner Solar System at ~4.0 billion years ago (half a billion years after meteorite ages have established the origin of the Solar System at 4.565 billion years). The Late Heavy Bombardment is thought to have occurred approximately 4.1 to 3.8 billion years ago when an especially large number of asteroids collided with the early terrestrial planets after they had formed and had accreted most of their mass. Evidence for this late bombardment comes from dating the Apollo samples: impact melts were found to have occurred in a rather narrow time interval. So, Wasserburg, Walker, Arnold and Gast came to be known as The Four Horsemen (of the lunar cataclysm). There are other interpretations of the apparent spike in the flux of impactors and no consensus has been reached.

Besides the rock and soil samples, the most fundamental data were those acquired over the years by the Apollo seismometer network and that have provided direct information about the Moon's internal structure. More than 1,700 meteoroid impacts were recorded – some involving quite large impacting bodies (up to 5,000 kg). Small Moonquakes were also recorded that take place at depths of 800-1,000 km. These occur at monthly intervals indicating that they are caused by tidal stresses resulting from the Moon's eccentric orbit about the Earth.

The mean density of the Moon – 3.346 gm/cc – is considerably less than that of the Earth (5.513 gm/cc) indicating a relative absence of heavy metals in its makeup. The passive seismic experiment of Apollo 15 demonstrated that the Moon has an internal structure – crust, mantle and core – quite like that of the Earth.

In Brief, What We Have Learned about the Moon

The Lunar and Planetary Institute, in Houston summarizes:

The lunar crust is rich in the mineral plagioclase and has an average crustal thickness of 60-70 kilometers, which is about 3 times the average crustal thickness on Earth. The lunar mantle lies between the crust and the core and consists mostly of the minerals olivine and pyroxene. The core is probably composed mostly of iron and sulfur and extends from the center of the Moon out to a radius of no more than 450 kilometers; i.e., the core radius is less than 25% of the Moon's radius, which is quite small. In comparison, Earth's core radius is 54% of Earth's radius. However, the size of the lunar core is not well constrained by existing seismic observations. Better constraints come from the laser ranging retro-reflector and magnetometer experiments.

The Moon's internal structure evidently has resulted from the precipitation of minerals from a global magma ocean soon after the Moon formed about 4.5 billion years ago giving rise to a mantle of ferromagnesian silicate minerals (such as pyroxene and olivine) and a crust of silicate minerals. Partial melting within the mantle in time gave rise to the eruption of mare basalts (fine-grained igneous rock common on Earth and mostly made of silicate minerals) on the lunar surface. The ages of these mare basalts were determined by radiometric dating of returned samples and range from 4.2 to 3.16 billion years. Some whose age was estimated from counting the areal density of craters are as young as 1.2 billion years indicating that the Moon was geologically active until quite recently. However, the majority of the basalts erupted between 3 and 3.5 billion years ago and they are found, with minor exception, in basins on the near side of the Moon. It remained unclear why there are fewer giant impacts on the far side (NASA's precision gravity mapping mission GRAIL of 2011 has shed some light on this). That side does, however, have the South Pole–Aitken Basin that, incidentally, is not filled with flood basalts.[41]

The connection between the basin-forming impacts and the later process of basaltic flooding remains unclear. Certainly, lunar volcanism is quite different from what we mostly experience on Earth where conical mountains are the most familiar expression of volcanism – the Earth's 'Ring of Fire' results from the collision and subduction of crustal plates or, in the case of the Hawaiian chain of shield volcanoes, the result of plate movement over a mantle 'hot spot'. The Moon lacks plate tectonics. The Earth has, however, also experienced lunar-like flood volcanism that coats huge areas of land with basaltic lava and, erupting from fissures, does not create mountains. The Deccan Traps in India, dated at 66 million years ago are one example among many others. The flood volcanism that filled the lunar basins with basalts would appear to be the result of the diminished crustal thickness that resulted from the giant impacts that formed the basins. The dark, low maria, a naked-eye feature of the lunar near side, are very lightly cratered in comparison to the lunar highlands. From this observation alone it is clear that the maria are younger than the highlands; the latter evidently accumulated their impact craters early in lunar history soon after the magma ocean had cooled and a crust had formed. The absolute age dating that was made possible by the return of the Apollo samples has established that both highland and maria are ancient. The highlands were evidently formed before 4.2 billion years ago and the maria, as noted, from about 4.2 to 3.16 billion years. This is interpreted to mean that, in a quite short interval, the rate of crater formation, initially very intense, fell off rapidly as the inner planets accreted from the available supply of asteroids/planetesimals. However, as noted, the interpretation is somewhat controversial.

Potentially, this finding from the Apollo missions, if confirmed, can be applied to the other inner planets including Earth and Mars. Of course, in the case of our own planet, plate tectonics and rapid resurfacing by erosion has eliminated the evidence of our early history. The counting of craters to establish their areal density is invaluable in determining the relative ages of different terrains on the Moon and other planets. However, among other things it is complicated by secondary crater formation. The interpretation of crater statistics is a quite complex subject; Michael Carr's book *The Surface of Mars* is recommended for more enlightenment.[42]

An impact by an asteroid, large or small, creates a shockwave that spreads out, excavating the cavity and, on the airless Moon, ejecting a substantial amount of material, some molten, to form radial patterns of debris – characteristic rays – and secondary craters. Lunar craters come in all sizes from simple bowl-shaped craters to large, complex craters with terraces, central peaks and multiple rings. *Copernicus* for example is a notable lunar crater 93 km across. The inner walls of the crater have collapsed to form a series of step-like terraces together with a central peak.

The NASA History Office in a 2015 update of its 1975 special publication, SP-350, *Expeditions to the Moon* summarizes the history of the Moon resulting from the Apollo science results:

---

[41] https://www.lpi.usra.edu/lunar/missions/apollo/apollo_14/experiments/pse/
[42] https://www.cambridge.org/core/books/surface-of-mars

During the melted shell phase from about 4.6 to 4.4 billion years ago, at least the outer 200 miles of the Moon was molten or partially molten. As this shell cooled, the formation and settling of crystals of differing composition resulted in the creation of major chemical differences between various layers tens to hundreds of miles thick. A crust, mantle, and core apparently were formed at this time. The crust consisted of light-colored minerals rich in calcium and aluminum (largely the mineral plagioclase); the mantle contained dark minerals rich in magnesium and iron (largely the minerals pyroxene and olivine); and the core probably was composed of dense, molten material rich in iron and sulfur.

The cratered highland phase that followed was extremely, almost inconceivably violent. The debris left over from the creation of the planets bombarded the light-colored crust. These highland surfaces have survived as the bright portions of the full Moon we see today. They were pulverized, re-melted, re-aggregated, and, finally, saturated with craters at least 30 to 60 miles in diameter. The sheer violence of those times is difficult to comprehend. The large basin phase was the time when very large basins were formed. This appears to have been the result of a distinctly more massive scale of bombardment than that which preceded their formation. These large basins dominate the surface character of the front side of the Moon and are responsible for the major chemical differences we have measured between various large surface regions. The light-colored plains phase that followed was a brief, still controversial period in which most old basins appear to have been partially filled with debris largely derived from the surrounding light-colored crust. The events that created these plains are poorly understood partly because several different processes related to both meteor impact and internal volcanism may have produced similar plains.

The basaltic maria phase was the main period during which the accumulation of heat from radioactive elements within the Moon produced melting and volcanic eruptions. Those eruptions filled all of the large basins with thick masses of dark-colored basalt called the maria. (These sea-like regions are the dark portions of the full Moon). The lunar basalts are very different from basalts on Earth; they contain much less sodium, carbon, and water and commonly have much more titanium, iron, and heavy elements. At least the upper parts of the maria are ancient lava flows up to 300 feet thick. Many flows differ significantly from each other in chemical and mineral characteristics, differences that vary with both the age and the region.

The quiet crust phase from about 3.0 billion years ago to the present was largely just that - quiet. Compared to the past, very little happened except for the formation of scattered, very bright craters like Tycho and Copernicus, the creation of regional fault systems like the Hyginus Rille, and the appearance of mysterious light-colored swirls like Reiner Gamma. Eruptions of basaltic maria also seem to have continued along a ridge and volcanic system that stretches for 1,200 miles along the north-south axis of Mare Procellarum. Some of the events may be indications of continuing internal activity and stress beneath a now strong crust, such as the slow, solid convection of the lunar mantle. For the most part, the surface of the Moon appears to have completed recording its history about three billion years ago. It has been largely unchanged except for the continued eroding rain of small meteors and now by the first primitive probing of men.

Lunar geoscientists have been able to reconstruct the geologic history of the Moon in considerable detail but the question of how the Moon came into being remains less than certain. The research by George Wetherill (1925-2006) of the Carnegie Institute in Washington DC (who was a major contributor to the art and science of high-precision geochronology) has provided much of the basis for the model of a giant-impact origin. In such a scenario, it might be expected that, even following a violently hot creation event, the Moon would have a component of water in its composition. Sample analysis indicated the Moon to be bone-dry. However, continuing post-Apollo research now suggests that the Moon does in fact have water, in the form of water-bearing minerals in an amount equivalent to a volume of water that "could exceed the amount of water in the Great Lakes here on Earth".[43]

Scientists at the Carnegie Institution's Geophysical Laboratory in Washington DC, along with other scientists across the nation, determined that the water was likely present very early in the Moon's formation history as hot magma started to

---

[43] https://www.nasa.gov/topics/moonmars/features/lunar_water.html

cool and crystallize. Isotopic analysis of the Apollo samples has, however, introduced a further problem because it was found that the relative abundances of the three different oxygen isotopes (O-16, O-17 & O-18) of Earth and Moon are the same whereas they differ for all the other objects in the Solar System. The oxygen isotopes of the smaller impacting planet (Theia) would have been expected to lead to a distinctive difference between Earth and Moon. However, it did not. Kevin Pahlevan and David Stevenson of Caltech in 2005 proposed "mixing between the Earth and the circum-terrestrial disk in the aftermath of the giant impact would homogenize the terrestrial and proto-lunar material".[44]

A new analysis by Washington University in St. Louis geochemist Kun Wang added the measurement of the relative abundances of the isotopes of potassium in lunar and in 8 terrestrial rocks "representative of the Earth's mantle". The lunar rocks were found to be enriched by about 0.4 parts per thousand in K-41 the heavier isotope. "The only high temperature process that could separate the potassium isotopes in this way is incomplete condensation of the potassium from the vapor phase during the Moon's formation. Compared to the lighter isotope, the heavier isotope would preferentially fall out of the vapor and condense".[45]

From all this it appears that the best explanation is that the proto-Earth experienced an extremely high-energy impact rather than a low-energy grazing impact as had once been supposed. The impact evidently vaporized both bodies leading to a massive, well-mixed magma cloud of melted rock from which the Earth and the Moon both condensed.

The Apollo missions have led to a detailed and satisfying understanding of the history of the Moon. Not surprisingly, however, the lunar science community was not slow to point to issues needing further exploration. In time, a series of robotic lunar missions – as described in Chapter 11 – were carried out to fill these gaps. The task goes on. Also, looking further ahead, radio astronomers see the Moon as a potential location for observations shielded from interference from terrestrial transmitters.

The Apollo program had been carried out under enormous time pressure by thousands of the most talented engineers, scientists and managers in the US under Administrator Webb and his successor Thomas Paine. Besides Webb, von Braun and the astronauts, the following names include some of the most famous ones within NASA and, in some cases, famous to the American public at large: Kurt Drebus (1908-1983), Hugh Dryden (1898-1965), Max Faget (1921-2004), Eilene Galloway (1906-2009), Robert Gilruth (1913-2000), Brainerd Holmes, (1921-2013), John Houbolt (1919-2014), Chris Kraft (b 1924), Gene Kranz (b 1933), George Low (1926-1984), George Mueller (1918-2015), Rocco Petrone (1926-2006), Sam Phillips (1921-1990), Robert Seamans (1918-2008), Joe Shea (1925-1999), Abe Silverstein (1908-2001) and Floyd Thompson (1898-1976). Biographical information about each can be found at NASA's History Office.

---

[44] https://www.lpi.usra.edu/meetings/lpsc2005/pdf/2382.pdf
[45] https://www.futurity.org/moon-potassium-isotopes-1246082-2/

CHAPTER 4

MISSIONS TO THE INNERMOST PLANETS: MERCURY AND VENUS

Mariner 1, 2, 5 & 10, Goldstone & Arecibo, Venera 3, 4, 5, 6, 7, 8, 9, 10, 11 &12

Pioneer Venus Orbiter & Pioneer Venus Multiprobe, VEGA 1 & 2

Our inner neighbor Venus has been a special challenge to astronomers for centuries because the brightest object in the night sky (other than the Moon) has no features that can be recorded by optical telescopes – indicating global cloud cover. All that was known till the dawning of the space age is that, in size, Venus is similar to our own planet and that its closer distance from the Sun, orbiting it in about 225 days, means that it must be significantly hotter than Earth. There was much speculation regarding the nature of the surface that lies beneath the clouds, with arguments being made for both ocean, desert and swamp. Venus thereby became the setting of many fanciful science fiction stories; the author, growing up in the 1950s UK, recalls with amusement following the adventures of *Dan Dare, Pilot of the Future* in the weekly *Eagle* comic as Dan battled his archenemy the *Mekon* ruler of the *Treens* of northern Venus.

In the early 20th Century, Venus was assumed by scientists to be relatively Earth-like, possibly even harboring life. In 1955 it was argued by astronomers Donald Menzel and Fred Whipple, on largely theoretical grounds, that a global ocean would cover the Venus surface. Much progress was subsequently provided by observations of Venus at radio wavelengths. These were made using the giant antennas that had been built in increasing size by radio astronomers since the 1930s. In 1956 Cornell Mayer (1921-2005), T.P. McCullough and R.M. Sloanaker of the US Naval Research Laboratory recorded intense 3.15 cm microwave emission from Venus. This implied a temperature of about 600K. In Russia A.D. Kuzmin and A.E. Salomonovich observing at wavelengths of 0.4 to 10.3 cm, estimated a similar temperature. The possibility of a global ocean or of any life on Venus seemed to be foreclosed although it was recognized that high in the atmosphere at an altitude of about 50 km conditions of temperature and pressure were rather like surface conditions on Earth.

Among the scientists in the US who were engaged in studies of the planets – with a particular interest in the possibility that life may have emerged elsewhere in our Solar System – was young Carl Sagan. Harold Morowitz and Carl in 1967 opened the discussion about the possibility that life forms might have evolved on Venus to inhabit the temperate layers of the clouds (at about 50 km altitude). They imagined "an isopycnic organism constructed as a float bladder".[46] Keay Davidson in his 1999 biography also describes Carl's participation in the study of Venus and his conclusion that the extreme recorded temperature was the result of greenhouse physics where sunlight incident on an atmosphere of carbon dioxide that is mostly transparent to visible light, heats a planet's surface and then, like a blanket, absorbs much of the IR radiation emitted by that hot surface. Having been studied in 1895 by the Swedish chemist Svante Arrhenius, this same physical effect applies to our own planet (as we are now all anxiously aware). In fact, we now know that on Venus very little sunlight actually reaches the surface.

It is somewhat surprising that in his famous book *COSMOS*, published in 1980, where Carl writes about Venus, he does not discuss the possibility of life there. He had by then concluded that the extreme temperatures on Venus represented "a

---

[46] Morowitz, H, & Sagan, C., 'Life in the Clouds of Venus?' *Nature* 215, 1259-1260 (1967)

planet-wide catastrophe", "a thoroughly nasty place", where "It is always raining sulfuric acid on Venus, all over the planet and not a drop ever reaches the surface". Others, however, including Charlie Cockell of Edinburgh University continued to pursue this possibility. Carl Sagan, who, alas, died two-dozen years ago would surely, like the rest of us, have been intrigued to learn of the announced spectroscopic detection (by a Cardiff University & MIT team) of phosphine in the microwave emission from Venus. Because the only known ways in which phosphine in nature is created (setting aside in the hydrogen atmosphere of Jupiter where Reinhard Beer and Fred Taylor reliably detected phosphine in 1975) involve biology, the implication could be that some form of life has evolved in the upper Venus atmosphere where conditions of temperature and pressure are similar to those on Earth. However, the observation and its interpretation are subject to much debate. Planetary spectroscopy experts generally look for more evidence than the identification of a single absorption line and now the phosphine detection is considered only tentative.

The first robotic deep space missions carried out by NASA were lunar pathfinders for the Apollo missions as described in a previous chapter. The technology developed was then applied to science missions targeted to our inner neighbors: Venus and Mercury. Spacecraft missions to them were thus among the earliest ventures into deep space beginning in the early 1960s; they continued in subsequent decades and in the present century a European-Japanese mission is in transit on its way to orbit Mercury in 2025. The Soviets made Venus a focus of their space program while NASA also carried out a number of JPL-led Mariner missions and NASA Ames-led Pioneer missions to Venus. There is a lot to learn – especially about Venus, the ugly sister planet to Earth.

Because Mercury and Venus orbit the Sun inside the Earth's orbit, this makes them difficult to study using telescopes – for when Earth is at inferior conjunction with respect to the planet (i.e., at closest approach) the observer is looking at the dark side of the planet. At superior conjunction when the planet is fully illuminated it lies at its furthest distance from Earth on the other side of the Sun.

In the late 1950s Gordon Pettengill and colleagues R.B. Dyce and Irwin Shapiro at MIT's Lincoln Lab (using the giant Arecibo dish) and Dick Goldstein and colleagues at JPL (using NASA's Deep Space Network antenna at Goldstone) had pioneered the effort to establish the rotation rate of Venus (and also of Mercury) by attempting to record radar echoes from Venus when it was closest to Earth. The first confirmed echoes from Venus were achieved in March 1961. These and later observations served to measure the rotation rate of Venus – found, remarkably, to be both very slow (243 days) and retrograde i.e., when viewed looking down at the north pole, Venus rotates in a clockwise sense. This remains a bit of a puzzle.

The first radar images taken from the Earth showed very bright (radar-reflective) highlands christened Alpha Regio, Beta Regio and Maxwell Montes. Given the impenetrable cloud and the limitations of available radar observatories, there was obviously much that could only be learned about Venus by high-resolution radar studies from a spacecraft.

The Soviet Union gave the exploration of Venus a very high priority and, after some learning experiences, in the 1970s enjoyed a great deal of success. The first two Venera spacecraft were launched in 1961 and were intended to fly by Venus. They did get underway but experienced telemetry failures.

The innermost terrestrial planets, Venus especially, were also priority targets for the US science community and the first spacecraft to fly there were designed and built at the Jet Propulsion Lab for launch as early as 1962. It would, inevitably, be a learning experience.

NASA's Deep Space Mission Selection Process

Before describing the first planetary missions that were undertaken – starting with the less-challenging inner planets – it may be helpful to explain the process by which those missions were chosen and the process that attempted to deal with the great inherent risk. For the first several decades of NASA's deep space missions, inception began with a Conceptual Study – *Pre-Phase A* – of how to meet an agreed science goal (i.e., supported by the National Academy of Sciences). The process has evolved but is still followed for costly 'flagship' missions. A Science Working Group was formed to prioritize the science goals and requirements. This led to an Announcement of Opportunity (AO) that the Program Office at NASA HQ issues, seeking proposals for instruments/experiments to be carried out by Principal Investigator-led science teams at universities, NASA Centers, and science organizations around the world. The proposed mission was assigned, *Phase A*, to a Center, most often JPL, for implementation – providing further study is judged successful and affordable. The preliminary design and project plan included the spacecraft design required to carry out the experiments, the required launch vehicle and launch window, the trajectory design, and operations *enroute* and at the target(s).

Then the effort moved into *Phase B* – the Definition Phase. Experience has shown that the more effort and resources that are applied in Phase B the more likely it is that the mission will not experience the kind of unacceptable cost growth that can potentially lead to the cancelation of a promising mission or, more often, delay the new start in favor of other Space Science missions waiting in the queue. Resources that are expended carrying out Phase B are always well spent because the early identification of potential showstoppers avoids expensive fixes downstream. Phase B creates a baseline technical solution with documented requirements including schedules and the specifications to initiate system design and development. In the case of JPL that, unusually for the NASA Centers, has the capability of building spacecraft in-house, the decision is made whether or not to contract out the spacecraft and, if so, to seek proposals. Responses by the science community to the AO receive in-depth review by a NASA peer group that grades them in order for NASA's Science Mission Directorate to make the selection of experiments; this takes into account scientific value, cost, management, engineering, and safety. In an effort that takes at least a year, major reviews are carried out in Phase B: System Requirements Review, System Design Review, and Non-Advocate Review.

At this point NASA has enough solid information to include (or not!) the mission in its proposed program and to meet with the Administration's Office of Management and Budget (OMB) seeking agreement to include a *new start* for the mission in the Agency's next budget request. If approved by the OMB and then by Congress the Project can get underway into the Design and Development Phase – *Phase C/D* – at the start of the next fiscal year (which begins each October). This phase (where nearly all the money is spent) is when the many subsystems, including instruments, are built and integrated into the spacecraft for test in a solar-thermal-vacuum chamber and finally transported to the Cape for launch. During the several years this takes, the Project carries out a series of reviews – Preliminary Design Review, Critical Design Review, Test Readiness Review, and Flight Readiness Review – before HQ Program Office managers and a Review Board made up of a dozen or more senior members. The latter are typically past and present NASA and Industry managers of comparable projects – individuals who have experienced the many varieties of problems that can bedevil space missions. They invariably raise many issues and submit them in written form to the Project Manager who investigates and provides a formal written response to each – and takes whatever action may be required.

The required multi-year budget profile to carry out the mission has been determined by now. This profile for the Project Manager includes a reserve to deal with unforeseen problems that inevitably arise. The HQ Program Manager also has reserve funds called Allowance for Program Adjustment (APA) intended to cover the cost of delays such as a schedule slip resulting from a launch vehicle problem. Because the Project Manager is aware that this HQ reserve exists, it isn't surprising that he or she will advance convincing reasons to release some APA in order to reduce the various kinds of risks that surface once development is underway. Cost overruns are not uncommon for "flagship" missions that push on the boundaries of what is possible.

Two-launch Insurance Policy

In the 1960s and on into the 70s missions were often lost in the first minutes after liftoff and so it was prudent to increase the chances of mission success by duplicating the spacecraft and the launchers. Over time the success rate of NASA's deep space missions – including their launch vehicles – improved to the point in the 1980s where it was no longer considered necessary to build and launch two spacecraft (albeit with the purchase of some critical spares, just in case). Inevitably, this costly approach is expensive as NASA seeks to find the optimum balance between mission performance, cost and risk. Critics like NASA Administrator Goldin in the 1990s have argued for taking more risk to reduce costs. Because a deep space mission typically takes the best part of a decade from conception to launch and then years to reach its destination, the consequences of a failure – which may only occur at the target – are severe, especially for the scientists but also for the reputation of the Agency and the implementing Center.

1962 Mariner 1 & 2

Venus was still virgin territory when Mariner 2 flew by at a distance of 35,000 km on 14 December 1962 (Mariner 1 having had a launch vehicle failure). Mariner 1 and 2 were three-axis-stable spacecraft based on the Lunar Ranger design – a spacecraft known at the time for a series of failures in attempting to return images as they impacted the Moon. The design would, nevertheless, serve as the progenitor of a remarkable series of Mariner and Voyager missions that, within three decades, encountered all the planets out as far as Neptune. Up to this point JPL had been a US Army Lab focused on missile development. Following the success of Explorer 1, and now that it was part of NASA, JPL was intent on establishing a new business: interplanetary exploration. Four more Explorer spacecraft had been launched, two of which experienced launch vehicle failures. Two lunar flyby probes, Pioneer 3 & 4 followed; of these the first had another launch vehicle failure while the second did fly by the Moon but was only a partial success. Next for JPL came the impactor Lunar Rangers and the Surveyor landers. Only now did the Mariners emerge. They required a powerful upper stage for the launch vehicle if they were to reach the planets; a USAF development – Centaur – was already underway to meet this need.

Development began at JPL for a 1,250 lb three-axis stable spacecraft derived from Ranger; Mariner A was to encounter Venus with a launch in 1962 while Mariner B would be launched to Mars in 1964. In fact, both of the missions would involve dual launches of two Mariner spacecraft – an insurance measure given the unreliability of both launch vehicles and spacecraft, something that, with an exception, continued to the Voyager launches in 1977. Jack James (1920-2001) was appointed to be the Mariner Project Manager and he reported to JPL's planetary program director Bob Parks (1922-2011). As would be the case with many of the missions that followed, there would be numerous challenges during the development, beginning with a delay in the availability of the Centaur; this required the first Mariner Venus mission to use the less powerful Agena upper stage and, in turn, led to a significant required reduction in the weight of the spacecraft. A second big challenge was the development of a sufficiently reliable system to make the accurate midcourse maneuvers (a roll-turn sequence, followed by a pitch-turn sequence and finally a motor-burn sequence) needed to make a close encounter with Venus. The project schedule was very tight.

The science plans for the mission were inspired by the discoveries of the radio astronomers and their interpretation by a young University of Chicago (later Cornell) physicist, Carl Sagan (1934-1996) and his student Jim Pollack (1938-1994): that the remarkably high temperatures observed might be evidence of a 'greenhouse' atmosphere of infrared-absorbing gases. Carl would later become famous as an author, television personality and space science visionary. The project science plans also included study of the interplanetary medium between Earth and Venus and the 'solar wind' that Caltech (later U of Chicago) astrophysicist Eugene Parker had hypothesized; this was the task that JPL scientist Marcia Neugebauer took on with colleague Conway Snyder (1918-2011). Ed Smith led the effort to characterize any planetary magnetic field. Lewis Kaplan, supported by Carl Sagan and Gerry Neugebauer (1932-2014), was responsible for the infrared and microwave radiometers; other science team members included James Van Allen (1914-2006), Hugh Anderson, Victor Neher (1916-1994) and Doug Jones. A camera was notably absent from a science payload that had to be squeezed into an 18 kg package. There was no scan platform: the science experiments were simply mounted on top of the spacecraft base.

The launch of Mariner 1 from the Cape took place on 22 July 1962 but not long after take-off the Atlas appeared to be in trouble and was destroyed by the range safety officer just before the Agena was about to separate. The launch of Mariner 2 on 27 August was successful, though not without some stressful moments – as was the cruise to Venus: "the mission endured one seemingly show-stopper crisis after another, only to recover and soldier on. It barely worked, recalls one JPL engineer who worked on Mariner 2 early in his lab career".[47]

---

[47] https://www.jpl.nasa.gov/mariner2

Valuable science results were acquired long before Mariner 2 reached Venus including confirmation of Eugene Parker's model of the solar wind – supersonic plasma streaming from the Sun. The cosmic dust collector only registered one hit during the journey indicating that meteoritic dust was not a hazard in the innermost Solar System. However, one of the two solar panels failed on the way to Venus and Mariner 2 was "heated to within an inch of its life". After its closest approach to the Sun on 27 December contact was lost.

As something of a surprise, Venus was found to lack a detectable magnetic field and, so, no radiation belts. The radiometer science team found the surface temperature of Venus to be between 300 and 400°C, confirming the terrestrial radio telescope observations. An accurate estimate of the Venus surface pressure was still lacking but these results confirmed the powerful greenhouse effect.

Shortly after, Jim Pollack by now at NASA Ames, based on an idea that was conceived in 1963 by Thomas Gold (1920-2004), proposed that Venus had once had oceans like Earth and that these had gradually evaporated into the atmosphere leading to positive feedback and a consequent runaway greenhouse (CO_2 and methane, CH_4, are not the only greenhouse gases, water vapor is too). At that time, in the early 1960s, terrestrial climate change was not yet a cause for existential alarm.

The Mariner 2 encounter had also served to make a very accurate measurement of the mass of Venus – 0.81485 times that of Earth – based on tracking the radio signal from the spacecraft. This first result demonstrated a critically important capability of interplanetary spacecraft in making an inventory of the planets, moons, asteroids and comets. Tracking the radio signal also provided an improved value for the Astronomical Unit (AU) – the average distance of Earth from the Sun – of 92,956,200 miles, plus or minus 300 miles.

Mariner 2's particles-and-fields instrumentation (a magnetometer, a cosmic ray detector, two other particle detectors and a cosmic dust detector) served to make an early characterization of the interplanetary medium. At Venus scans made by the microwave radiometer on the planet's dark side, on the lighted side and near the day/night boundary led to the conclusion that there were no significant temperature differences across Venus. Measurements made by the IR radiometer reached the same conclusion.

The success of Mariner 2 came in time to steady the reputation of JPL, one that had suffered because of the failure of the first five Lunar Rangers and the consequent suspension of that program for more than a year. The first successful Ranger, Ranger 7, was still in the future. The numerous problems like these, encountered early on in the Space Age, have served to require NASA and JPL to establish a very systematic, phased approach to developing and carrying out missions, one that would significantly reduce the failures (albeit at some expense) in an exploration program of unprecedented complexity. It is worth taking a diversion to explain briefly how the process has evolved and has been in place now for decades. (Note: besides the time-honored approach described above, NASA also undertakes cost-constrained *Discovery* missions proposed by planetary scientists in academia in response to announcements of opportunities by the Agency.)

Several researchers had developed models of the circulation of the slowly rotating Venus atmosphere including Richard Goody and Alan Robinson (1933-2009) at Harvard who concluded that the diurnal temperature difference would be small and that the circulation would be between equator and poles rather than sub-solar and anti-solar points. It was a complex problem with relatively little hard data to inform it: the highly reflective clouds and dense atmosphere meant that very little (a few percent) of the sunlight falling on Venus contributes to heating the surface and the slow rotation rate reduces the Coriolis force. The one other piece of observational evidence was provided by the UV telescope images that suggested that the upper atmosphere is in a state of 'super rotation' such that the atmosphere rushed around the planet much more rapidly than the surface was rotating. There was a clear need for more data of the kind that could only be acquired by long-lived orbiters.

The composition of the clouds that hide the surface of Venus was a puzzle that in 1972 was solved by astronomer Godfrey Sill of the University of Arizona and confirmed in 1973 by Louise & Andy Young of JPL who identified sulfuric acid as the constituent; they had based their confirmatory analysis on data provided by James Hansen who, later as head of the NASA Goddard Institute for Space Studies in New York City, would be at the forefront of warnings about the direction in which Earth's climate is changing as a result of mankind's continuing build-up of CO_2 in our atmosphere.

1964 Goldstone and Arecibo Radio Telescopes

The first radar images of Venus taken from the Earth showed very bright (radar-reflective) highlands. Dick Goldstein christened them Alpha Regio, Beta Regio and Maxwell Montes. Given the impenetrable cloud and the limitations of available radar observatories, there was obviously much that could only be learned about Venus by high-resolution radar studies from orbit.

1965 Venera 3 & 1967 Venera 4

The Soviet Union's next mission to Venus was Venera 3, designed to enter the atmosphere and land on the surface. This it apparently did in March 1966, but the probe was lacking a working communication system at that point so we shall never know for sure. Launched on 12 June 1967, Venera 4 was a large, complex spacecraft again designed to enter the atmosphere and make measurements of temperature, pressure and composition all the way to the surface. The atmospheric entry was successful in the face of heat-shield temperatures of up to 11,000°C and deceleration of 300g. The descent lasted 93 minutes before the signal was lost, some distance from the surface. The probe did provide important information about atmospheric composition: 90-93% carbon dioxide, 7% nitrogen, 0.4-0.8% oxygen, and a trace of water vapor (c.f. present values: 96.5 % carbon dioxide, 3.5 % nitrogen, 150 ppm sulfur dioxide, 70 ppm argon, 20 ppm water vapor, carbon monoxide, 17 ppm, helium 12 ppm, neon 7 ppm).

1967 Mariner 5

NASA also returned to the Venus action with the launch of Mariner 5 on 14 June 1967, a spacecraft whose original role had been as a backup to the Mariner 4 Mars spacecraft (Mariner 3 having failed at launch). Dan Schneiderman (1922-2007) and Timothy Parker served as Project Managers with Conway Snyder as the Project Scientist. Modifications to deal with the challenging thermal environment at Venus included adding thermal insulation and reducing the size of the solar panels. The spacecraft carried only basic instrumentation – UV photometer, solar plasma probe, magnetometer, trapped radiation detector and radio science – and did not include an imaging system. Flyby of Venus took place in October 1967 just one day after the Venera 4 probe entered the Venus atmosphere. The radio science occultation experiment (the time history of the frequency) was limited to measurements down to an altitude of about 32 km where the atmosphere is critically refractive, making it inaccessible to radio occultations. Mariner 5 provided a measurement of the atmospheric pressure and temperature profile, confirming the extreme conditions observed by Venera 4.

1969 Venera 5 & 6

Little information is available about Venera 5 and 6, apparently beefed-up versions of Venera 4 that were launched in January 1969, entered the atmosphere of Venus in May 1969 and landed near the equator. Both probes entered the Venus atmosphere on the night-side, deployed their parachutes and were tracked as they descended, each for nearly an hour sending back compositional data before contact was lost at altitudes of about 25 km (Venera 5) and about 10 km (Venera 6).

1970 Venera 7

The Soviets continued to develop the Venera spacecraft design with the goal of making a soft landing on Venus and making direct observations from the surface. Venera 7 comprised a bus with a spherical lander probe that was designed to withstand higher pressures and temperatures. The pressure vessel was made of titanium and lined with shock-absorbent material. In order to reach the surface more quickly a smaller parachute was used. Venera 7 launched to Venus in August

1970 and entered the Venus night-side atmosphere on 15 December, descending under a parachute from an altitude of 60 km. Evidently the parachute experienced a failure at some point so that the probe was in freefall for about 30 minutes before making an equatorial landing that was much harder than planned – probably leading the spacecraft to bounce onto its side and misdirecting the radio antenna. Nevertheless, 23 minutes of very weak recorded data were received, only providing information about the surface temperature: 475°C.

1972 Venera 8

Two years later, in March 1972, the Soviets launched Venera 8, a modified design of atmospheric probe and lander. A second mission failed to leave Earth orbit. Venera 8 employed a refrigeration unit to chill the descent capsule in an attempt to increase its surface life. Before reaching Venus the probe interior was cooled to -15°C. Everything worked as planned. The parachute opened in reefed mode at an altitude of 60 km, and at 30 km altitude the parachute was fully opened.

At the surface where the lifetime had increased to 63 minutes Venera 8 confirmed the extreme harshness of surface conditions – 470°C and 90 atmospheres pressure – but also determining that lighting at the surface ("like on Earth on an overcast day") would make surface imaging practical. The spacecraft confirmed that little sunlight makes it to the surface of Venus so that direct surface heating by the Sun does not account for the very high temperature. Other important new information from photometry concerned the high altitude of the clouds and relative clarity of the atmosphere down to the surface.

An analysis by NASA Goddard's JE Ainsworth and JR Herman of the Venera 7 and 8 determined that "the lower boundary of the horizontal retrograde '4-day' wind is defined by a 50–60% decrease in wind speed in the vicinity of 44 km, and there exists a retrograde wind 'plateau' of 15 to 40 m/s winds extending from 40 km down to the vicinity of 18 km, where the winds decrease rapidly to the order of 0.1 m/s near the surface".[48]

Measurements by a gamma ray spectrometer indicated that the surface rocks were similar in composition to granite.

1973 Mariner 10

At this point NASA returned with another mission to Venus, Mariner 10, but one that was, in fact, primarily designed to explore Mercury where it was planned to make a series of flyby encounters. Gene Giberson (1923-1999), later JPL's Assistant Director for Flight Programs, was the Project Manager and Jim Dunne was the Project Scientist. Bruce Murray (1931-2013) – not yet JPL Director, still a Caltech professor – was Principal Investigator for the imaging science team; there were two vidicon cameras, one wide and one narrow angle. Other PIs were Stillman Chase of the Santa Barbara Research Center, the infrared radiometer; Lyle Broadfoot of the University of Arizona, the UV spectrometers; H. Taylor Howard (1932-2002) of Stanford University the celestial mechanics and radio science experiment; Norm Ness of the University of Delaware, the magnetometers; Herb Bridge of M.I.T. (1919-1995), the scanning electrostatic analyzer and electron spectrometer; and John Simpson of the University of Chicago (1916-2000), the charged particle telescope.

Mariner 10 was launched in November 1973 on its two-planet mission. This was to be the first mission to Mercury – its primary target (in fact, Mariner 10 was the *only* Mercury mission until MESSENGER was launched 30 years later). Getting to the innermost planet is quite a launch energy challenge as also is coping with the extreme solar heating that is inevitable. Travelling inward to Venus and Mercury from Earth, the upper-stage rocket must direct its thrust so that the spacecraft loses a significant amount of orbital energy about the Sun; the spacecraft's orbital transfer to Venus is along an elliptical path that has its aphelion at Earth and its perihelion at Venus. Travelling directly to Mercury is even more challenging from an energy point of view. And, obviously, all this must be done at exactly the right time so that arrival at Venus's orbit and Mercury's coincides with the target planet being there. What made possible the multiple flybys of Mercury was the celestial mechanics insights of Italian scientist Giuseppe (Bepi Colombo (1920-1984). Launched on an Atlas Centaur Mariner 10 flew

---

[48] https://agupubs.onlinelibrary.wiley.com/doi/abs/10.1029/JA080i001p00173

by Venus on 5 February 1974 and used its gravity-assist to reach Mercury on 29 March 1974 after a 146-day journey. On the way there were numerous technical malfunctions to challenge the operations team. Telescopic observations of Venus over decades had determined that the global cloud cover is essentially featureless at visible wavelengths.

However, in the 1960s observers had found indications of structure detectable at UV wavelengths. A very large-scale feature – shaped like a Y tipped on its side – had been observed, apparently rotating about Venus with a period of 4 to 5 days. This implied winds moving at 100 meters/sec – something of a puzzle give the slow planetary rotation period of 243 days.

Mariner 10's cameras were therefore equipped with UV filters and a plan to continue imaging Venus for eight days as it approached and swung by on its way to Mercury.

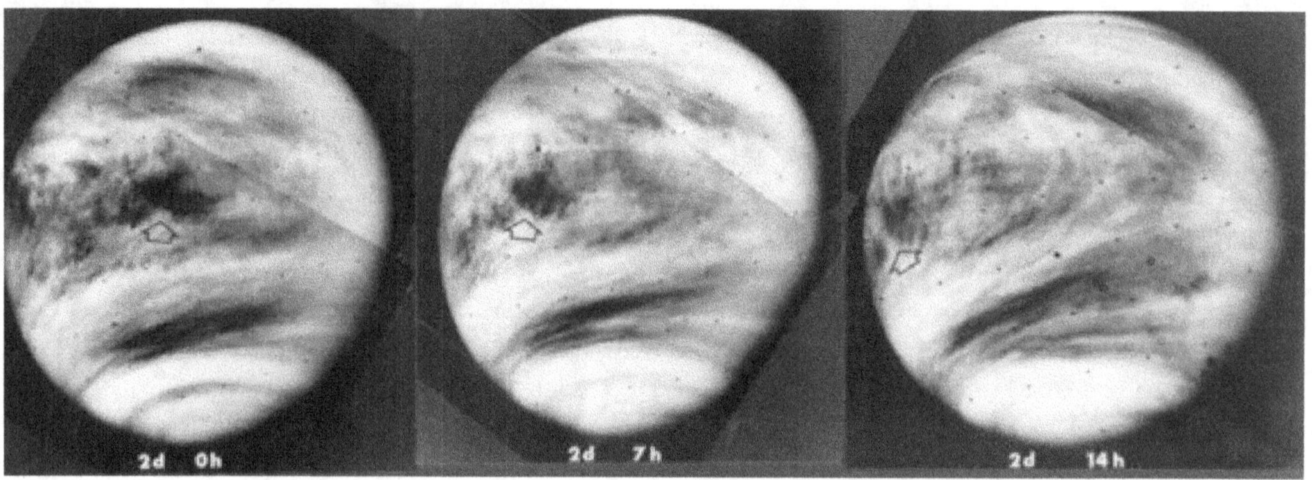

The observations fully confirmed the earlier puzzling interpretation of high-velocity cloud-top motion. The contrast observed in UV light is evidently the result of the sulfur in clouds that had been discovered earlier by astronomers, clouds that are composed of dilute sulfuric acid. Understanding the general atmospheric circulation of Venus would become an important goal of the next NASA missions to Venus – Pioneer Venus Orbiter and Pioneer Venus Multiprobe. These were managed by the NASA Ames Research Center in Mountain View California that inherited the Pioneer name not only for Venus missions but also missions to the outer planets.

Decades later, in June 2020, JPL released a newly-processed image of Venus that showed more colorful detail in the cloud tops of Venus along with the comment:

"These cloud particles are mostly white in appearance; however, patches of red-tinted clouds also can be seen. This is due to the presence of a mysterious material that absorbs light at blue and ultraviolet wavelengths. Many chemicals have been suggested for this mystery component, from sulphur compounds to even biological materials, but a consensus has yet to be reached among researchers."[49]

Mercury's orbit is slightly more eccentric – eccentricity 0.21 – even than that of Mars. The first flyby of Mercury by Mariner 10 took place with closest approach on its night side so the highest resolution images of the half-illuminated disc were obtained about 30 minutes before. Radio tracking provided a determination of the mass of Mercury and from this its density was found to be 5.5 gm/cc – similar to that of Earth.

Given its much smaller size and, therefore, lesser internal self-compression, the density measurement implies that the size of Mercury's core is proportionately larger than that of Earth. But Mercury's magnetic field is only 1% as strong as Earth's and is evidently created by the convection of electrically conductive iron in the outer core – not yet solidified as some had

---

[49] http://spaceref.com/venus/newly-processed-views-of-venus-from-mariner-10

anticipated. Mercury rotates slowly – it is locked in a 3:2 resonance with the Sun, rotating once every 59 days as it orbits the Sun every 88 days. A good understanding of why the magnetic field is so weak remains to be generally established. It is, however, strong enough to create a magnetosphere, slowing the solar wind.

The second and third encounters with Mercury enjoyed better viewing geometry and coverage enabling the imaging resolution to improve to ~1 km. However, the same side of the planet was illuminated each time, so Mariner 10 could only map about 45% of the surface. Mercury was found, in appearance, to be much like the Moon with heavily cratered terrain similar to the lunar highlands and, also, large basins filled with smooth plains. An important difference between the bodies is the presence of lobate scarps (cliffs) on Mercury that are kilometers high and run for hundreds of kilometers – suggestive of thrust faults arising from crustal stresses that are perhaps the result of cooling and shrinkage of an iron-rich core. Analysis of phase changes in the radio signal as Mariner 10 passed behind Mercury allowed the planetary radius to be accurately measured as well as a sensitive measurement of the density profile of the tenuous planetary atmosphere.

1975 Venera 9 and 10

Interkosmos's next Venus mission followed in June 1975: Venera 9 and 10 were significantly more ambitious and effective. Each consisted of an orbiter and lander. With an orbiter (the first at Venus) as a communication link to Earth the lander could return much more data – including images – while the orbiter could carry out its atmospheric investigations for an extended period (from 26 October to 25 December 1975). The descent system had been modified to use three main parachutes that opened at an altitude of 62 km with the lander making a slow 20-minute descent through the clouds while collecting atmospheric data.

The lander comprised a hemispherical insulated pressure vessel mounted with shock absorbers on a deformable landing ring. On top of the pressure vessel was a disc-shaped aero-brake; this served as a reflector for the cylindrical fixed communications antenna above it. The pre-cooled Venera 9 lander made it successfully to the surface and touched down on its doughnut-shaped landing cushion on 22 October. There it functioned for almost an hour. The lander was tilted at a 30-degree angle and its cameras could only see as far as few dozen meters. The landing site at 31°N, 292°E is covered with angular boulders and partly weathered rocks about 30 to 40 cm across as revealed by the first images returned from the surface of Venus. The scene is consistent with that of a volcanic landscape.

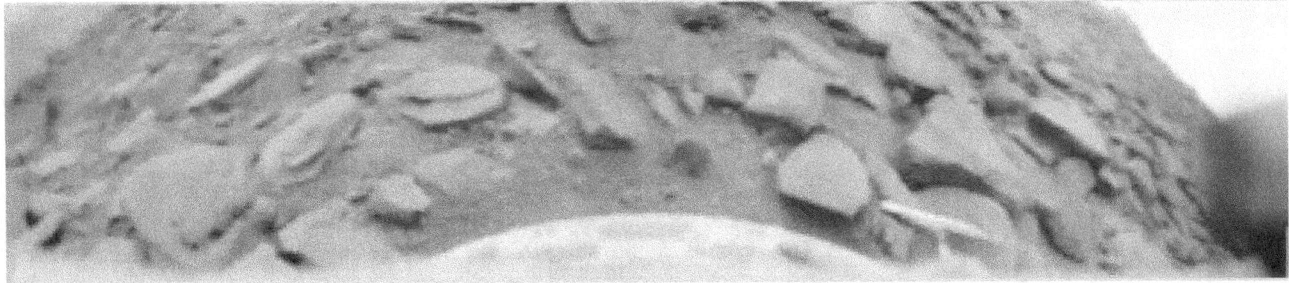

The lander acquired temperature, pressure and composition data all the way down and determined that it had passed through a 30-40 km thick layer of clouds with a cloud base at 30-35 km altitude. The surface pressure at the landing site was 90 atmospheres and the temperature 485°C.

Surface light levels were comparable to those of a cloudy summer's day on Earth and the atmosphere was clear. So imaging was not a problem – other than removing one of the camera lens covers! Only one 180-degree panorama could be returned. It took about 30 minutes to transmit all the data. The Venera 10 mission was similarly successful with the lander touching down about 2,200 km from its sibling. Again, only one 180-degree panorama was returned – a rocky and bleak-looking site. The image shows flat slabs of rock, partly covered by fine-grained material, not unlike a volcanic area on Earth. The large slab in the foreground extends over 2 meters across. The lander was equipped with a 'density sensor' that assessed that of the rock on which the lander touched down recording about 2.8 gm/cc. The density and gamma ray spectroscopic measurements are consistent with basalts.

1978 Pioneer Venus Orbiter

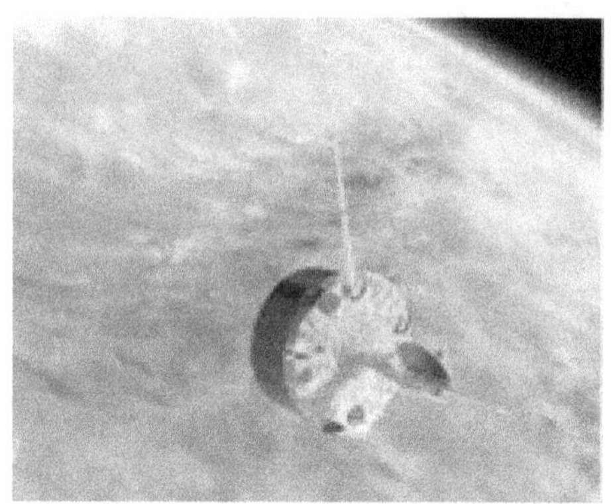

NASA's Space Science Program Offices do not lack for advice from a well-organized community of eminent scientists. According to the NASA History Office "The official history of Pioneer Venus dates the beginning of the project to October 1967, shortly after the Venera 4 and Mariner 5 spacecraft visited Venus. Three scientists, Richard M. Goody (Harvard University), Donald M. Hunten (University of Arizona; Kitt Peak National Observatory), and Nelson W. Spencer (Goddard Space Flight Center) formed a group to consider the feasibility of exploring the Cytherean atmosphere from a spacecraft. The group's formation led to a study published in January 1969 by the Goddard Space Flight Center".[50] The resulting Pioneer Venus missions were carried out by NASA Ames Research Center.

NASA's previous missions to Venus, Mariner 2 and 10, had made only a superficial start toward reaching an understanding of the planet under the clouds. So, as NASA History Office records: "In June 1970, to address the lack of a serious Venus mission, the NASA Lunar and Planetary Missions Board and the Space Science Board brought together 21 scientists to study the scientific potential of a mission to Venus. Richard Goody and Don Hunten (1925-2010), who had helped start the Goddard study (mentioned above), co-chaired the meeting. Their report, known as the Purple Book (the color of its cover) recommended that exploration of Venus should be prominent in the NASA program for the 1970s and 1980s. The group presented its recommendations to NASA management, and the Space Science Board endorsed them".

The US science community eager to explore Venus included two score of the big hitters in the community: Richard Goody and Don Hunten, Don Anderson. Ian Axford, Alan Barrett, Leverett Davis, Tom Donahue, John Gille, Seymour Hess, Garry Hunt, Robert Knollenberg, John Lewis, Gordon Pettengill, Bob Phinney, Keith Runcorn, Vern Suomi, Patrick Thaddeus, Len Tyler, James Weiman and George Wetherill. It was no surprise that the Space Science Board endorsed the report that, besides continuing to push study of the Venus atmosphere also stressed the important role of radar in studying the surface. One radar approach that was considered would have had Earth transmit the signal and a receiver on the spacecraft collect the echoes. Such a spacecraft system would have been less massive but probably less effective than a radar altimeter – an instrument that was finally included in the Pioneer Orbiter's science payload on the urging of Danny Herman, head of the Advanced Studies Branch in HQ's Planetary Exploration Office. The altimeter proved to be a spectacular success and laid the groundwork for a Venus orbiter mission (Magellan Chapter 8) that would use side-looking synthetic aperture radar to map the surface at much higher resolution.

In May 1978 NASA launched two Pioneer Venus spacecraft that were managed by the NASA's Ames Research Center; Charlie Hall and Richard Fimmel managed the Orbiter and Multiprobe Projects. Larry Colin the Project Scientist. Unlike the 3-axis-stable Mariners, these were spin stabilized – a configuration not well-suited to imaging. The Pioneer Venus Orbiter, a cylindrical bus covered in solar cells, was heading to a highly elliptical 24-hour period orbit about Venus. The orbit was of near-polar inclination (actually 105 degrees) so that science coverage could be acquired over the entire planet. Periapsis was close to the equator and it was from this low altitude part of the orbit that the radar altimeter acquired data; apoapsis was 67,000 km. Orbit insertion took place in early December 1978.

---

[50] Andrew J. Butrica, *To See the Unseen: A History of Planetary Radar Astronomy*

There were 17 experiments carried by the orbiter intended to address numerous aspects of atmospheric and ionospheric questions while the radar altimeter (Principal Investigator Gordon Pettengill) would provide a first opportunity to map the hidden surface and learn of its geologic evolution. Over the 243 days of Venus' slow rotation underneath the Pioneer orbiter an elevation map was gradually built up.

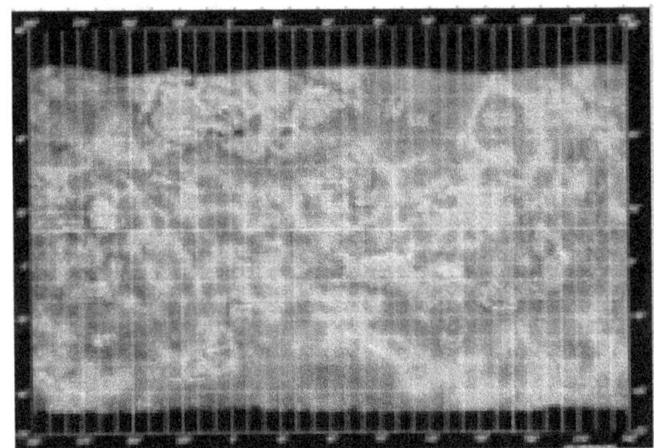

The radar's resolving power at closest approach was 23x7 km and the surface map – showing a considerable range of elevations – provided a tantalizing picture of what Venus is like. The map revealed that about 70% of Venus is made up of rolling plains, about 10% are highlands and 20% lowlands. The northern highlands, named Ishtar Terra, are about the size of Australia and rise steeply from the plains. In the middle of Ishtar are found mountains – Maxwell Montes – that rise 11 km above the mean. Far to the south east of Ishtar and stretching 10,000 km just south of the equator is Aphrodite Terra made up of very rough terrain and including a roughly circular feature along with several gigantic trenches. The Pioneer Venus Orbiter data served as powerful motivation to return to Venus with the capability for much higher resolution surface mapping.

The payload included a cloud photo-polarimeter (PI James Hansen & Co-I Larry Travis of the Goddard Institute of Space Studies) that would acquire spin-scanned global images of the clouds at UV wavelengths and track the evolution of the Y feature. They concluded that there are at least three different cloud materials that contribute to the images: a thin haze layer, sulfuric acid clouds and an unknown ultraviolet absorber below the sulfuric acid cloud layer. The linear polarization data confirmed that the low- and mid-level clouds were sulfuric acid with radius of about 1 micrometer. Above the cloud layer was a layer of sub-micrometer haze.

Circulation of the Venus atmosphere was observed mainly through the global UV imaging coverage and the atmospheric soundings provided by the multi-channel IR Radiometer (PI Fred Taylor of JPL and later Oxford University). The former instrument acquired its coverage from the high-altitude part of the orbit while the radiometer, like the altimeter, mainly operated at low altitudes. The Y feature appears to be a planetary scale Kelvin Wave, the kind that in the Earth's oceans can be created as a long wavelength swell by tides, an abrupt change in the wind field or, occasionally, by an earthquake. The source that excites the wave on Venus and leads to its rapid encirclement of the planet is uncertain; its Y shape has been tentatively attributed to the variation with latitude of wind speeds – winds further from the equator having less distance to travel to circle the globe.

Oxford University and JPL collaborated in providing the Pioneer Orbiter's multi-channel infrared radiometer (Fred Taylor PI) and this produced some of the most interesting atmospheric results; The radiometer measured the atmospheric thermal emission at seven levels above the clouds to derive the vertical temperature structure and it also had a channel to study the structure of the clouds and one sensitive to water vapor. The general circulation of the Venus atmosphere – deduced from the sounding data – is in many ways simpler than on Earth because of the slow planetary rotation. Hadley Cell circulation – air rising at the equator and being transported pole-ward to descend and return nearer the surface – is the main feature. The rising air makes it most of the way to the poles, descending at about 60° latitude. There are two holes in the north polar cloud cover, located about 1,500 km from, and on opposite sides of, the pole and rotating around it every four days. The descending air evidently leads the cloud droplets to evaporate and produces the observed holes.

Fred Taylor observes: "Probably the air is rising slowly everywhere on Venus and descending only near the poles, thus supporting a basic planet-wide cloud. This is in marked contrast to the other cloudy planets we know well, where the tendency is for more localized circulation cells to form with clear regions in the descending parts. The basic difference between Venus and Earth, Mars, Jupiter and Saturn in this respect is the slow rotation of Venus which allows the monocellular (Hadley) regime to be stable – not withstanding the rapid zonal winds superimposed upon it".

Fred Taylor also reports the discovery of another important consequence of the atmospheric circulation: "Greatly enhanced amounts of infrared flux were found to be emerging poleward of 80° latitude, in a localised, elliptical region which evidently is a clearing in the cloud cover forced by the descending air in the return branch of the Hadley cell". "The circumpolar collar is a current of very cold air which surrounds the pole at a radial distance of 2500 km. It is nearly 1000 km across, but only about 10 km thick in the vertical direction. Temperatures inside the collar are about 30°C colder than at the same altitude outside, so the feature generates pressure differences that would cause it to dissipate rapidly were it not continually forced by some unknown mechanism. Inside the collar lies the region of reduced cloud cover caused by the descending branch of the Hadley cell. Because of the zonal momentum transported from lower latitudes, the descending air is also rotating rapidly forming a polar vortex analogous to the eye of a hurricane or whirlpool but much larger and more permanent."

Climate Change on Planet Earth

Modelling of the thick, hot atmosphere of Venus has aroused considerable interest and concern among scientists who turned their modelling to illuminate existential questions about the warming of our own planet's atmosphere:

> Since the mid-1970s, James Hansen has focused on studies and computer simulations of the Earth's climate, for the purpose of understanding the single planet-wide cloud. He is best known for his testimony to Congress on climate change in the 1980s that helped raise broad awareness of the global warming issue. In recent years Dr Hansen has drawn attention to the danger of passing climate-tipping points, producing irreversible climate impacts that would yield a different planet from the one on which civilization developed. Dr Hansen disputes the contention, of fossil fuel interests and governments that support them, that it is an almost god-given fact that all fossil fuels must be burned with their combustion products discharged into the atmosphere. Instead Dr Hansen has outlined steps that are needed to stabilize climate, with a cleaner atmosphere and ocean, and he emphasizes the need for the public to influence government and industry policies.[51]

In the 29 June 2020 issue of *The New Yorker* there is a lengthy article by Elizabeth Kolbert about James Hansen titled *The Climate Expert Who Delivered News No One Wanted to Hear* and from which this alarming statement is taken: "Carbon dioxide isn't just approaching dangerous levels; it is already there. Unless immediate action is taken – including the shutdown of all the world's coal plants within the next two decades – the planet will be committed to a change on a scale society won't be able to cope with. This particular problem has become an emergency." The alternative, as the books of Ecclesiastes and Isaiah propose, would seem to be that we should eat, drink and be merry for tomorrow we die.

1978 Pioneer Venus Multiprobe

The Pioneer Venus Multiprobe spacecraft was launched on an Atlas Centaur from the Cape on 8 August 1978. It carried one large and three small entry probes whose purpose was to determine the structure of the atmosphere as they descended at four widely separated locations on Venus. Only the large probe was equipped with instruments to measure composition. Wind speeds and turbulence were inferred from the radio signals as they descended. The small probes were not equipped with parachutes and none were designed to survive at the surface.

Each probe consisted of a spherical titanium pressure vessel surrounded by a conical ablative heat shield. Each was filled with high-pressure inert gas to prevent crushing. The

---

[51] https://theyearsproject.com/science-advisor/james-hansen-ph-d/

large probe and one of the three small probes were aimed at the dayside of Venus while two small probes were targeted to the night side, one at high northern latitude. They were released on approach about 3 weeks before reaching Venus on 9 December 1978, spreading out, each with its different target.

In the case of the Large Probe there were 14 sealed penetrations in the pressure vessel: one for the antenna, four for electrical cables, two access hatches and seven for instruments. A 40 amp-hour silver-zinc battery provided power. This probe had a comprehensive payload: a neutral mass spectrometer, a solar flux radiometer, an atmospheric structure experiment, a nephelometer (to measure the density of cloud particles), a cloud particle size spectrometer, a gas chromatograph, an infrared radiometer, and radio science experiments. The large probe jettisoned its heat shield at 67 km altitude and began to make measurements all the way down to the surface. Its parachute was jettisoned at 47 km altitude and 36 minutes later the probe reached the surface in a hard landing.

The identical small probes designated North Probe, Night Probe and Day Probe were of a simpler design in which the aero-shell was permanently attached to the pressure vessel and they carried fewer instruments – an atmospheric structure experiment, a nephelometer that measures the size and concentration of suspended particles, a net-flux radiometer, and radio science experiments. The small probes reached the surface after a 50+ minute descent and, one, the Day Probe continued to transmit for more than an hour, presumably because of a fortuitous post-impact orientation of its radio antenna.

The probes entered haze at an altitude of about 75 km and the main cloud cover at about 50 km altitude, exiting the clouds at about 30 km altitude. Wind speeds gradually fell from 100 meters/sec to almost calm conditions at the surface just as the earlier Venera landers had found. The Large Probe confirmed the 1970s observations of Godfrey Sill and the Youngs that the clouds contain sulfuric acid droplets and elemental sulfur. Based on net radiation measurements it was concluded that the atmosphere experiences radiative cooling at high latitudes and radiative heating at low latitudes. Cloud structure was significantly different at the three sites sampled by the small probes. "A feature of unusually large opacity found near 60 kilometers at the north probe site is probably related to the unique circulation regime revealed by ultraviolet and infrared imagery".[52]

1978 Venera 11 and 12

In the same Venus launch window as the Pioneers, the Soviets launched two of their latest Venus probes: Venera 11 on 9 September 1978 and Venera 12 5 days later. Independent of their planetary science objectives, the Venera 11 and 12 carrier spacecraft, together with Prognoz 7, were part of a French-Soviet collaboration SIGNE 2 in which gamma ray detectors provided by France formed an interplanetary gamma ray burst triangulation network. The landers were separated at Venus and the carrier spacecraft flew on by. They provided a communication link for the landers for only about 95 minutes. 143 gamma ray bursts were observed and led to the 'most plausible model' of their origin to be 'a binary with a neutron star'. This kind of interdisciplinary collaboration is something that had not been contemplated by NASA.

The basic lander design was similar to that of Venera 9 and 10. Separating from its flight platform bus (which flew on by the planet and acted as a data relay) on December 23 the Venera 11 lander entered the Venus atmosphere two days later and after an hour-long descent (on a parachute to an altitude of 50 km where it was jettisoned) made a soft landing on 25 December. This was 4 days after Venera 12. The landers were instrumented to measure the temperature during descent and at the surface along with atmospheric and soil composition. Lightning and thunder were also recorded. Unfortunately, most of the instrumentation failed to work as planned: especially, there was much disappointment when the imaging system lens covers on both spacecraft failed to separate.

---

[52] *To See the Unseen: A History of Planetary Radar Astronomy* By Andrew J. Butrica

NASA's Late 1970s Plans for Radar Mapping of Venus: VOIR

The small planetary radar science community and the larger planetary geology community, excited by the Pioneer Venus Orbiter's radar altimetry results, had been eager to using synthetic aperture radar imaging from Venus orbit to map the planet at a resolution comparable to what Mariner 9 had achieved at Mars (see Chapter 5). This was to be done by means of the Venus Orbital Imaging Radar (an apt acronym: VOIR) mission – the highest priority new start for the Solar System Exploration Program. The mission, to be launched on the Space Shuttle (then in final development) received approval by the outgoing Carter Administration for inclusion in the President's last budget submission to Congress: the Fiscal Year 1982 budget.

The White House's interest in deep space exploration was such that Laurel Wilkening (1944-2019), supporting the Solar System Exploration Program Office on leave from the University of Arizona (where she was Department Head of Planetary Sciences and Director of the Lunar & Planetary Laboratory) was invited to join President Carter in watching coverage of Voyager 1's encounter with Saturn in November 1980. Also in 1980 the Public Broadcasting System (PBS) began airing the 13 part television series *COSMOS: A Personal Voyage* presented by Carl Sagan (one of the most eloquent of astronomers and planetary scientists). It was the most widely watched series in the history of American public television and has since been broadcast in more than 60 countries and seen by over 500 million people. A book was also published to accompany the series.

So, President Carter's interest was evidently widely shared by the American people. Alas, in Washington DC the political winds can change direction quite rapidly just as on Venus, as will be discussed in Chapter 8 (Exploration Program Cancellation & Recovery). And did so when Ronald Reagan became the 40th President of the USA in January 1981.

1981 Venera 13 and 14

The Soviet team was ready to go again three years after the Venera 11 and 12 launches and with essentially the same equipment. Venera 13 and 14 were launched on Proton rockets from the Baikonur Cosmodrome 5 days apart: Venera 13 on 30 October 1981 and Venera 14 on 4 November 1981. As before, after a four-month cruise, the landers were separated from their flight platform (which also served as their data relay) on approach to Venus and entered the atmosphere. They deployed their parachutes at 63 km altitude, released them at 47 km and descended by aero-braking down to the surface. Three distinct cloud layers were detected on the way down.

Venera 13 landed east of Phoebe Regio at 7.5°S, 303°E on 1 March 1982 about 950 km of where Venera 14 would touch down four days later. This time the camera lens covers were successfully detached and each spacecraft acquired a color panoramic view of what turned out to be their colorless surroundings. A color scale was included in each image and this allowed the yellow/orange atmospheric color effect produced by the thick atmosphere to be removed. The resulting images reveal dark scenes of bedrock slabs and fine-grained soil.

Besides the camera system, the landers were instrumented with an X-ray fluorescence spectrometer, a drilling arm and surface sampler, a dynamic penetrometer, and a seismometer to conduct investigations on the surface. Samples acquired by a drilling arm were deposited in a sealed chamber and their composition – found to be that of an igneous rock similar to basalt – determined by an X-ray fluorescence spectrometer. Venera 13 survived at the surface for a remarkable 127 minutes while Venera 14 survived for 57 minutes in an environment of 470°C and a pressure of 94 atmospheres. Silicon chips won't work after they reach 250°C.

1983 Soviet Venera 15 and 16

The Soviet Union pressed on with their comprehensive program of missions to Venus and in 1983 followed their series of landers with two spacecraft, Venera 15 and 16, that would begin the high-resolution radar mapping of Venus from orbit. They were instrumented with side-looking radar and the goal of mapping the northern hemisphere from 30°N to the pole at

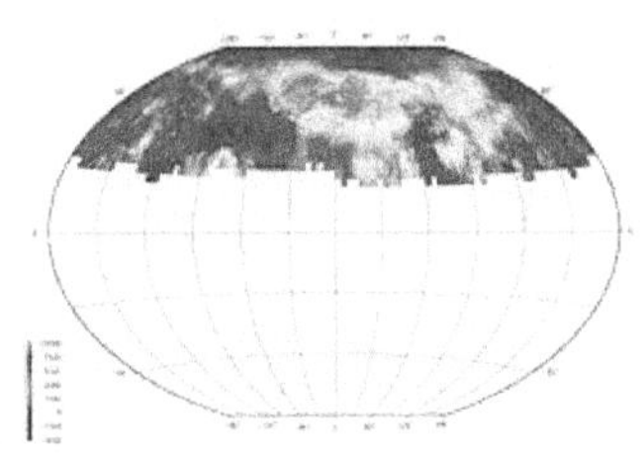

Venera topography by Venera 15/16

a resolution of 1-2 km. This would involve the acquisition of 150 strips of coverage 7,000 km in length. The spacecraft were launched in June 1983 on Protons out of the Baikonur Cosmodrome and arrived at Venus a day apart four months later. They were inserted into near-polar eccentric orbits with periapsis at 62°N and 24-hour period. The periapsis altitude was 1,000 km and an apoapsis 65,000 km. The radar mapping and altimetry were carried out from the low-altitude part of the orbit over 8 months.

The decision to map from 30°N and up to the pole meant that the coverage did not include most of the landing sites of the earlier Venera missions. With a useful fraction of Venus now mapped at high resolution, Valerie Barsukov, Alexander Basilevsky and colleagues were able to describe the geology and geomorphology of Venus thus:

"The predominant type of terrain of the studied area is a plain that was tentatively subdivided into five morphological types: ridge-and-band, patchy rolling plain, dome-and-butte plain, smooth plain, and high smooth plain. Stratigraphically, the ridge and band plains are the oldest and the smooth plains are the youngest. The stratigraphic position of the other types is yet to be determined. Large sections of the plains show similarities to the mare-type basaltic plains of the moon, Mercury, and Mars. Other types of terrain are combinations of ridges and grooves in various patterns: linear parallel, orthogonal, diagonal or chevron-like, and chaotic. In some places the ridge-and-groove terrain is stratigraphically below the plain material, but in other places it appears to be plains material that has been subsequently deformed. Near the eastern end the ridge and band plains are the oldest and the smooth plains are the youngest.

The stratigraphic position of the other types is yet to be determined. Large sections of the plains show similarities to the mare-type basaltic plains of the moon, Mercury, and Mars. Other types of terrain are combinations of ridges and grooves in various patterns: linear parallel, orthogonal, diagonal or chevron-like, and chaotic. In some places the ridge-and-groove terrain is stratigraphically below the plain material, but in other places it appears to be plains material that has been subsequently deformed. Near the eastern and western boundaries of Ishtar Terra large (several hundred kilometers in diameter) ring-like features can be seen that are named coronae or ovoids. Evidence of tectonic deformation and the presence of flow-like patterns support their designation as volcano-tectonic features. Beta Regio seems to be an uplifted plain showing evidence of rifting and volcanism. All types of terrain are sparsely peppered with craters of obvious impact morphology. Their average density gives the plain an age range of 0.5 to 1x10 Byr. The fact that many impact craters are still in the pristine state indicates a very low rate of surface reworking, at least for the last 0.5 to 1x10 Byr. No evidence for water-erosion-sedimentation processes has been found. The tectonic activity of Venus has no equivalent on the Moon, Mercury, or even Mars, and can be compared only with that of the Earth. Intensive horizontal deformation, previously known only on Earth, occurs on Venus, but in a characteristic Venusian style".[53]

Based on the areal density of the craters that had been mapped, Basilevsky and colleagues estimated that the surface is about one billion years old – just yesterday! Jim Head (another Bellcomm alumnus) of Brown University had been invited to be a guest investigator on the Venera 15 and 16 missions. As NASA was planning a similar radar-mapping mission – VOIR – the willingness of the Russian team to share their analyses of the data with US colleagues proved very helpful. Alexander Basilevsky and Valery L. Barsukov, director of the Vernadsky Institute, presented Venera 15 and 16 results at the 1984 Lunar and Planetary Science Conference held in Houston and a year later they described their

---

[53] http://adsabs.harvard.edu/full/1986LPSC...16..378B

ongoing analyses at the first Brown-Vernadsky 'microsymposium' held at Brown University. These planetary science microsymposia have been annual events since then with the location alternating between the US (Providence & now Houston) and Moscow.

The Venera radar mapping data also allowed US scientists Robert Grimm and Sean Solomon to assess the history of the crust of Venus given the very high surface temperature that might be expected to lead to solid-state creep of the crustal material. They concluded that the preserved relief of the largest craters constrained the crustal thickness to be less than or equal to 10-20 km and that the average rate of crustal generation has been much smaller on Venus than on Earth – or that some kind of crustal recycling has occurred.

1984 Vega 1 and 2

The last two Soviet spacecraft to visit Venus were Vega 1 and Vega 2 (Vega is a contraction of Venera and Gallei (Russian words for Venus and Halley)). The new names came about because the Soviets, unlike the US, planned to take advantage of the 1986 Comet Halley apparition by launching a spacecraft to the comet by way of a gravity assist at Venus. This was to be a mission of considerable ambition: balloons to be deployed in the Venus atmosphere, Venera landers, and flybys of Comet Halley! Jacques Blamont, (1926-2020) a leading scientist at the French Space Agency CNES, was the principal advocate of using balloons to explore planetary atmospheres. The thick Venus atmosphere is, perhaps, the most obvious target for such an experiment and the Soviets were carrying out regular missions to the planet. Blamont had recognized that, as the comet-bound spacecraft acquired their gravity assists at Venus, each could drop a balloon probe into the Venus atmosphere and he persuaded his Soviet colleagues to support the plan.

The spacecraft were launched on December 15 and 21, 1984. Vega 1 arrived at Venus on June 11, 1985 and Vega 2 four days later. Each released a 1,500 kg, 240 cm diameter spherical descent unit a few days before arrival at Venus. These probes each contained a lander and a balloon, similar to a terrestrial weather balloon. The balloon deployment was a choreographed sequence of events that followed the entry of the Lander/Balloon capsule into the atmosphere on 11 June 1985: a parachute attached to the cap on the landing craft opened at 64 km altitude and the balloon was pulled out by another parachute at 61 km altitude. 200 seconds after entry, at 55 km altitude, the furled balloon was extracted and inflated 100 seconds later at 54 km when the parachute and inflation system were jettisoned. Ballast was jettisoned when the balloon reached roughly 50 km and the balloon floated back to a stable height between 53 and 54 km. The 3.4-meter diameter balloon was made of a Teflon cloth matrix and was filled with helium to an overpressure of 30 mbar. The battery-powered communication equipment was suspended on a 13-meter line. Total mass was 21.5 kg – 12.5 kg for the balloon and tether, 6.9 kg for the gondola, and 2.1 kg of helium in the balloon. Science instrumentation consisted of pressure, illumination and temperature sensors plus an anemometer to measure wind speed. "The mean stable height was 53.6 km, with a pressure of 535 mbar and a temperature of 300-310K in the middle, most active layer of the Venus three-tiered cloud system. The Vega 1 balloon drifted westward in the zonal wind flow with an average speed of about 69 meters/sec at nearly constant latitude. The balloon measured downward gusts of 1 meter/sec and horizontal wind velocities up to 240 km/hour. The probe crossed the terminator from night to day after traversing 8500 km. The probe continued to operate in the daytime until the final transmission was received on 13 June after a total traverse distance of 11,600 km". [54]

---

[54] https://nssdc.gsfc.nasa.gov/nmc/spacecraft/display.action?id=1984-125F

CHAPTER 5

RACE TO MARS: MARINERS AND VIKINGS

Mariner 4, 6, 7 Flybys & Mariner 9 Orbiter,

Viking Orbiter & Lander 1 & 2

The National Academy's 2019 report on the *Search for Life in the Universe* reminds us how the new science discipline of Astrobiology came into being: The first international conference on the origins of life took place in Moscow in 1957 … astronomers, geologists, biologists, and others rapidly saw the potential of access to space as a new venue for research. The founding of the Committee on Space Research (COSPAR), of NASA and of the National Academy's Space Studies Board (SSB) in 1958 accelerated the creation of a multidisciplinary space science community: in 1959 Gerard Kuiper went to the University of Arizona to found what became the Lunar and Planetary Laboratory/Department of Planetary Science. Thomas Gold went to Cornell University to establish the Center for Radio Physics and Space Research separate from an astronomy department. Robert Sharp at Caltech transformed the Division of Geological Sciences into the Division of Geological and Planetary Sciences. Joshua Lederberg and others interested in the search for life beyond Earth banded together to establish the new scientific discipline of exobiology. NASA funded its first exobiology project in 1959. In a response to a request by NASA, the SSB published an extensive series of reports on life in the universe. The first concluded:

> The biological exploration of Mars is a scientific undertaking of the greatest validity and significance. Its realization will be a milestone in the history of human achievement. Its importance and the consequences for biology justify the highest priority among scientific objectives in space – indeed in the space program as a whole.

Given the outstanding importance of exploring Mars in search for evidence of extra-terrestrial life, the US-Soviet space competition had an ambitious dimension besides landing men on the Moon – first, to establish by means of robotic spacecraft whether life may have evolved on Mars and second, to follow-up by landing a human crew on that planet. As in the lunar case, the robot mission precursors to manned missions would again involve flyby spacecraft and orbiters to be followed by highly instrumented landers. Minimum energy Mars launch windows open at intervals of 26 months, the synodic period (the time between two successive oppositions) with respect to Earth. The Soviet efforts again ran into misfortunes while NASA, after disappointing results from Mariners 4 (1965) and 6 & 7 (1969), was much more fortunate with both the Mariner 9 orbiter in 1971 and the two subsequent Viking – both orbiter & lander – missions in 1976.

In the late 1960s NASA seriously considered, and then abandoned, a very ambitious and expensive plan to use a single Saturn V to send two orbiters and two landers to Mars. This plan would have served to keep the great launch vehicle in production and maintain the possibility of moving on to a manned Mars mission. Even the proposed robotic mission was, however, judged to bust the bank and was not approved. Not to be discouraged, this plan evolved into the still-very-ambitious orbiter & lander Viking program that was managed overall by NASA's Langley Research Center with JPL managing the orbiters. The program began with Mariner flyby missions and then Mariner & Viking orbiters. Like all planetary missions of that era the uncertainty of mission success was hedged by duplicating the Titan-Centaur launchers and spacecraft – a policy that served well and continued in most cases up to and including the launches of two Voyager spacecraft in 1977.

1964 NASA Mariner 4

NASA's first Mars mission was a flyby, Mariner 4, launched in 1964. JPL's Mariner and, later, Voyager spacecraft were all 3-axis-stabilized spacecraft with a scan platform that was used to point cameras and other instruments at targets designated by the science teams. As such they were ideal for acquiring images and spectrometer data whereas a spin-stable spacecraft like the NASA Ames Pioneers were a less expensive match for magnetometer experiments and instruments that measure the flux of energetic particles. Radio science experiments are the key to measuring gravity fields and also support determination of atmospheric temperature/pressure profiles (accomplished by tracking the radio signal as a spacecraft disappears behind the planet). Radio science experiments perform well with both kinds of spacecraft.

Mariner 5 was built as a backup to Mariner 4 to be launched 26 months later if necessary. Because Mariner 4 was successful Mariner 5 was saved and modified for a 1967 Venus mission.

NASA's Mariner Mars program experienced most of its misfortunes at launch: Mariner 3 (in 1964) and Mariner 8 (in 1971) were both lost at this point. The latter loss was a formative experience for the author. The Mariner 4 spacecraft did make it all the way to Mars and flew by the planet at the end of July 1965 at a distance of ~10,000 km. It had an extensive science payload but, not surprisingly, it was the images of Mars that received the most attention. Disappointingly, they revealed little more than lunar-like impact craters and a somewhat obscuring atmosphere. Edward Smith's magnetometer team made the most important discovery of the mission, namely, that Mars lacks a magnetic field like that of Earth. The implications of this for the evolution of Mars are still being understood. In fact, as was learned decades later, Mars does have a very weak magnetic field limited to parts of the southern hemisphere. This absence of a detectable field placed important constraints on its present-day interior and on the interaction of Mars with the solar wind. Without a significant magnetosphere the solar wind impinges directly on the top of the atmosphere and, adding to the thermal escape mechanism (Jeans escape), erodes it over time.

1969 NASA Mariners 6 and 7

Two more Mars flyby missions had been planned before the first orbiters that were due to be launched in 1971. Mariners 6 and 7 were launched to Mars in 1969 and successfully flew by Mars in late July and early August 1969, returning data from their cameras, UV and IR spectrometers and radiometers. As always, radio occultation measurements (observing the attenuation of the signal causing small changes in frequency as the spacecraft vanished behind the planet and later reappeared) were used to sound the atmosphere. The radio science team led by Arvydas (Arv) Kliore (1935-2014) of JPL reported the surface pressure at Meridiani Sinus to be about 6.6 mbar (compared with Earth's 1 bar) along with a temperature of about 276K. The re-emerging radio beam occurred at high northern latitudes indicating a surface pressure of about 6.4-mb and temperature of about 158K. Thus, the thin atmosphere of frigid Mars, while capable of creating global scale dust storms, was determined not to be able to support surface water in its liquid phase except transiently (the triple point of water is 6.1-mb and 273K).

Measurement of the full range of atmospheric pressure from the top of the highest point to the lowest and throughout the seasons would await the observations made by an orbiter. In the present era the dramatic Mars dust storm season begins just after its perihelion passage and can be global in scale. Over geologic time the obliquity changes and the orbital eccentricity changes also – leading to important geologic consequences discussed later. Mariner 6 & 7 had been programmed to image areas in the equatorial region and the region around the South Pole. By chance both succeeded in missing Mars' most characteristic features – huge canyons, channels cut by floods of water, huge ancient volcanoes, giant impact basins, and layered sediments at the poles that together are the most spectacular and characteristic features of the planet. There was a moment of excitement when data from the infrared spectrometer of Principal Investigator George Pimentel (1922-1989) appeared to have identified methane in the atmosphere but this finding was quickly withdrawn.

1971 Soviet Mars 2 and 3

The Soviets had been first off the mark in attempting robotic Mars missions, launching their first attempt in 1960 and trying again several times before, in 1971, they succeeded in inserting their Mars 2 spacecraft into orbit. Mars 2 had released a descent module some hours before. This evidently entered the atmosphere too steeply and crashed. The Mars 2 orbiter in its 18-hour orbit began its planned observations of Mars. The Mars 3 spacecraft was identical to Mars 2 and its descent module made it successfully to a soft landing in the southern highlands of Mars using parachutes and retrorockets – a remarkable achievement. Alas, it only operated for a matter of seconds and then, for an unknown reason, ceased to communicate. So, the tiny rover that the lander carried did not have a chance to achieve a historic 'first'. The accompanying Mars 3 orbiter was not as successful in achieving its desired orbit as Mars 2 and was only able to achieve a highly elliptical orbit of nearly 13 days period. As it happened neither orbiter was able to return useful images because the planet was experiencing a dust storm of global scale and unprecedented duration. Images acquired by Chick Capen at the Lowell Observatory in Arizona recorded just how enveloping the 1971 dust storm was.

The Soviet orbiters had pre-programmed science data acquisition so their recorders were filled with mostly useless image data before the dust had settled out. Other instrument data only provided information about the upper atmosphere and the range of surface pressure and temperatures. This kind of misfortune must have been enough to make grown men weep and likely the Soviet team did so.

1971 NASA Mariner 8 and 9

The Mariner 8 and 9 spacecraft, planned to be the first two US Mars orbiters, were similar in design to the flyby Mariners, mainly requiring only the addition of the fuel tanks and engine of a sizeable retro-propulsion module. The principal contribution of the two orbiters was expected to be Mariner 8: global high-resolution mapping of the entire planet along with Mariner 9: an investigation of the seasonal changes that astronomers had observed over previous centuries. The imaging science team was led by Hal Masursky (1922-1990) and included several of his USGS colleagues: Jack McCauley (1932-2012), fellow Brit Michael Carr, Larry Soderblom & Danny Milton. Merton Davies (1917-2001) was the team geodesist. Also on the team were astronomers Brad Smith (1931-2018) and Carl Sagan (1934-1996) Nobel Prize winning molecular biologist Joshua Lederberg (1925-2008), atmospheric scientists Conway Leovy (1933-2011) and Jim Pollack (1938-1994), geologist Bruce Murray (1931-2013), planetary scientists Bill Hartmann, Joe Veverka and Jim Cutts (another Brit), engineer Elliot Levinthal (1922-2012) and the author who was tasked to plan image acquisition to support all of the team's science goals. Several of the team became great personal friends of the author and those that have moved on are greatly missed.

A small team of Bellcomm scientists that included the author had been recruited by the Viking Project Office at NASA Langley to participate in the Mariner Orbiter imaging team to better understand the threat to landers posed by the global dust storms that astronomers had observed. The team was led by Bill Thompson and included mathematician Curt Greer who wrote software for the author to complete and use to plan Mariner's approach and orbital imaging sequences. Those were early, enormously stimulating days! The reader is referred to Keay Davidson's biographic *A Life of Carl Sagan to* better understand the spirit of those times.

On 9 May 1971 Mariner 8 was launched on an Atlas Centaur from the Cape. The Atlas functioned as intended but the Centaur tumbled out of control and Mariner 8 returned to Earth splashing down some 560 km north of Puerto Rica. Mariner 9 was launched successfully on 30 May – a big relief – and reached Mars on 14 November 1971 to become the first US spacecraft to orbit another planet. Because of the Mariner 8 launch failure, the mission's science objectives – global geologic mapping and surface & atmospheric dynamics – that had been shared between the two spacecraft were rapidly combined into the plans for the remaining mission. Global mapping became the obvious priority.

Mariner 9 entered a 64° inclination, highly elliptical orbit of 12-hour period. Periapsis altitude (closest approach) was 1,650 km while aphelion was ten times further. The closest approach was at 64°S so that mapping of the southern hemisphere was greatly favored in terms of spatial resolution. The two cameras and spectrometers on the scan platform were normally pointed along the ground track but off-nadir views of high latitudes could be obtained as well as oblique images toward the planet's limb in order to study the atmosphere. One camera provided wide-angle coverage with ~1 km resolution from the lowest point in the orbit. The second had ten times higher resolution. The wide-angle camera was used to systematically map the entire planet providing 1–2 km resolution coverage of the southern hemisphere with much less good resolution in the north. The narrow-angle camera acquired images in the overlap between the wide-angle frames so that, in this way, both global coverage and an enhanced ability to interpret the mapping data were provided.

By contrast to the great misfortune of the Soviet effort, Mariner 9 was able to wait out the dust storm until early January 1972, at which time it began its systematic mapping of Mars with cameras and spectrometers. During those tense early months of waiting, the Mariner 9 radio-science team was able to measure the gravity field and shape of Mars and begin to assess its internal structure. Meanwhile the radio-science team continued to observe the change in their received signal as, each orbit, the spacecraft passed behind and, later, re-emerged from behind the limb of the planet. From these

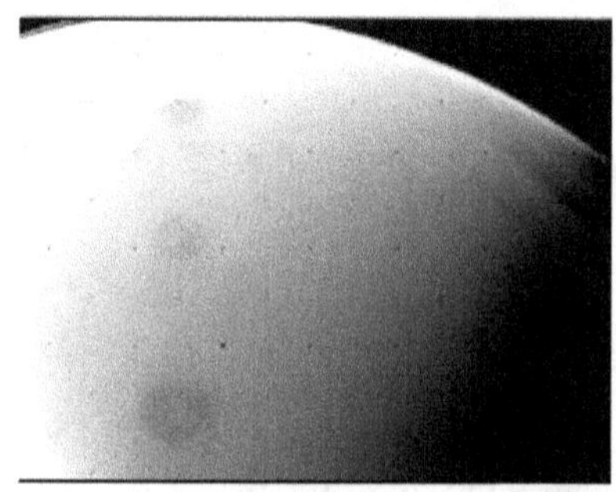

observations the changing vertical temperature/pressure structure of the dust-filled CO_2 atmosphere was recorded. These measurements, carried out throughout the mission, provided essential data for the modelling of the atmospheric circulation by Conway Leovy and his colleagues at the University of Washington. For week after week the dust storm persisted making everyone highly nervous. Curiously, the first images that began to convey information about the surface showed giant dark spots, three of which were in a line across the equator and looked somewhat like craters. The fourth dark region was in the location of what astronomers had called Nix Olympica – the 'Snows of Olympus'. Also in the equatorial region was a pattern of brightening that had linear features – all of which was tantalizing and frustrating in equal measure to a team that had spent many years preparing for the first global view of Mars. By January the dust finally began to settle out and the dark crater-like features resolved themselves into the enormous calderas of giant volcanoes.

If NASA had been planning a first mission to land astronauts on Mars then the Mariner experience with the global dust storm that lasted for many months would have provided the Agency with much food for thought (decades later it also provided the plot for a Hollywood movie called *The Martian*). When the dust storm finally began to clear, the Mariner 9 science teams were privileged to be the first to find out what Mars is really like. As the cameras systematically mapped the planet, starting in the southern hemisphere, the imaging team were somewhat in awe as they grasped the highly evolved nature of the planet in contrast to the impression gained by the previous flyby Mariners. It soon became clear what the earlier flybys had missed. Short of discovering astronomer Percival Lowell's imaginary canals and Little Green Martians, Mars was about as exciting as anyone could have imagined for it is, in its alien way, strikingly beautiful and is dynamic with a complex history that will keep geologists, atmospheric scientists, glaciologists and astrobiologists busy for many decades to come.

Mariner 9's repeated imaging of selected surface sites established that the seasonal changes that astronomers had recorded as the 'wave of darkening' (and that had captured Lowell's imagination) were simply the result of the transport of dust from one place to another by fierce winds. Just a few years later the Viking orbiters provided much improved global coverage. It's worth mentioning that all the Mariner 9 images were of low intrinsic contrast – because of the still dusty atmosphere – and required extensive computer processing to enable their interpretation. As such, JPL image processing advances led to some of the software that we take for granted today. The ability to mosaic images digitally did not yet exist and the science team was provided with hard-copy photographs that were cut and pasted together by hand.

The southern hemisphere of Mars is heavily cratered like the lunar highlands and includes three giant basins identified on old maps as *Hellas, Argyre* and *Isidis*. Unlike the lunar basin maria, they are not filled with lava. The Hellas basin is about 7 km deep. There are no mascons on Mars so the histories of the Moon and Mars have significant differences.

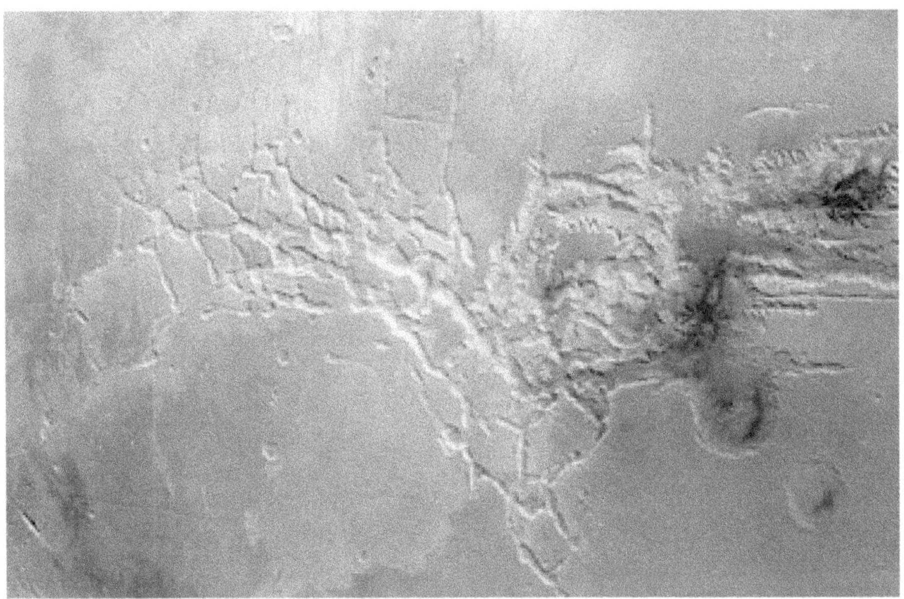

In the equatorial region, the brightening was found to be associated with a labyrinthine maze of steep-walled valleys – Noctis Labyrinthus – at the western end of a huge canyon system – Valles Marineris (also up to 7km deep) – that runs just south of the equator, a quarter of the way across Mars. Arizona's Grand Canyon is not even tiny by comparison.

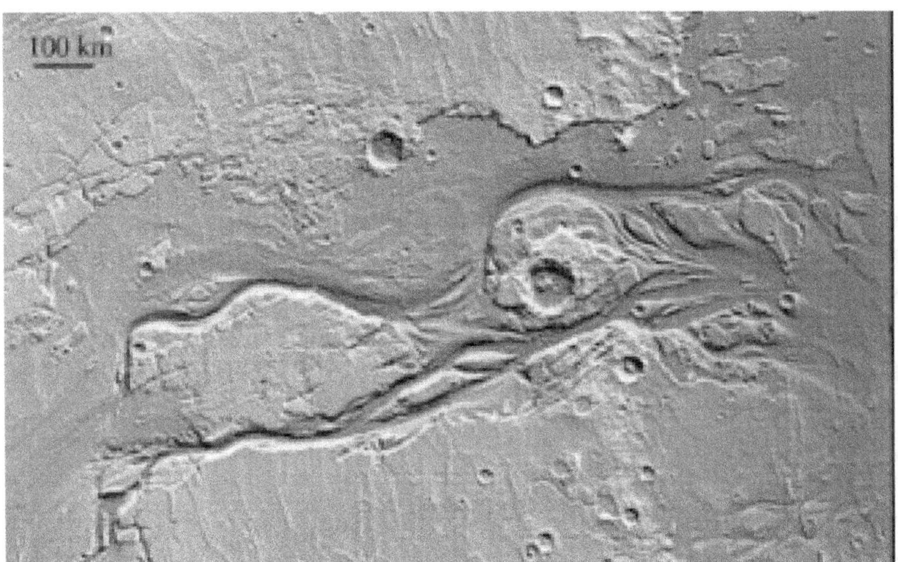

In addition to the giant Valles Marineris, there are numerous enormous channel features arising from a region of extensive collapse – chaotic terrain – that were compelling evidence that catastrophic flooding by water had taken place in the past: another breakthrough in understanding Mars and its many differences from the Moon.

Mariner 9 observed two quite different kinds of channels. The most remarkable are streamlined outflow channels that extend for hundreds of kilometers. *Kasei Vallis* is the largest of these stretching for 3,500 km, 400 km wide and cutting 2.5 km into the surface It begins in a canyon, *Echus Chasma* close to Valles Marineris, runs northward and empties into a smooth circular plain in the northern equatorial region. This region, named *Chryse Planitia*, is 1,600 km in diameter and has a floor 2.5 km below the Mars datum. These amazing channels appear to have been cut by immense floods similar to those that created the Channeled Scablands on the eastern side of Washington State in the US – the result of the breaching of a large glacial lake (Lake Missoula) as proposed in the 1920s by J Harlen Bretz (1882-1981) and J T Pardee (1871-1960). The canyons of Valles Marineris might have been lakes that were subsequently breached catastrophically. Some of the channels begin full-bore in regions of chaotic collapse suggestive of the release of ground water under enormous pressure, perhaps as a result of an earthquake or an impact.

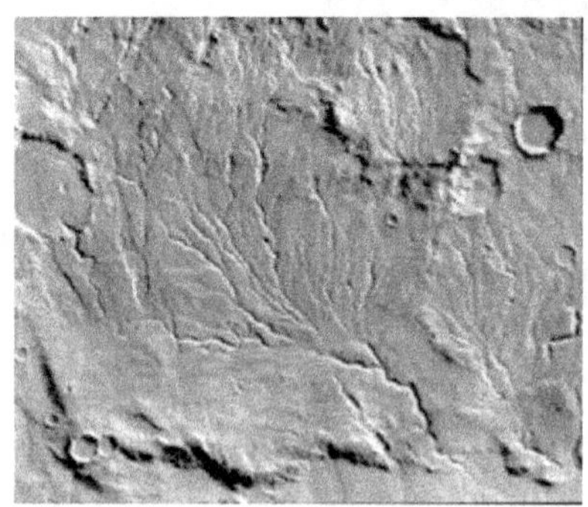

Other channel features that were resolved by the Mariner 9 cameras are branching networks of valleys in the ancient cratered southern highlands that appear much like terrestrial river drainage basins. An individual valley is less than 5 km wide but the network may extend for hundreds of km evidently draining into topographic lows. The higher resolution images and laser altimetry acquired by subsequent orbiters would establish the distribution, ages and nature of these channels. Could it be that once it had rained on Mars? There were yet other kinds of channels that were judged to be evidence of ground-water sapping erosion by groundwater that emerges as seeps and springs.

No one had anticipated these discoveries because the present thin atmosphere and its cold temperature make surface liquid water incompatible with present conditions. The discoveries provided considerable encouragement to astrobiology members of the imaging science team like Carl Sagan and Joshua Lederberg that Mars might well have once been hospitable to the evolution of life. Indeed, the evidence of surface water in the past on Mars is among the most important findings of Solar System exploration to date. Using later orbiters to identify past environments in which water was freely available at the surface, and using landed spacecraft to study them, would become the principal thrust of subsequent missions to Mars whose direction has been to *Follow the Water*!

When the northern hemisphere had been mapped (from higher orbital altitude with less good spatial resolution) it became clear that Mars exhibits a distinct crustal dichotomy – the elevation of the much-less-heavily-cratered north is 1 to 3 km lower than the southern highlands. The origin and consequences of this massive planetary scale feature was a major puzzle and has been the focus of much subsequent research including the search for evidence that the northern lowlands might even have been the location of an ocean – see Mars Reconnaissance Orbiter findings in Chapter 14. The earlier uninspiring Mariner 6 & 7 images had made Mars appear to be not much different from the Moon but Mariner 9 changed that.

For the several years preceding the Viking launches in 1975 there followed intensive analysis of the Mariner 9 data to better understand this amazing planet and to prepare to choose landing sites for the Viking spacecraft in their search for evidence of extra-terrestrial life. The imaging coverage provided the data that allowed geologic maps of Mars to be produced by the US Geological Survey – maps that have been fundamental to our understanding of the planet. Hal Masursky and colleagues at the US Geological Survey in Flagstaff took on this task. Even with the limited Mariner 9 resolution available, David Scott and Michael Carr were able to create the first global map of Mars at 1 to 25 million scale.

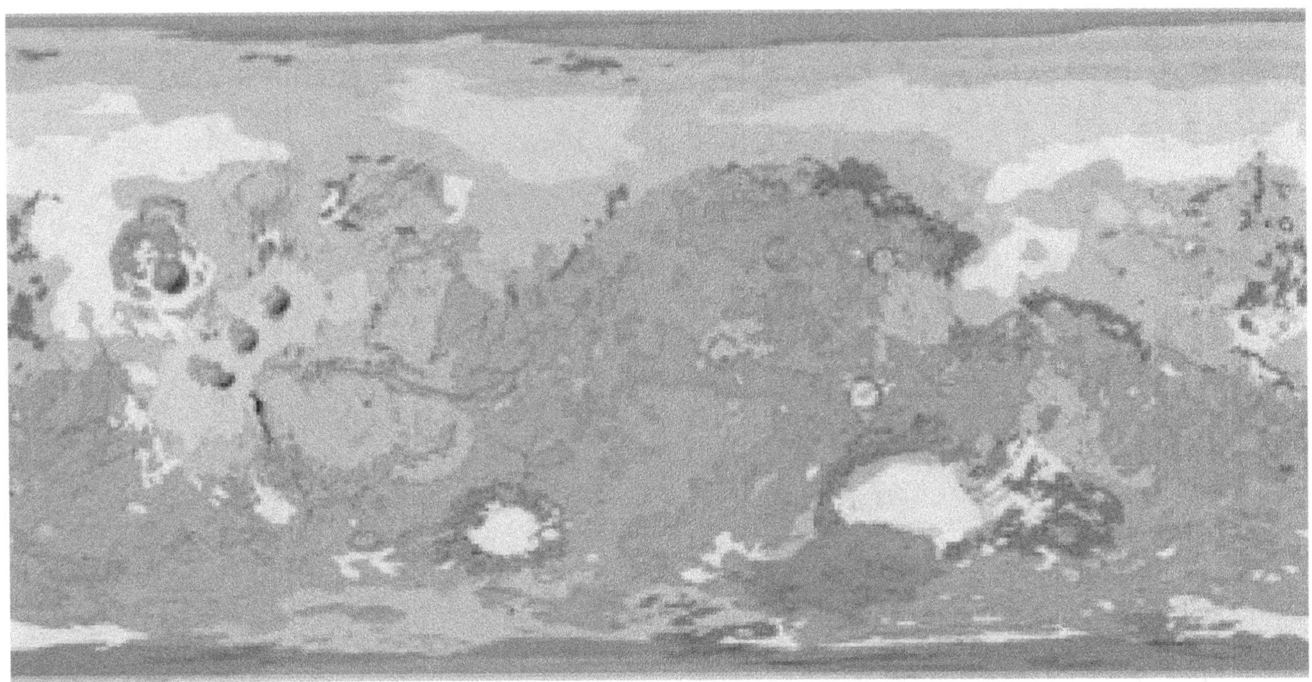

Mike Carr would lead the imaging team of the Viking Orbiter mission that would follow Mariner 9. The latitude/longitude grid of the geologic maps was – and is – based on the coordinate system that the Lowell Observatory had used long before but, following Mariner 9, this is now precisely controlled by the geodetic grid that Merton Davies (1917-2001) of the RAND Corporation developed using the wide-angle Mariner images. He had already established a preliminary control net for Mars back in 1970 and, years before, had played a leading role in the creation of the US space reconnaissance satellites. Appropriately, the 48 km diameter crater *Davies* – located on the prime meridian at 46°N – is named for him. He was involved in a number of other deep space missions and extended his work to Mercury, Venus and the moons of the outer planets.

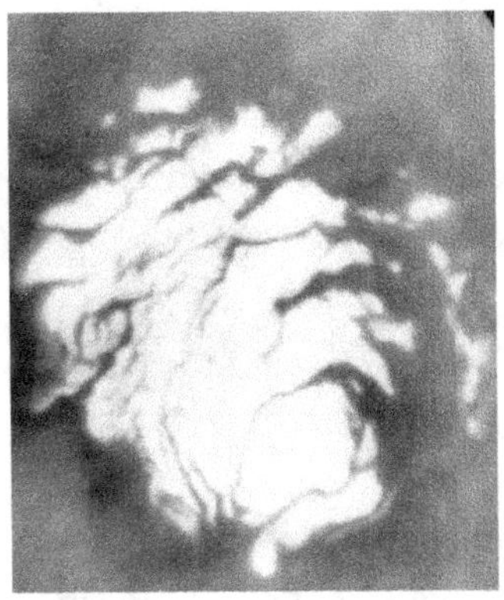

As astronomers had monitored, the two polar caps grow during the lengthy winter when the carbon dioxide atmosphere condenses as dry ice on the surface and retreats during the summer, in both cases to reveal permanent ice caps. Mariner 9 made the first orbital observations of the residual south polar cap during the southern summer combining images, infrared spectroscopy and infrared radiometry to assess the temperature and composition of the bright CO_2 ice and – a big surprise – the thick sequence of layered deposits that make up dark surrounding terrain. These deposits lack craters and are evidently the youngest terrains on Mars.

In winter, the polar caps grow underneath cloud cover (the 'polar hood' that had been recorded by astronomers). The north cap grows to ~60°N with more than a meter of dry ice at the pole. As much as a third of the atmospheric CO_2 is deposited as the cap grows and this is released back during the spring leading to a significant annual cycle of atmospheric pressure. This had been predicted by Robert Leighton (1919-1997) & Bruce Murray and by Conway Leovy before the first space missions took place.

Because in the present era southern winters are longer and colder than the northern winters, the southern polar cap grows larger than in the north – to ~50°S – during winter but the permanent summer ice cap in the south is smaller (350 km across) than the one in the north (1,000 km). Spiral troughs cut through both residual caps. More unexpected was the observation that the permanent caps sit upon layered sedimentary deposits characterized by lighter and darker material in layered stacks hundreds of meters thick. The individual layers appeared to be tens of meters in thickness and the deposits extend laterally for hundreds of km. The sediments are interpreted to be the result of dust and ice falling to the surface and accumulating there. Evidently the layers are the result of cyclically changing climate similar to the Milankovitch Cycle that

Earth experiences as its orbital eccentricity and obliquity change systematically with time due to the gravitational attraction of neighboring planets (particularly the large ones – Jupiter and Saturn). In 1938 Milutin Milankovitch proposed that such cycles may partly explain the Ice Ages. William Ward, a recent Caltech PhD, soon published an analysis of the phenomenon in terms of variations in the insolation of Mars resulting from oscillations of the orbital eccentricity (varying from 0.005 to 0.141) and of the obliquity (the tilt of the rotation axis) that varies from 14.9 degrees to 35.5. The obliquity variation has the much larger effect: Ward calculated that the yearly insolation at the poles of Mars varies by over 100% between the extremes of the obliquity range. The orbital periapsis pole changes from north to south and back again on a 25,000-year cycle.

A huge sea of sand dunes – forming a 'collar' up to 500 km in extent surrounds the layered polar terrain in the north. Given that dust storms are so common it is no great surprise to find that wind action on the surface is pervasive. Dunes are not formed simply by the deposit of dust particles picked up by the wind and suspended in the air but rather by the process of saltation that had been studied by Ralph Bagnold (1896-1973) who had been commander of the British Army's Long Range Desert Group during World War II: sand-sized particles are lifted by the wind and, falling back, knock up others that bounce and jump along.

Because NASA is in the aeronautics business as well as space exploration it has just the kind of capabilities needed to understand how sand dunes form on Mars. Ron Greeley (1939-2011) of Arizona State University founded the Planetary Aeolian Laboratory at Ames with a Mars Surface Wind Tunnel and a Titan Wind Tunnel also. Tests carried out in this facility indicate that the optimum size for particle movement on Mars is similar to that on Earth – about 0.1 mm – which is somewhat surprising given how thin the atmosphere is. In addition to the north polar collar, sand dunes are also found inside craters where the saltating sand accumulates.

The wind also creates erosion and deposition of dust in the down-wind wake of those craters that have elevated rims, the result of vortices that are formed. This feature provides a handy indicator of the prevailing wind direction. The Viking orbiters carried out more detailed mapping of the wind-created geology of Mars a few years later while the landers provided close-up surface images in two locations.

Conway Leovy and Jim Pollack were the members of the imaging team most focused on atmospheric dynamics. Leovy and Yale Mintz (1916-1991) had created the first Mars general circulation model in the 1960s by modifying a UCLA model for Martian conditions. Using Mariner 9 data Leovy, Mintz, Pollack and Paul Greiman analyzed the effect of topography on the general circulation during southern winter and found that it forces very large amplitude quasi-stationary planetary waves. The Mariner 9 observations became the first constraints to be incorporated in a general circulation model and led to increasing improvements – in particular taking into account the effect of dust – in collaborations ever since (at NASA Ames, Oxford University, NOAA's Geophysical Fluid Dynamics Laboratory, Le Laboratoire de Météorologie Dynamique and at JPL). In time, by combining insights from a broad range of interdisciplinary fields it should be possible to reliably learn how the planet has evolved over 4+ billion years from a more Earth-like state to today's desert Mars, how its interior cooled with a loss of its magnetic field, how it lost its early atmosphere and surface water – perhaps having passed like Earth through one or more Snowball Planet episodes.

Liquid water was present on the Earth's surface early in its history at a time when the Sun was (theory tells us) significantly less luminous than at present. This circumstance is the 'faint young Sun paradox' pointed out by Carl Sagan in 1972 and that remains poorly understood. If the Mars valley networks are indeed evidence of precipitation and drainage then a powerful greenhouse effect caused by an early thick atmosphere of IR absorbing gases including CO_2 and water vapor would have been needed. Many decades later we are still compiling the evidence. A quite different interpretation of how the valleys formed has emerged recently. A paper by Anna Grau Galofre and colleagues in *Nature Geoscience 3 August 2020* argues that early Mars was still covered in ice when the valleys were formed mainly by sub-glacial and fluvial erosion. "The inference of sub-glacial channels among the valley networks supports the presence of ice sheets that covered the southern highlands during the time of valley network emplacement." Their approach to the problem "departs from the common approach to studying individual valley networks to present a global statistical picture of the variability in valley morphology as well as a physical interpretation of their origin".

Given the stark nature of Earth's Moon and the apparently similar nature of Mars revealed to three earlier spacecraft, Mariner 9 recharged the imagination and reinvigorated an anxious science community. Ironically, however, NASA's following Mars mission – the Vikings of 1976 – had a different effect as astrobiology expectations for that mission were, in retrospect, too high. Those Viking landers have, in fact, carried out the only Mars biology experiments thus far.

Although planning for the first robotic Mars missions had been undertaken in the expectation that crewed missions would quite soon follow, by the time the Viking missions were carried out in the 1970s, decisions taken by the Nixon administrations had ensured that this would not be the case. The Mariner 9 Orbiter and the Viking Orbiters and Landers were therefore carried out as purely scientific endeavors.

Mariner 9 did not get close to the two small moons of Mars but did acquire complete longitudinal imaging coverage by observations of the moons at several points in their equatorial orbits – taking 80 images in all. This first close-up view of the two small moons of Mars was intriguing in part because a few years before, in 1966, it had been wisely proposed by Fred Singer (1924-2020), physics professor at the University of Virginia, that they would be a logical destination for the first manned missions to Mars – orbital rendezvous being much less energetically challenging than landing on Mars while providing the opportunity for the crew to control landed robots in real time. This idea has returned for renewed consideration decades later. Both moons are irregular in shape and dark like C-type asteroids (the source of carbonaceous chondrite meteorites). Phobos (above) is about 22 km across and orbits Mars at an altitude of about 9,000 km while Deimos is about 13 km in size and orbits at an altitude of about 23,500 km. The period of Phobos' orbit is 7.66 hours while that of Deimos is 30.35 hours. Both are in synchronous rotation. They are dark, irregular bodies with crater densities close to saturation. It was concluded that both satellites consist of consolidated, though highly fractured material. Their origin and

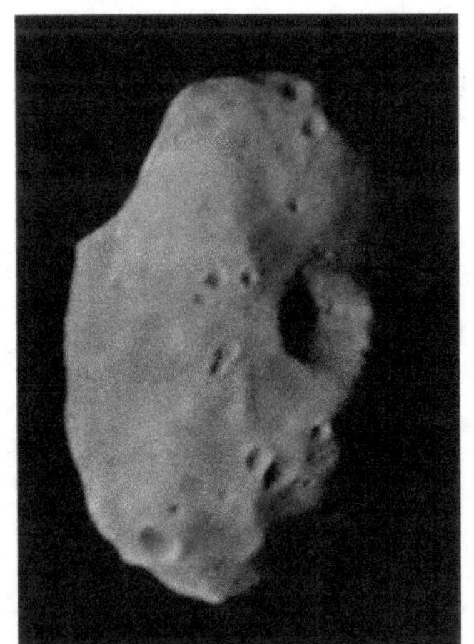

history remains quite uncertain and controversial.

One of many craters on Phobos is disproportionately huge, ~9 km in diameter, and its formation must have come close to destroying the body. It has been given the name Stickney after the wife of the moon's discoverer Asaph Hall who in 1878 wrote that he "might have abandoned the search [for Martian satellites] had it not been for the encouragement of [his] wife".[55] The moons are very much like the small asteroids that have been encountered by a number of spacecraft in recent years; their capture by Mars is probable, but by no means certain, since both capture and then equatorial circularization of their orbits about Mars would be required.

The Mariner 9 results had far exceeded anyone's expectations given the discouragement of the observations made by the earlier Mars Mariners. Indeed, there was a certain euphoria among planetary scientists in anticipation of space endeavors that might follow. Imaging team member and influential planetary scientist Carl Sagan in 1973 published his prize winning book, *The Cosmic Connection,* one that became a classic work of popular science and in which he looked forward not only to the continuing scientific exploration of the Solar System but to the eventual "time in our future history when the Solar System will be explored and inhabited". This perception that mankind was on the verge of an unbounded future of exploration and colonization of our Solar System was part of the national mood in the 1970s but was not to endure among policy makers after the results came back from the Viking Mars lander missions that soon followed.

One part of the vision that *has* endured has been of a different kind – the search for evidence of *intelligent* life elsewhere in our Milky Way Galaxy – not by means of spacecraft but by radio astronomers. The *Search for Extraterrestrial Intelligence (SETI)* was initially supported by NASA but subsequently by private funding, and has now for decades been searching the sky for signals from advanced civilizations elsewhere in our galaxy. This subject will be returned to in a second volume in the context of the ongoing (since the early 1990s) astronomical discovery of planets orbiting stars other than our Sun.

---

[55] https://en.wikipedia.org/wiki/Stickney_(crater)

US and Soviet Plans for First Landings on Mars

NASA's plans had originally called for launching two Mars landers, each combined with an orbiter, to Mars in the coming 1973 launch opportunity but, not surprisingly, this challenging mission – *Viking* – slipped to the 1975 window. The Soviets, meanwhile, stayed on track to take advantage of that opportunity and recover leadership in exploring Mars. Their *Mars 4 & 5* orbiters and *6 & 7* landers were launched in July and August of 1973 but without success.

In August 1972, after ten months in orbit, Mariner 9 was handed over to the Viking science team (there was much overlap of membership including that of the author) to begin to identify sites for the landing of two spacecraft in July 1976.

The Viking Orbiters

The Viking 1 Orbiter & Lander were together inserted into Mars orbit on 19 June 1976 and trimmed to a 1,513 x 33,000 km, 24.66-hour site certification orbit on 21 June. Imaging of candidate sites had begun immediately and the landing sites would be selected based on these pictures. Subsequently, the orbiters were fully occupied in mapping Mars once the landers were safely down on the surface with results described below.

The Viking Orbiter cameras were similar in design to those flown on the Mariner spacecraft but each spacecraft had two identical cameras where Mariner 9 had a wide-angle and a narrow-angle camera. The cameras were mounted on a scan platform along with instruments to measure surface temperature and the amount of water vapor in the atmosphere. The 475 mm focal length lens covered an area of 40 x 44 km from an orbital altitude of 1,500 km. Each image was made up of 1,056 x 1,182 picture elements so, at closest approach, resolution (2+ pixels) was ~100 meters. There were six selectable filters. Acquisition of each image and its subsequent readout to the tape recorder required ~9 seconds, so a frame was taken on alternate cameras every 4.5 seconds and produced a side-by-side swath of images – a big step up in resolution from the Mariner 9 coverage.

The Viking 1 orbiter's periapsis was reduced to 300 km in March 1977 and thereby increased the resolution of the images. The mission thereafter concentrated on maximizing the mapping coverage and acquiring repeated coverage of dynamic phenomena – polar phenomena, transient condensate clouds and dramatic dust storms. The Viking orbiters observed more than 20 local dust clouds and two global dust storms during 4 years. Most of the dust activity occurred during southern hemisphere spring and early summer when Mars was near perihelion and insolation was near maximum. About half the local dust clouds occurred near the edge of the southern polar cap where winds are enhanced by a steep regional temperature gradient. The other half occurred mainly in the southern hemisphere near regions where atmospheric circulation models incorporating topography predict positive vertical velocities. Although dust clouds observed from Earth show a similar partial correlation with models, some ambiguity exists concerning interpretation of regions near Hellespontus that have spawned the most spectacular dust storms on record.

On 7 August 1980, because the Viking 1 orbiter was running low on attitude control gas, its orbit was raised to 320 x 56,000 km thereby avoiding possible contamination of the surface until about the year 2019. Operations ended on 17 August 1980 after 1,485 orbits. The Viking 2 Orbiter periapsis was lowered to 778 km in December 1976 and the inclination increased to 80 degrees to improve visibility of high latitudes. Orbiter 2 developed a leak in its propulsion system and its mission ended on 25 July 1978 after returning 16,000 images during 706 orbits around Mars.

Viking Landers 1 and 2

The landers were designed to slow down using heat shields and parachutes with touchdown using rocket engines; there was no terminal guidance to avoid any hazards. Potential landing sites would be ones that were associated with evidence of the past (or even recent) presence of liquid water and, most importantly, sites that promised landing safety. The biggest concerns were steep slopes and rough terrain with boulders and debris from impact craters – the latter information well beyond the resolution of the Mariner 9 narrow-angle camera. Also, the Viking science team still had to resolve debates about where on Mars one might possibly find transient liquid water – perhaps near the edge of a retreating polar cap. So, the final narrowing-down of Viking landing site selection was necessarily postponed until after orbit insertion when the Viking orbiters could contribute their observations. After the first Viking combined orbiter & lander spacecraft reached Mars in June 1976, a month-long process of data acquisition and analysis by the investigators took place at the US Geological Survey in Flagstaff, Arizona along with continuing meetings of an anxious team at JPL.

As previously noted, the NASA Langley Research Center managed the overall Viking program with the support of JPL with its responsibility for the orbiters. And it was from JPL that mission operations were carried out. James Martin (1920-2002) was the Project Manager and Thomas Young the Mission Director; they both had the highly successful Lunar Orbiter experience behind them. Gerry Soffen (1926-2000) was the Project Scientist. Gus Guastaferro who would, years later, follow Tom Young as Director of NASA's Solar System Exploration Program, was the Viking Deputy Project Manager. Martin Marietta had been selected in 1969 and tasked with designing and building the two landers along with the entry, descent and landing systems. The orbiter design was based on the Mariner 9 spacecraft.

Tim Mutch of Brown University was the Viking lander imaging team's Principal Investigator and was a major science contributor during the planning and execution of the mission. He subsequently served as Associate Administrator for Space Science at NASA HQ while, at the same time, Tom Young was appointed to head the Solar System Exploration Division. The author served for six years at NASA HQ as Deputy Director – first to Tom Young, then to Gus Guastaferro (another Viking manager) and, lastly, to Jesse Moore before being appointed Director himself. The Vikings were everywhere at that time including Al Diaz who served as the author's Deputy for a couple of years before his further promotion. Frank Carr then took over as Deputy Director. Diaz and Carr were both important contributors to the management of the program.

Landing Sites

The two selected landing sites for Viking were both in the northern, summer hemisphere. The first site was chosen to be at low elevation where the atmospheric pressure would be high and the terrain as flat and lacking in obstacles as could be judged. The chosen site was in a plains region at ~23°N called Chryse Planitia, chosen because it appeared relatively smooth & flat and, importantly, because the site is at the terminus of the outflow channel Kasai Valles originating in the equatorial highlands.

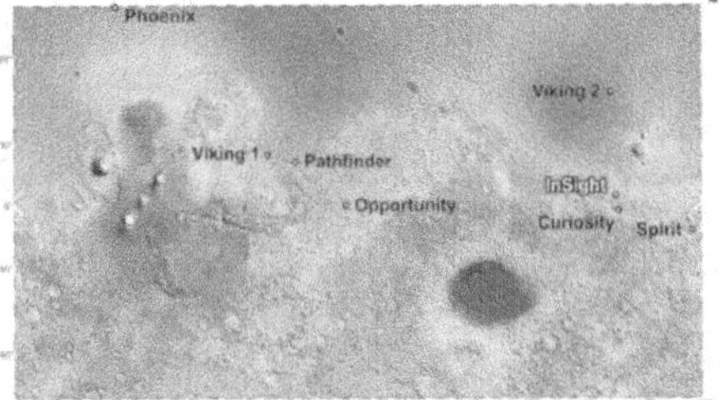

Viking 1 (the combined orbiter and encapsulated lander) was launched on a Titan III Centaur from the Cape on 20 August 1975. Insertion into a highly elliptical orbit took place on 19 June 1976 after which there followed an intensive effort to analyze the proposed landing site. This took place at the USGS in Flagstaff (an overnight train ride from Pasadena) and led to the conclusion that the pre-selected site was too rough and risky. So, attention moved to other sites in Chryse Planitia that were more acceptable.

The Viking Landers

Each 576 kg triangular-shaped Viking lander is about 3 meters across and 2 meters high. Carl Sagan serves as size comparison. The lander has three legs with circular footpads and provided just 22 cm of ground clearance. Three terminal descent engines provided attitude control and reduced the lander's velocity after parachute separation. Two 35-watt plutonium-238 radioisotope thermoelectric generators (RTGs) developed by the U.S. Department of Energy provided electrical power. Rechargeable batteries served to augment power when needed. The landings were the first that were to be carried out by NASA on a planet with an atmosphere and this made for much more complication than landing on the Moon. The landers had been sterilized as rigorously as possible and were packaged in an aero-shell with an ablative heat shield that provided the critical initial deceleration. At 6 km altitude parachutes were deployed and a few seconds later the aero-shell was jettisoned followed by the extension of the three legs. After the parachutes were cut, retro-rockets were fired beginning at 1.5 km altitude and continued to do so for the last 40 seconds all the way to the surface. It was necessary to determine and deal with any horizontal velocity that the lander acquired from the wind so the terminal descent radar was turned on at 12 km altitude. It could measure velocity to +/-1 meter/sec. The vertical landing velocity was about 2.4 meter/sec.

Touchdown of Viking 1 on Chryse Planitia (22.3°N, 48.0°E, ~2.7 km altitude) took place with a giant sigh of relief at JPL (ten minutes of light-time later) on 20 July 1976 a few weeks after what would have been highly symbolic 4 July 1976. NASA is understandably risk-averse when it comes to decisions affecting the success of multi-billion dollar missions (and the repercussions of failures). And, by the way, 20 July is also a key date in NASA's own history.

Viking 2 had launched from the Cape on 9 September 1975, arriving at Mars on 7 August 1976 where it was inserted into a 24.6-hour period orbit with 1,500 km periapsis and 55.2 degrees inclination. The Viking 2 landing site had also been selected with a desire to land as far north as practical, preferably close to where the northern polar cap – mostly made of carbon dioxide ice but also trapping water ice – had recently retreated. 73°N was considered to be the latitude where liquid water might be briefly stable at the season in question but, given the still frigid environment so close to the pole, the biologists agreed to settle for a northern site at a middle latitude. Again, there were weeks devoted to landing site certification. On 3 September 1976 the Viking 2 lander was directed to a site at 48.3°N, 226°W (altitude -4.2km) in Utopia Planitia about 200 km to the west of the 104 km diameter crater Mie (named in 1973 for German physicist Gustav Mie – of Mie Scattering and Nazi-opposition fame). So, both landers were targeted to the lowland north of Mars.

Some important measurements were able to be made by the landers on the way down to the surface as the aero-shells were instrumented with mass spectrometers that measured the abundance of the atmospheric constituents and, in particular, the ratio of the heavy and light isotopes of nitrogen. Similar measurements were made at the surface by mass spectrometers on the landers – providing the following result: N14/N15 = 170/168. From this it became possible to make a first estimate of the degree of out-gassing that Mars has experienced – in particular how much water once found its way onto the surface and into the atmosphere. Inevitably the modelling and calculation are complicated, full of assumptions and having to take into account various loss mechanisms including Jeans escape, hydrodynamic blow-off, photochemical reactions and solar wind scouring at the uppermost top of the atmosphere. These losses will be experienced preferentially by the lighter isotopes of gases and a comparison of ratios of nitrogen isotope abundances with terrestrial values can lead to an estimate of the loss. Not all loss of atmospheric water and carbon dioxide is to space: water and carbon dioxide (as ice, hydrated minerals and carbonate minerals) are locked up in the polar caps and regolith. Michael McElroy and his colleague Yuk Yung used the Viking measurements to calculate that an amount of water equivalent to 130 meters deep was out-gassed, on average, across the surface of Mars.

It is not a great stretch of the imagination to suppose that an ocean, hundreds of meters deep, once covered the northern lowlands and persisted for a billion years or more. However, convincing evidence of this early state has been difficult to find. Orbiter imaging teams had been seeking to identify a paleo-shoreline around such an ancient ocean with limited success until recently when the Mars Reconnaissance Orbiter has provided new insight as discussed later in Chapter 13.

The mass spectrometer measurement of the atmospheric composition would later turn out to be the key to identifying (through their content of trapped gases) a number of oddball meteorites – the SNCs (Shergottites, Nakhlites, & Chassignites) and Allan Hills 84001 (collected in Antarctica) – as having their origin on Mars as discussed in Chapter 10.

The ways in which the atmospheres of planets and moons have been lost over billions of years continues to be researched. David Catling & Kevin Zahnle point out that in addition to the loss mechanisms mentioned above, impacting asteroids early in Solar System history might have accounted for the loss of the entire early atmosphere but for "simple chance" given that the frequency of large impacts diminished rapidly 3.8 billion years ago. They also point out that the red color of Mars would be the result of the loss of hydrogen from an original ocean of water – leaving behind oxygen that oxidized the original grey volcanic rocks (a similar rusting process that they believe happened on Earth and Venus). As discussed in Chapter 14, NASA launched the MAVEN spacecraft to Mars in 2013 to measure the rate of escaping ions and neutral atom to provide more data from which to construct the atmospheric history.

Both Viking landers touched down successfully six weeks apart ending some knuckle-biting suspense for the team. In the case of Viking 2 it could have ended badly as one leg did land on a rock – but only a fairly small one – so that the lander sits at an angle of about 8 degrees. They communicated directly with Earth as well as via the orbiters overhead. On each lander, scientific instrumentation included two 360-degree cylindrical scan cameras, three biology experiments in an interior

environmentally controlled compartment, chemical composition experiments, meteorology, seismology, and magnetic properties instruments. All the experiments except one were carried out as planned and acquired reams of data. Unfortunately, the Viking 1 seismometer failed to un-cage, limiting the analysis possible by the Viking 2 instrument to the measurement of the seismicity of Mars and preventing direct determination of interior structure – a big hole that persists to today and is trying to be plugged by NASA's InSight lander since its landing in the Elysium Planitia in November 2018 – Chapter 14.

The first panoramic view returned by Viking 1 from the surface of Mars showed rocks ranging in size from about 10 to 20 cm across scattered across a small dune field. A dark rock at the center of the frame, nicknamed "Big Joe", was about 2 meters across and only 8 meters from the lander. Touchdown on Mars is risky.

The first color image of Utopia Planitia returned by the Viking 2 Lander indicated that the lander was tilted at an angle of 8 degrees. The colors of the rocks – scattered throughout the scene – and soil were similar to those at the Viking 1 site. In one direction, to the northeast, drifts of fine-grained material dominated the view while to the southeast was a saturated field of small angular blocks stretching to the horizon.

The surface material at both landing sites was best characterized as iron-rich clay. Seasonal dust storms, pressure changes, and transport of atmospheric gases between the polar caps were observed through the several years they were operational (1,824 days, 1,775 sols for Viking 1, 1,316 days, 1,281 sols for Viking 2).

The X-Ray Fluorescence Spectrometer successfully measured elemental abundances of nine soils at Viking 1 and eight soils at Viking 2. They are broadly alike, with an iron-rich chemistry similar to that of palagonite (an alteration product from the interaction of water with volcanic glass of chemical composition similar to basalt). The result of the Viking 2 analysis of the soil was consistent with the weathering products of basaltic lavas – abundant silicon and iron plus significant amounts of magnesium, aluminum, sulfur, calcium and titanium. The sulfur and the chlorine in the soil, presumed present as sulfides and chlorides, might have been the result of the evaporation of seawater. All the soil samples that were heated in the gas chromatograph-mass spectrometer gave off water. The soil was determined to be between 3% and 7% magnetic materials by weight – magnetite (Fe_3O_4) and maghemite (Fe_2O_3) – the products of basalt weathering. These measurements were not inconsistent with the notion that the landing had taken place on the floor of an ancient ocean but by no means provided confirmation.

Life Science Soil Sample Analysis

The landers scooped up soil samples at both sites. Each carried out four different types of analyses with results that the biology science team led by Harold Klein (1921-2001) concluded did not indicate the presence of life. There was, and is, one dissenter: the Principal Investigator of the Labelled Release experiment, Gilbert Levin – a contrary position he still holds today. The life science experiments (the only ones ever conducted on Mars) were:

The *Gas Chromatograph-Mass Spectrometer* of PI Klaus Biemann analyzed the components of untreated soil released as it was heated. It could measure molecules present at a level of a few parts per billion. The soils were found to contain less carbon than lifeless lunar soils.

The *Gas Exchange* instrument of PI Vance Oyama (1922-1998) incubated soil samples and measured any gases given off after first replacing the atmosphere with inert helium and then adding liquid nutrients. A gas chromatograph was used to measure the concentration of various gases in the incubation chamber (including oxygen, carbon dioxide, nitrogen, hydrogen and methane) on the assumption that metabolizing organisms would consume or release at least one of these. No such signature was observed.

The *Labelled Release* experiment of PI Gilbert Levin treated a soil sample with a drop of a very dilute aqueous solution containing seven different nutrients, each tagged with radioactive carbon-14. If microorganisms in the soil had metabolized one or more of these nutrients, then the air above the soil would reveal this by monitoring for radioactive carbon dioxide gas. Levin's team observed a steady stream of radioactive gases being given off by the soil immediately following the first injection of solution. Both Viking landers carried out the experiment with the same result first using a sample from the surface exposed to sunlight and the second probe taking the sample from underneath a rock. However, injections of the nutrient solution a week later did not produce the same result.

The *Pyrolytic Release* experiment of Caltech's Norman Horowitz incubated its sample for several days in an environment of light, water, and a Mars-like atmosphere of carbon monoxide and carbon dioxide. These gases were made with radioactive carbon-14 that would have been incorporated into any photosynthetic organism and revealed to a radiation counter when the sample was subsequently baked at 650°C. No such luck!

Taken together, the biology experiments results were judged by the team in the absence of any carbon-containing organic molecules to indicate the absence of life. This absence of organic molecules was a bit of a surprise because the surface of Mars is, like ours, subject to meteoritic in-fall that would have included organics. The negative analysis result of the Viking landers contributed to the lack of any follow-on Mars missions for two decades. It would await the results of the 2007 NASA Phoenix lander mission (Chapter 13) to better understand the absence of organics in the soil.

In subsequent missions to Mars there have been no experiments to search for evidence of life but, rather, experiments to identify Martian surface environments that might have been hospitable to the evolution of life earlier in the history of the Red Planet.

In a 10 October 2019 *Scientific American* blog PI Levin argues the case that his experiment did discover positive evidence of Martian life, pointing out that, among a longer list of other considerations:

- Ultraviolet (UV) activation of the Martian surface material did not, as initially proposed, cause the LR reaction: a sample taken from under a UV-shielding rock was as LR-active as surface samples;
- Complex organics, have been reported on Mars by Curiosity's (Chapter 13) scientists, possibly including kerogen, which could be of biological origin;
- The excess of carbon-13 over carbon-12 in the Martian atmosphere is indicative of biological activity, which prefers ingesting the latter;
- Terrestrial microorganisms have survived in outer space outside the ISS;

- Methane has been measured in the Martian atmosphere; microbial methanogens could be the source;
- The rapid disappearance of methane from the Martian atmosphere requires a sink, possibly supplied by methanotrophs that could co-exist with methanogens on the Martian surface;
- Large structures resembling terrestrial stromatolites (formed by microorganisms) were found by Curiosity (the mobile lander mission reported in Chapter 13)[56]

This is clearly a critical area where other astrobiologists remain unconvinced. The author is not qualified to comment further except to say that, as has always been NASA policy, future sample return missions to Mars will be carried out to ensure that Planet Earth is fully protected from possible contamination.

A Co-Investigator of the Labelled Release experiment team, Patricia Ann Straat has recently published a book, *To Mars With Love*, in which she describes the Viking Biology Experiments in detail and discusses the evidence that leads her to conclude that the experiment may, indeed, have discovered evidence of life on Mars. Noted astrobiologist Chris McKay of NASA Ames has reviewed Dr. Straat's book in the journal *Astrobiology* vol. 19, April 2019, and concludes "Whether biology or nor, this is a remarkable book, by a remarkable woman, about a remarkable instrument on the most remarkable mission ever to go to Mars". The jury is still out.

The Seismometer of PI Don Anderson (1933-2014) on the Viking 2 lander at the Utopia landing site suffered from wind noise during the daytime hours and no large seismic events were recorded. Mars is much less seismically active than Earth. A possible local seismic event was detected on sol 80 – interpreted as a natural event it had a magnitude of 3 and was distant 110 km. Preliminary interpretation of later arrivals in the signal suggested a crustal thickness of 15 km.

Post-Mission Viking Data Analysis

Arguably, the subsequent hiatus until the 1990s in missions to Mars was in fact beneficial as it allowed the enormous data returned from the landers and orbiters (especially) to be thoroughly digested. A flood of science papers followed including those by USGS geologist Michael Carr, leader of the orbiter imaging team. He had contributed many papers on the geologic history of Mars and had digested the others to produce his magisterial books: *The Surface of Mars* and *Water on Mars*.

The greatly improved resolution and coverage of the Viking orbiter images led to a surge in the geologic mapping of the planet (carried out as before by the USGS Astrogeology Branch) and ability of the science community to further understand the evolution of Mars. The Viking orbiters returned 4,600 surface images at a better resolution than Mariner 9 had achieved and led to the creation of a classic series of geologic maps and other products by the USGS: global mosaics, north and south polar stereographic maps, globes, and hemispheric views.

Deciphering the history of Mars is helped greatly by the fact that the planet has not experienced plate tectonic recycling of the crust – though there is ample evidence of large-scale tectonic stress leading to fractures and the formation of spectacular canyons. Much of the effort by geologists has been involved in classifying different terrains according to superposition relationships and differing areal densities of craters – allowing them to determine the relative ages of the geological units. Eventually it is expected that the time history of the events that have shaped Mars will be as complete and confident as that determined for the Moon – progress that will require the return of samples from selected locations for in-depth mineralogical analysis and radiometric age dating. In the absence of such Mars samples, geologists have been taking advantage of the known lunar impact history by assuming that the varying rate of impacts (large and small) on Mars would have been similar to that of the Moon i.e., that the early asteroid bombardment was intense and then rapidly declined at about 3.9 billion years ago. It is still necessary to make allowance for 1) the different impact velocities of the asteroids given the different gravity fields of Mars and Moon and 2) the different rate of secondary craters formation and preservation on the two bodies. The calculations and logic involved tends to be complicated and the reader is referred to Michael's book *The Surface of Mars* for the detail and for the different results at which researchers arrive.

---

[56] https://blogs.scientificamerican.com/observations/im-convinced-we-found-evidence-of-life-on-mars-in-the-1970s/

The earliest stratigraphic time period in the history of Mars – 4,100 to 3,700 million years ago – has been given the name *Noachian* and is dated from the time the giant impact basins (Hellas, Argyre, Chryse and Isidis) and the cratered southern highlands were formed. The few hundred million years during which Mars accreted – the pre-Noachian – is, of course, not directly observable. The periods are each named after the region of Mars that best typifies the areal density of craters that define each. Noachis Terra lies to the west of Hellas Planitia where the heavily cratered terrain resembles the lunar highlands. Terrain of Noachian age is observed on about 50% of Mars. For a few hundred million years the whole planet suffered a similar intense bombardment to that experienced by the highlands of our Moon. The cooling of the core and loss of the magnetic field may already have taken place. The composition and density of the atmosphere during the Noachian when the valley networks were formed is an important area of research.

The next stratigraphic time period is the *Hesperian* – ~3,700 to ~3,000 million years ago – when volcanism was widespread and the outflow channels were carved. Flood basalts covered much of the planet, solidifying and accumulating craters at a much-reduced rate. These lava flows evidently erupted at a very high rate and with low viscosity from many sources. The Hesperian terrain resurfaced about 30% of Mars. Hesperia Planum lies to the northeast of Hellas Planitia. (The plains on Mars may be named planitia or planum, depending on elevation.) The continent-sized (~5,000 km across) Tharsis 'bulge' further developed during this time, eventually creating the three huge Tharsis volcanoes (Arsia Mons, Pavonis Mons and Ascreaus Mons) as well as, to the west, giant Olympus Mons and, to the north, Alba Patera, a volcano that in areal extent exceeds even that of Olympus Mons. The growing crustal stresses of the Tharsis bulge led to the formation of a vast system of rifts radial to the bulge. The Valles Marineris themselves are a dramatic consequence of this crustal stress. They may have filled with water to become huge lakes – water that was subsequently released as catastrophic floods. Water under great pressure released from beneath deep permafrost is an evident second source of the water that created the outflow channels. At the eastern end of the Valles Marineris is a huge area of collapsed 'chaotic' terrain from which one of the extraordinary runoff channels, Tiu Valles, emerges full bore to flow to the north, creating massive streamlined islands on its way to debouch into Chryse Planitia where Viking 1 landed.

Volcanism on Mars has continued until quite recent times and so the third Martian time period – the *Amazonian* – is in fact the longest: ~3,000 million years ago to the present time. It is named for Amazonis Planitia to the west of Olympus Mons – a region that has accumulated the least areal density of craters. It is judged that a thin atmosphere and associated bleak climate have persisted during this period such that liquid water would not be an element of the erosive and depositional forces that continued to slowly shape the surface. Water in the form of deep permafrost, however, has surely been an important part of the environment during the last three billion years and, below the permafrost seal, there may be liquid water. Back in the Hesperian period a sizeable ocean may have occupied the northern lowlands. Extensive surface melting to create a magma ocean together with out-gassing to create a thick greenhouse atmosphere would have characterized Mars very early in its formation; this would have been at about the same time that the Moon and the other inner planets were forming (about 4.5 billion years ago). Orbiting the Sun at a significantly greater distance than Earth it is hard to see how Mars would not have included a substantial amount of water ice and other volatiles in its makeup. So, a residual atmosphere would likely have survived the severe losses resulting from the asteroid bombardment. After the crust solidified, Mars accumulated craters at a rapid rate including a handful of huge impact basins. The valley networks are pervasive on surfaces that date from 3.9 billion years ago. At that time these were most probably created by liquid water

and would be consistent with an atmosphere much thicker than today's. The surface of Mars might have been quite hospitable to life for an extended time. Then the impact rate relatively abruptly diminished. As Mars lacked a magnetic field by this time, the solar wind would have begun the erosion of the atmosphere. The planetary inventory of water would have partitioned between the polar caps, the cryosphere and, deep down, a liquid water table. Volcanism was evidently widespread across Mars throughout its history but at a diminishing rate. The smaller Martian craters – less than 5 km in diameter – have ejecta patterns that resemble those of lunar craters. Above that size there is a striking difference indicative of impacts that were made into a surface containing ground ice – a pattern that indicates that the ejecta flowed from the impact site rather than were blasted out on ballistic trajectories. As Michael Carr describes: "Craters in the 5 to 15 km size range generally have a single ejecta sheet that extends about one crater radius from the rim. The edge of the sheet has either a low ridge or an escarpment. It is usually nearly circular, with a jagged or slightly lobate margin". [57]

The Absence of Martian Organics

As noted, no matter that the Viking landers returned quantities of essential information about Mars, the failure to produce convincing evidence of past or present life on Mars effectively took the steam out of the Mars element of NASA's planetary exploration program – especially given the many other priorities of the planetary science community. A high-latitude NASA lander mission years later – the *2007 Phoenix* – would throw more light on this negative finding (Chapter 13). Consideration of sending spacecraft or astronauts to land on Mars to continue its exploration would not emerge again until much later. It would be a dozen or more years before NASA would restart its robotic Mars exploration missions. There have been no further attempts to carry out biology experiments with soil samples. The Viking engineering team did propose to create mobile Vikings by replacing the landing pads with tracks but did not find an interested audience. A study was also carried out by a JPL team led by Viking PI Tim Mutch in which the author participated that proposed a six-wheeled, RTG-powered Mars rover (very much like the *Curiosity* rover that is today exploring Gale Crater) for a launch in the 1980s but the time was not ripe for such an ambitious mission to get beyond what NASA calls *Pre-Phase A* study.

Viking Lander 2 mission ended after 3 years and 7 months in April 1980 following a failure of its battery. Viking Lander 1 soldiered on for another two years – till November 1982 and, unusually, was only shut down after a software update error that caused the lander's antenna to go down, terminating power and communication. By now the author, exercising questionable judgement, had since 1977 moved from his position at JPL on the Viking Orbiter and Voyager imaging teams to NASA HQ in Washington DC.

Stormy Weather Ahead for Solar System Explorers

In the fall of 1977 with two Voyager Grand Tour spacecraft being prepared for launch, with the two Pioneer Venus missions preparing for launch in 1978, with the Galileo Jupiter Orbiter and Probe mission in early development, and with the Venus Orbiting Imaging Radar mission closing in on an approved new start, NASA's Planetary Exploration Program looked very healthy but, as it happened, this was an illusion. After the Pioneers were launched there would be no more deep space launches for more than a decade. This was in part because NASA had phased out its use of expendable launchers and now relied solely on the Space Shuttle – which proved expensive and fragile. The 'lost decade', as discussed in Chapter 7, was also the result of decisions made by the Ronald Reagan administration following his election in 1980.

A personal note: at this point in late 1977, the author, out of the blue, was asked by Tom Young to become his Deputy as part of the Planetary Program management at NASA HQ in Washington DC (a different culture entirely to that of JPL!). At that time Tom, previously Deputy Viking Project Manager, was the new Planetary Program Director. He, in turn, reported to Tim Mutch, the new Associate Administrator for Space Science. And a tragic note: Tim (1931-1980) died in a climbing accident on Mount Nun in the Himalayas just a few years after the completion of the Viking lander missions. A large crater on the equator of Mars at 55°W was named in his honor and the Viking 1 lander was formally renamed *Thomas A. Mutch Memorial Station* on Jan 7, 1981 by Administrator Frosch.

---

[57] Michael Carr, *The Surface of Mars*, p 48, Yale University Press, 1981

CHAPTER 6

ASTRONAUT SPACEFLIGHT AFTER APOLLO

Apollo-Soyuz Test Project, Boeing Mars Study, Skylab, Space Shuttle, Buran, Mir, Shuttle-Mir, International Space Station, NASA Mars Reference Mission, Constellation Program, Private Sector Plans, Artemis

A Space Task Force headed by Nixon's Vice President Agnew had been created early in 1969 (before the Apollo 11 landing in late July of that year) to recommend what next for the manned program; it provided three ambitious alternatives, one extremely so: a 50-man Earth-orbiting space station and a lunar base, to culminate with a Mars landing in the mid-1980s. This was a time of seemingly unlimited optimism (indeed intoxication) by space exploration enthusiasts: planning for the 1980s was to include space stations not only in low Earth orbit but also in geosynchronous orbit and in lunar orbit. A Space Shuttle and a Nuclear Shuttle would augment Saturn V-derived launchers. Lunar and Mars bases would be established.

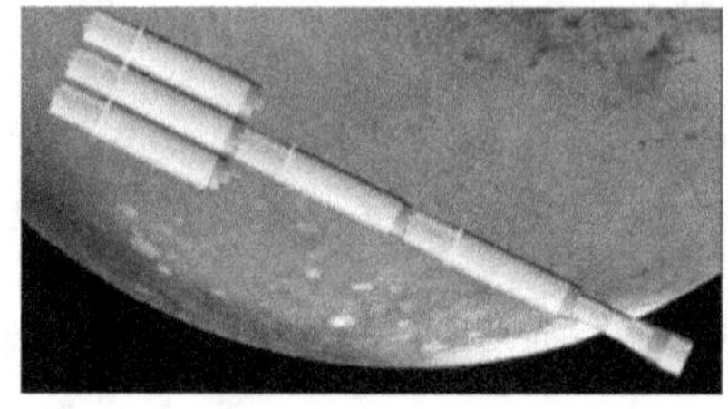

In January 1968 Boeing had issued a manned Mars mission report that was the result of a 14-month study. The spacecraft for the Mars mission used five modular nuclear-thermal rocket stages to provide the necessary performance to transfer to Mars several unmanned probes, a manned Mars lander, a crew compartment module, and a bi-conic Earth Entry vehicle. A Nuclear Thermal Rocket heats and expands a working fluid, such as hydrogen in a nuclear reactor. It has twice the efficiency of the best chemical engines because of the ultra-high temperature of the exhaust. Boeing proposed a Mars orbit rendezvous approach as developed at NASA's Lewis Research Center. Total development cost through the first two missions to Mars was estimated to be 50% more than the Apollo program (this when Apollo was already accounting for about 4% of the total Federal budget). The required launch vehicles were much bigger versions of the Saturn V with strap-on solids. If the start of full-scale development began in 1976, the first landing on Mars could be made in 1985-1986! Given all the events in world history since that time, this

plan may yet turn out to be the closest that we get to landing astronauts to Mars.

In a lengthy article in the March 2000 *Air & Space Magazine* David Portree described the ambitions of Von Braun and associates within NASA and in the aerospace companies who shared his thinking about how near-term manned exploration of Mars might build on and magnify the legacy of Apollo. As noted, the giant Saturn V launcher was an essential element of that grand plan – along with the nuclear-thermal upper stage (NERVA) that had been under development by the US Atomic Energy Commission & US Air Force since 1956 and that had undergone testing at the Jackass Flats (!) in Nevada. Other enabling technologies included a reusable Shuttle launcher and a Space Station where refueling of the Mars-bound stages would be carried out. Two crewed spacecraft powered by clusters of these NERVA engines were to journey together to Mars each with a crew of three.

All of the crew would descend to the surface from the orbiting mother craft, land close together to provide mutual support, and carry out up to sixty days of scientific investigations. If necessary all six crew could return to Mars orbit in one of

the two ascent vehicles, thereby providing an important measure of safety. The crewed mission would be supported by a number of robotic sample collection landers that would ascend to orbit and deliver their samples to the crewed mother craft.

A less ambitious approach offered as an alternative was a program that contemplated undertaking a robotic Mars landing before a date that would be set for the manned mission. A third option was a 'minimum' program that would develop a space station and a shuttle vehicle but would delay the Mars landing to late in the century. Costs were estimated at between $8 billion and $10 billion per year by 1980 for the most ambitious option and from $4 billion to $5.7 billion annually by 1976 for the least.

James Webb's successor as NASA Administrator, Thomas Paine, had seriously misread the intentions of President Nixon as had Vice President Agnew. As the NASA History Office records, the general reaction to these recommendations were negative. Many scientists shared the point of view that such plans were "the utmost folly." Even Congressional supporters of manned space flight were opposed. A strongly dissenting point of view in favor of early astronaut missions to Mars emerged from a number of individuals and organizations in the 1980s.

Although NASA, with encouragement from incoming Administrations, has carried out numerous studies of how to carry out crewed landings on Mars, this lack of wide support has led to a continuing failure to actually implement such an expensive and risky program.

Although the lunar landings are now in the distant past, it is worth recalling the successful Apollo-Soyuz Test Project (ASTP) that followed a few years later. Long before the International Space Station was gradually assembled in low Earth orbit the ASTP demonstrated that complex manned space missions can be completed successfully as international projects between erstwhile antagonists. Thus, in 1972, President Nixon and Premier Brezhnev signed the Agreement Concerning Cooperation in the Exploration and Use of Outer Space for Peaceful Purposes Act, committing themselves to a joint space mission in 1975.

This first US-Soviet space mission carried out in Earth orbit was conducted in July 1975 and marked the end of the Space Race, serving as a symbol of superpower détente. It was also the end of US manned spaceflight until the first Shuttle flight in 1981. Administrator Paine initiated the collaboration in an October 1970 letter to Mstislav Keldysh president of the Soviet Academy of Sciences. Technical meetings followed and the Nixon Administration confirmed its support. An agreement was signed in April 1972 and the Apollo-Soyuz mission proceeded with a Saturn IB launch on 15 July 1975. The three astronauts were Thomas Stafford, Vance Brand and Deke Slayton (1924-1993) while the Soviet cosmonauts were Alexey Leonov and Valeri Kubasov (1935-2014). The docking of the Apollo Command Module and the Soyuz spacecraft was of only two days' duration but this served its symbolic purpose.

After Apollo, the US and the Soviet Union both oriented their crewed space flight programs toward the building and operation of facilities in low Earth orbit in which to develop experience with long duration human spaceflight and to carry out experiments in microgravity. In April 1971, the Soviet Union launched its Salyut 1 space station into a 51.6-degree inclination orbit where it functioned for 175 days before it re-entered in October 1971. Salyuts 2 through 7 were launched at intervals through 1982 and laid the groundwork for the much-more-advanced, modular Mir space station. The Mir space station was assembled in low Earth orbit beginning in 1986 and had a 14-year operational lifetime during which time over 100 individuals – cosmonauts and astronauts – from a dozen different countries were part of the crew.

In 1973 NASA, using the Saturn V, launched the Skylab space station – an orbital workshop – into a 50-degree inclination orbit. Several three-astronaut crews occupied Skylab over the following six years. However, NASA did not follow this by launching a more advanced space station but instead used the Space Shuttle to make repeated flights (32 in all from 1983 to 1998) carrying a reusable laboratory – NASA's Spacelab – into low Earth orbit.

The von Braun Vision for Space Exploration

Looking back, one can see that NASA's leaders remain caught up in the logic expounded by von Braun for how space exploration should proceed. In von Braun's visionary thinking – formulated in the 1930s, 40s, and 50s – the intelligence, versatility and oversight of the human explorer were central to the enterprise; capable robot spacecraft were not even in the realm of science fiction. Logically, routine access to near-Earth space would be provided by a reusable launch vehicle. A space station in low Earth orbit would be the needed staging point for missions to the planets (recall Stanley Kubrick's 1968 epic '2001, A Space Odyssey'). Exploration of the Moon and Mars would follow after these capabilities had been put in place. What actually happened was the dramatic race to land humans on the Moon, after which, with a much reduced budget, there was an attempt by NASA to return to the von Braun scenario: scrap the expendable launch vehicles (Saturn Vs, Atlases, Titans), build a reusable launcher (Shuttle), build and operate a space station (Freedom) and then, maybe, return astronauts to the Moon or, instead, go on to Mars. In fact, by 2015 mankind's reconnaissance phase of the exploration of the Solar System had already been substantially realized – but by means of *robot* spacecraft of ever-greater capability, endurance and accomplishment. The reusable launch vehicle – the Space Shuttle – though brilliant in concept and an amazing technology, proved to be much more expensive to operate and more fragile than anticipated. The Russian space station Mir and the *International* Space Station have served to prove that long-duration zero gravity human space flight can be carried out successfully in a relatively protected radiation environment (low Earth orbit inside the magnetosphere) and, *equally important*, that large-scale international cooperation on space projects is possible and mutually beneficial.

1981-2011 Space Shuttle

The Shuttle was a partially reusable low-Earth-orbital spacecraft system that was operated from 1981 to 2011. The first orbital test flights took place in 1981 with operational flights beginning in 1982. A total of 135 missions were launched carrying to orbit numerous satellites, interplanetary probes, the Hubble Space Telescope, and construction & servicing of the International Space Station. The success was, however, temporary. As the Presidential enquiry reported: "On January 28, 1986, STS-51-L disintegrated 73 seconds after launch due to the failure of the right Solid Rocket Booster, killing all seven astronauts on board *Challenger*. The disaster was caused by low-temperature impairment of an O-ring, a mission critical seal used between segments of the SRB casing. Failure of the O-ring allowed hot combustion gases to escape from between the booster sections and burn through the adjacent ET, leading to a sequence of events which caused the orbiter to disintegrate."[58]

Dick Scobee   Michael J. Smith   Ellison Onizuka   Judith Resnik   Ronald McNair   Gregory Jarvis   Christa McAuliffe

---

[58] https://en.wikipedia.org/wiki/Space_Shuttle

After being brought back into service there was a second disaster. On February 1, 2003, *Columbia* disintegrated during re-entry, killing all seven of the STS-107 crew, because of damage to the carbon-carbon leading edge of the wing caused during launch.

| Rick Husband | William C. McCool | David M. Brown | Kalpana Chawla | Michael P. Anderson | Laurel Clark | Ilan Ramon |

From this time on until late May 2020 it had been necessary for NASA to rely on Russian Soyuz launchers to rotate US crew members of the ISS. This mutually beneficial US-Russia circumstance has been a rare element of cooperation during politically difficult times that have included Russian interference in the 2016 Presidential election.

2020 SpaceX Falcon

Now the SpaceX Falcon rocket (including first stage reuse) with its Dragon crew vehicle provides transport to and from the ISS. Two NASA astronauts, Robert Behnken and Douglas Hurley launched to the ISS on the Falcon 9 rocket on May 30, 2020. They were equipped with snazzy white and black SpaceX spacesuits to replace NASA's familiar orange baggy suits. They splashed down under parachutes in the Gulf of Mexico off the coast of Pensacola, Florida on Sunday 2 August. Given the intervening years of Space Shuttle and Soyuz launches, this was the first such splashdown for American astronauts since Thomas Stafford, Vance Brand, and Donald "Deke" Slayton landed in the Pacific Ocean off the coast of Hawaii on July 24, 1975, at the end of the Apollo Soyuz Test Project.

While this renewed US launch capability is a source of much satisfaction for NASA, the ending of the US-Russian cooperation in this transportation function also is a matter for some regret.

1994-1998 Shuttle-Mir

The remarkable (and little noted) *Shuttle-Mir* program carried out between 1994 and 1998 effectively served to ease the way into the development and construction of the International Space Station. There were 11 Shuttle flights to rendezvous with Mir. US astronauts were part of the Mir crew for a total of almost 1,000 days. Similarly, Russian cosmonauts joined the crew of the Shuttle.

NASA's History Office in a Special Publication (SP 2001-4225) provides an in-depth report on how this US-Russia cooperative program – managed in the US by astronaut Frank Culbertson and in Russia by cosmonaut Valery Ryumin – came into being. Culbertson observed that the success was due to "the unwavering support and guidance of key leaders such as George Abbey, Dan Goldin, and Yuri Koptev despite the most intense political pressure from outside the two space agencies". Also, "the Russians and Americans always found a way to meet each other, sometimes halfway, sometimes on totally different paths, but always striving to find that common place, always trying to learn and to teach at the same time".[59] The Russian contribution to building the International Space Station meant that resources were no longer available to maintain the Mir Station which ended its life in March 2001 by a controlled atmospheric re-entry.

Relations between space program representatives – program managers, scientists and engineers – from all of the agencies and countries involved in space exploration have always been very positive – perhaps not surprisingly so given the commonality of the scientific goals and the motivations of the individuals involved. In the mid-1980s during the Administration of President Ronald Reagan relations between the US and Soviet space science program managers were excellent. NASA delegations headed by Deputy Associate Administrator Sam Keller (1930-2014) and the author (ably supported by Diane Rausch and her colleagues of NASA's International Affairs Office) met with their counterparts regularly to exchange plans for planetary exploration missions and to see if there were areas of collaboration that would be of potential mutual benefit. Jim Head of Brown University was a regular participant on the US side and had particularly productive meetings with Alex Basilevsky of the Vernadsky Institute; the annual Brown-Vernadsky Microsymposia grew out of these meetings.

Where Ronald Reagan had proposed in 1981 to terminate NASA's Solar System Exploration Program (Chapter 9), his successor (and his vice-President) George HW Bush arrived in the White House with great enthusiasm for what came to be known as the *Space Exploration Initiative* – crewed supported by robotic exploration. On the twentieth anniversary of the Apollo 11 landing, July 20, 1989, President George HW Bush announced ambitious plans for NASA: construction of Space Station Freedom, sending humans back to the Moon to stay, and eventually sending astronauts to explore Mars in a continuing program unlike that of Apollo. It was humanity's destiny to explore and America's to lead. NASA was therefore tasked to carry out a *90 Day Study* to determine how this would be undertaken. The study was led by the Johnson Space Center's Director Aaron Cohen and led to the conclusion published in November 1989 that a budget of $500 billion spread over 20 to 30 years would be required. At that price there was no support in Congress and the proposed initiative quietly died. It would have been better to have carried out the NASA study before making the announcement. President George HW Bush served only one term and was succeeded by Bill Clinton in January 1993. Al Gore became Vice President. They were re-elected in 1996.

Regular Space Policy Recommendations to NASA

President Bush's proposed initiative represented just the latest effort to define a long-range space program for the US. Appointed by President Reagan and chaired by former NASA Administrator Thomas Paine, the National Commission on Space had been charged with recommending a civilian space program that would advance the broader goals of American society over the next century. The Commission's 1986 report was titled *Pioneering the Space Frontier* and recommended a new direction "to lead the exploration and development of the space frontier, advancing science, technology, and enterprise, and building institutions and systems that make accessible vast new resources and support human settlements beyond Earth orbit, from the highlands of the Moon to the plains of Mars."[60]

The recommendations did not lead to any action and the next serious consideration of mounting a crewed mission to Mars emerged in a 1987 study led within NASA by former astronaut Sally Ride (1951-2012). This study was intended to provide priorities and longer direction for the US space program. A distinguished aerospace executive, Norman Augustine, chaired the effort.

The parts of the August 1987 report (entitled *NASA Leadership and America's Future in Space*) that applied to the crewed exploration of space included the establishment of a permanent lunar outpost and, as early as 2010, the first landings of astronauts on Mars.[61]

---

[59] https://www.nasa.gov/mission_pages/shuttle-mir/
[60] https://history.nasa.gov/painerep/begin.html
[61] https://history.nasa.gov/riderep/cover.htm

Although the Ride Report was not acted upon, interest in pursuing a crewed Mars program did not go away and, in fact, began to resurface quite soon. In early 1988 in the last year of his Administration, President Reagan announced new national space policy goals, the last of which was to expand human presence and activity beyond Earth orbit into the Solar System. Plans to build Space Station Freedom had been announced by President Reagan in his 1984 State of the Union Address.

Space Station Freedom in June 1993 came very close to being cancelled by Congress. After several cutbacks Freedom eventually evolved to become the US component of the International Space Station (ISS). Happily, in September of that year the Administration of President Clinton (with NASA Administrator Dan Goldin) reached an agreement with Russia to partner in its construction. By the middle of the following year Norman Augustine was called on again to head an advisory panel to reconsider space exploration priorities, the outcome being a common-sense recommendation that NASA, while building the Space Station, should concentrate on both Space and Earth sciences while looking to make crewed space exploration a "go-as-you-pay" endeavor.

Given the fragility of the Space Shuttle as demonstrated by the 1986 Challenger disaster and the expensive complexity of the now-international development of the Space Station, 'go-as-you-pay' meant that human exploration beyond low Earth orbit was a non-starter throughout both the George HW Bush and Bill Clinton administrations (1989 to 2000). That, however, did not discourage Mars exploration enthusiasts from developing creative ideas for how astronaut Mars exploration might be carried out. A number of organizations in the US that advocate crewed exploration of Mars have formed since Apollo – beginning with the University of Colorado's *Case for Mars* conferences and including *The Planetary Society, The Mars Society* and *The Mars Institute*. All are still vigorously alive and well. Their enthusiasm and lobbying, along with that of the planetary science community, have helped NASA's Solar System Exploration program surmount obstacles but NASA's crewed spaceflight program has been limited to Earth orbit. The *Augustine Report* referred to above was issued in 1990 during the second year of George HW Bush's presidency and made the following recommendations:

> That the United States' future civil space program consist of a balanced set of five principal elements: a science program, which enjoys highest priority within the civil space program, and is maintained at or above the current fraction of the NASA budget; a *Mission to Planet Earth* (MTPE) focusing on environmental measurements; a *Mission from Planet Earth* (MFPE), with the long-term goal of human exploration of Mars, preceded by a modified Space Station which emphasizes life-sciences, an exploration base on the Moon, and robotic precursors to Mars; a significantly expanded technology development activity, closely coupled to space mission objectives, with particular attention devoted to engines + a robust space transportation system.[62]

NASA's next long-range planning document is known as the Stafford Report after the Chairman Thomas Stafford of the 'Synthesis Group'; it was charged by George HW Bush's Vice-President Quayle and NASA Administrator Richard Truly to "seek out the best and most innovative ideas in the country." The 1991 report described four alternative architectures: (1) Mars exploration, (2) science emphasis for the Moon and Mars, (3) the Moon to stay and Mars exploration, and 4) space resource utilization.[63]

Five Hazards of Human Spaceflight

Whoever proposes to take on the challenge of a human mission to Mars will need to deal with what NASA identifies as the 5 Hazards of Human Spaceflight: https://www.nasa.gov/hrp/5-hazards-of-human-spaceflight

A human journey to Mars, at first glance, offers an inexhaustible amount of complexities. To bring a mission to the Red Planet from fiction to fact, NASA's Human Research Program has organized hazards astronauts will encounter on a

---

[62] http://www.nss.org/resources/library/spacepolicy/AugustineReport1990.pdf
[63] https://history.nasa.gov/staffordrep/main_toc.PDF

continual basis into five classifications. Pooling the challenges into categories allows for an organized effort to overcome the obstacles that lay before such a mission. However, these hazards do not stand alone. They can feed off one another and exacerbate effects on the human body. These hazards are being studied using ground-based analogs, laboratories, and the International Space Station, which serves as a test bed to evaluate human performance and countermeasures required for the exploration of space.

Various research platforms give NASA valuable insight into how the human body and mind might respond during extended forays into space. The resulting data, technology and methods developed serve as valuable knowledge to extrapolate to multi-year interplanetary missions.[64]

1. Radiation

The first hazard of a human mission to Mars is also the most difficult to visualize because, well, space radiation is invisible to the human eye. Radiation is not only stealthy, but considered one of the most menacing of the five hazards.

Above Earth's natural protection, radiation exposure increases cancer risk, damages the central nervous system, can alter cognitive function, reduce motor function and prompt behavioral changes. To learn what can happen above low-Earth orbit, NASA studies how radiation affects biological samples using a ground-based research laboratory.

The space station sits just within Earth's protective magnetic field, so while our astronauts are exposed to ten-times higher radiation than on Earth, it's still a smaller dose than what deep space has in store.

To mitigate this hazard, deep space vehicles will have significant protective shielding, dosimetry, and alerts. Research is also being conducted in the field of medical countermeasures such as pharmaceuticals to help defend against radiation.

2. Isolation and confinement

Behavioral issues among groups of people crammed in a small space over a long period of time, no matter how well trained they are, are inevitable. Crews will be carefully chosen, trained and supported to ensure they can work effectively as a team for months or years in space.

On Earth we have the luxury of picking up our cell phones and instantly being connected with nearly everything and everyone around us. On a trip to Mars, astronauts will be more isolated and confined than we can imagine. Sleep loss, circadian desynchronization, and work overload compound this issue and may lead to performance decrements, adverse health outcomes, and compromised mission objectives.

To address this hazard, methods for monitoring behavioral health and adapting/refining various tools and technologies for use in the spaceflight environment are being developed to detect and treat early risk factors. Research is also being conducted in workload and performance, light therapy for circadian alignment, phase shifting and alertness.

3. Distance from Earth

The third and perhaps most apparent hazard is, quite simply, the distance. Mars is, on average, 140 million miles from Earth. Rather than a three-day lunar trip, astronauts would be leaving our planet for roughly three years. While International Space Station expeditions serve as a rough foundation for the expected impact on planning logistics for such a trip, the data isn't always comparable. If a medical event or emergency happens on the station, the crew can return home within hours. Additionally, cargo vehicles continual resupply the crews with fresh food, medical equipment, and other resources. Once you burn your engines for Mars, there is no turning back and no resupply.

---

[64] *The Reference Mission of the NASA Mars Exploration Study Team*, p3-39, Stephen Hoffman & David Kaplan Editors July 1997, NASA Special Publication 6107

Planning and self-sufficiency are essential keys to a successful Martian mission. Facing a communication delay of up to 20 minutes one way and the possibility of equipment failures or a medical emergency, astronauts must be capable of confronting an array of situations without support from their fellow team on Earth.

4. Gravity (or lack thereof)

The variance of gravity that astronauts will encounter is the fourth hazard of a human mission. On Mars, astronauts would need to live and work in three-eighths of Earth's gravitational pull for up to two years. Additionally, on the six-month trek between the planets, explorers will experience total weightlessness.

Besides Mars and deep space there is a third gravity field that must be considered. When astronauts finally return home they will need to readapt many of the systems in their bodies to Earth's gravity. Bones, muscles, cardiovascular system have all been impacted by years without standard gravity. To further complicate the problem, when astronauts transition from one gravity field to another, it's usually quite an intense experience. Blasting off from the surface of a planet or a hurdling descent through an atmosphere is many times the force of gravity.

Research is being conducted to ensure that astronauts stay healthy before, during and after their mission. NASA is identifying how current and future, FDA-approved osteoporosis treatments, and the optimal timing for such therapies could be employed to mitigate the risk for astronauts developing premature osteoporosis. Adaptability training programs and improving the ability to detect relevant sensory input are being investigated to mitigate balance control issues. Research is ongoing to characterize optimal exercise prescriptions for individual astronauts, as well as defining metabolic costs of critical mission tasks they would expect to encounter on a Mars mission.

5. Hostile/closed environments

A spacecraft is not only a home, it's also a machine. NASA understands that the ecosystem inside a vehicle plays a big role in everyday astronaut life. Important habitability factors include temperature, pressure, lighting, noise, and quantity of space. It's essential that astronauts are getting the requisite food, sleep and exercise needed to stay healthy and happy.

Technology, as often is the case with out-of-this-world exploration, comes to the rescue in creating a habitable home in a harsh environment. Everything is monitored, from air quality to possible microbial inhabitants. Microorganisms that naturally live on your body are transferred more easily from one person to another in a closed environment. Astronauts, too, contribute data points via urine and blood samples, and can reveal valuable information about possible stressors. The occupants are also asked to provide feedback about their living environment, including physical impressions and sensations so that the evolution of spacecraft can continue addressing the needs of humans in space. Extensive recycling of resources we take for granted is also imperative: oxygen, water, carbon dioxide, even our waste.

NASA Mars Design Reference Mission

In 1991 the NASA Centers combined to support a NASA JSC-led study to develop a detailed plan for sending the first crews to Mars early in the 21st Century. Michael Duke of JSC, who had played an important role during Apollo and was a long-time colleague and friend of the author, headed the study. Contributors to the study included several dozen senior scientists, engineers and managers from the NASA Centers and Industry. The author (now moved on from NASA HQ and heading the new Center for Mars Exploration at NASA Ames) led the Ames contribution. Even decades later the proposed approach has been kept up to date by NASA JSC. The plan, Mars Design Reference Mission 1, took a key element of its approach from the ideas of one enthusiast – Robert Zubrin, an engineer then working for the Martin Marietta Corporation; he had resurrected the concept of using Martian *in situ* resources (a proposal of the University of Colorado *Case for Mars* team) to greatly reduce the mass needed to be launched to carry out a human Mars mission. The core of the idea is to send an unmanned precursor vehicle to Mars with a small nuclear reactor and a tank of liquid hydrogen to combine with atmospheric carbon dioxide to create liquid methane and liquid oxygen propellants ($2H_2+CO_2 > CH_4+O_2$) for the return journey of a crew that would be launched 500 days later. This approach saves considerable launch mass (most of the mass saved is effectively replaced by the integrated energy provided by the nuclear reactor over many months). The use of Martian atmospheric carbon dioxide in this way might conceivably be applied to some future robotic sample return mission using an RTG (like that of the Viking landers) rather than a nuclear reactor.

Accordingly, the study – conducted through 1993 and made available as a NASA Special Publication (SP6107) in 1997 – adopted *in situ* fuel production for ascent to Mars orbit as an enabling element of the mission architecture. The study was aimed at launches beginning in 2007 using Saturn V-class rockets (or, it was hoped, a Russian *Energia* equivalent), three at each window 26 months apart; the payloads were all directed to Mars without the need for any in-orbit assembly at Earth. International cooperation with Russia was seriously contemplated but this was not vigorously pursued. Crew safety was the obvious major consideration – a risk many times greater than was the case for the Apollo astronauts – though the hazard of galactic cosmic rays was not fully understood at the time.

The interplanetary ionizing radiation of concern consists of two components: galactic cosmic radiation (GCR) and solar particle events. NASA policy establishes that exposure of crews to radiation in space shall not result in health effects exceeding acceptable risk levels. At present, acceptable risk levels are based on not exceeding long-term cancer risk by more than 3% above the natural cancer death probability (which is approximately 20% lifetime risk for the US population as a whole). At present the information required to calculate acceptable risk from radiation exposure during a Mars mission, especially for the GCR, is not available. Although doses (the average particle energy deposition by incident particles) can be calculated, the conversion of this information into a predicted radiation risk cannot be done accurately. The National Research Council recently issued a report estimating the uncertainty in risk prediction for GCR can be as much as 4-15 times greater than the actual risk, or as much as 4-15 times smaller.[65]

The report noted that once the crew was on the surface of Mars they would be shielded from the GCR by a factor of 75% due to the Martian atmosphere and the planet itself. The time spent traveling in space in both directions between Earth and Mars would be the most radiation-hazardous part of the mission so reducing the journey time would be very important.

In the plan, the first crew was to be launched only after a fueled Mars Ascent Vehicle was ready and checked out and after a fueled Earth Return Vehicle was checked out in low Mars orbit. A nuclear thermal stage of the kind already under development in the 1960s was part of the baseline. This high-performance stage allowed the crew to travel on fast trajectories to and from Mars (~200 days) while cargo missions would be launched on minimum energy trajectories (350 days) to maximize their mass.

It should be noted that at this time, in the early 1990s, the Viking missions of the mid 1970s had been the most recent Mars missions and so the science scope of the proposed crewed mission – involving long distance excursions in a pressurized rover plus, at the base, a capable sample analysis laboratory – was enormous as compared to the situation now, following decades of orbital and mobile surface exploration (Chapters 13-16).

The planned crew size for the mission was six. The many different skills that the crew would require were to be captured by each individual having a specialist skill plus at least one secondary skill. This preliminary assessment of necessary crew size and composition considered the tasks required of the crew in the context of safety, risk and the dynamics of an international crew. The study included the initial crewed mission and the establishment of infrastructure at a single site on Mars, one that would support a series of subsequent missions. Four launches would be needed to send the payload necessary for the first crew while three launches would be required to support the subsequent missions.

The approach adopted stands in contrast to an 'all-up' approach in which a single vehicle, assembled in low Earth orbit, is capable of landing the required assets in a single mission to the surface – the approach adopted in the previously

---

[65] Human_Exploration_of_Mars_5.0_Addendum.pdf

mentioned 1989 "90 Day Study". A key new assumption was that very precise landings could be carried out so that the required assets and the crew could rendezvous on the surface of Mars. This allowed these assets to be launched separately and to leave Earth orbit directly without complex assembly there. This capability would be demonstrated using precursor robotic spacecraft. So, the approach involved the following: no extended LEO operations; no rendezvous in Mars orbit prior to landing; short crew transit times to and from Mars (200 days or less); long surface stay-times (500-600 days); a heavy lift launch vehicle capable of delivering either crew or cargo directly to Mars; commonality between the crew's transit habitat (7.5 meters in diameter) and the Mars surface habitat.

The required launch vehicle capability was determined by the largest payload intended to land on Mars: 50 metric tons for a crew habitat sized for six people and that had to be transferred on a high-energy trajectory. This required a launch vehicle that could deliver 200-225 metric tons to LEO. The interplanetary transportation system comprised: a trans-Mars injection stage, a biconic aero-shell for Mars orbit capture and Mars entry, a descent stage for surface delivery, an ascent stage for crew return to Mars orbit, an Earth return stage and an Apollo Command Module for Earth entry and landing.

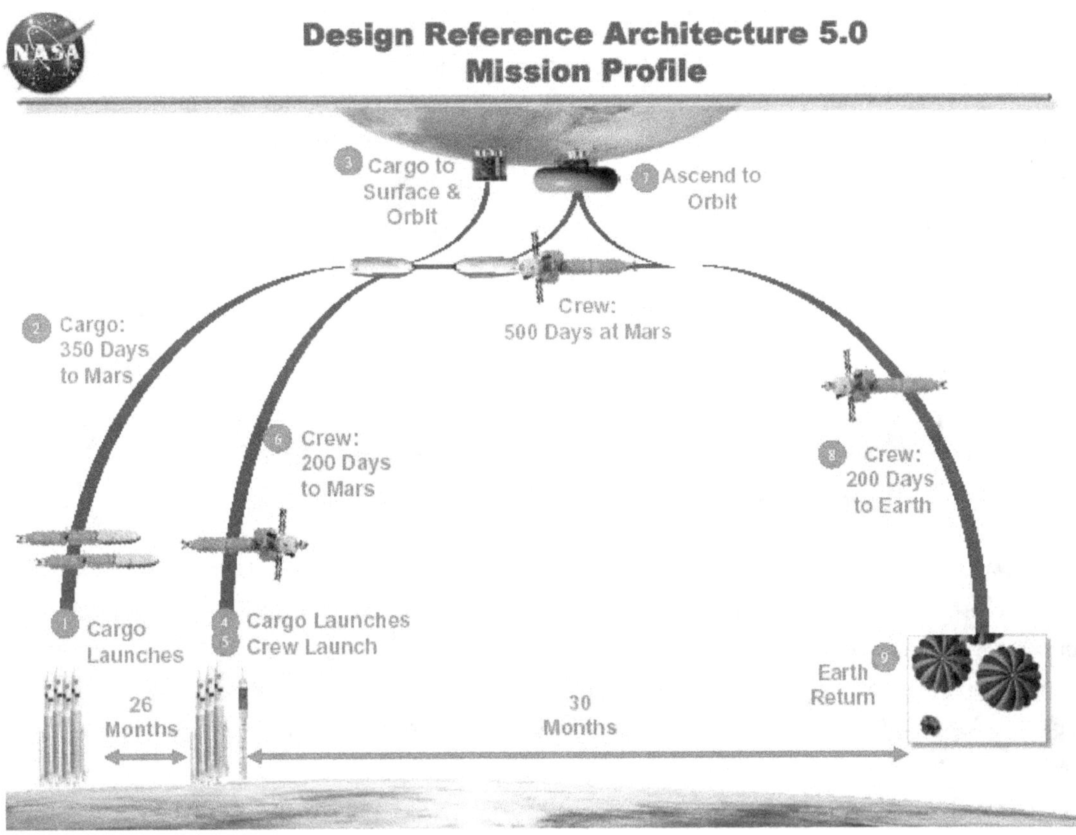

Adding up all the technology developments (including dependence on compact nuclear reactors) that would have been needed to carry out Mars Design Reference Mission 1, it is clear that their funding would have required a major augmentation of NASA's budget, perhaps a doubling. This was at a time when Space Shuttle operations and Space Station development costs were already causing headaches in Washington. Accordingly Mars DRM 1 was put on the shelf to be further refined as opportunity allowed. The Mars Design Reference Mission has in fact been greatly refined in studies over the years with the most recent DRM 5.0 carried out in 2009 with an in-depth addendum in July 2009, and a second addendum in March 2014. The study was the product of the NASA HQ Mars Architecture Steering Group and was edited by Bret Drake of NASA JSC. DRM 5 has served to identify and systematically assess the principal challenges that will confront the planning and execution of landed missions. The study considered a number of different landing sites chosen to meet well-defined science goals that would justify the role of exploration by astronauts.

Mars Design Reference Mission Role of Human Explorers

They would have greater access to the near subsurface of Mars, which would yield insights into climate and surface evolution, geophysics, and, potentially, life. Humans would be able to navigate more effectively through blocky ejecta deposits that would provide samples that were excavated from great depth and provide a window into the deeper subsurface. Humans could also trench in dozens of targeted locations and operate sophisticated drilling equipment that could drill to a depth of 500 to 1,000 meters below the surface (The drilling depth range of 500 to 1,000 meters below the surface represents the HEM-SAG team consensus depth, where it is believed that subsurface water may be found. Clearly, additional investigation will be needed to narrow the depth of drilling). Our current understanding of the crust of Mars is limited to the top meter of the surface, so drilling experiments would yield unprecedented and immediate data. Drilling in areas of gully formation could also test the groundwater model by searching for a confined aquifer at depth.[66]

Improved access to the deep subsurface might be an important contribution by astronauts supervising drilling on the surface of Mars – speeding up the day when we may know whether or not life ever evolved on Mars and, if so, whether life still persists there. Although robotic access to the top several meters of the surface will have many benefits, high levels of UV radiation and oxidizing perchlorates (Phoenix Mission Chapter 14) probably preclude finding evidence of past life in the near surface. Robotic access to the deep subsurface might not be more of a challenge than carrying out the operation with human oversight. An alternative to drilling in order to gain access to depth would be use of an electrically heated probe that would melt its way through the rock.

DRM 5.0 provided some caution with respect to radiation risk from solar storms but not from galactic cosmic rays. The latter are too energetic for any shielding to be effective. It is important to note that the risk assessment that is provided by the radiation discipline indicates that both the short-stay (Opposition Class) and long-stay (Conjunction Class) mission options pose a high risk that crew-members will exceed current permissible radiation exposure limits. The Medical Care analysis did not identify specific knowledge gaps for resolution through goal-directed research beyond those that were identified by the other disciplines; however, there are technology development needs to enable the capability that is required by the Spaceflight Health Standards Document.

The medical capability to manage a large, acute radiation exposure is currently beyond the scope of space flight medical care, especially if extreme measures such as bone marrow transplantation would be required. However technology development in this arena during the next 20 years may allow this treatment capability, or even development, of effective countermeasure performance to progress considerably. The acute solar exposure risk that is posed by the short-stay mission may be the highest unmitigated risk from a CHP Team perspective, aside from the accepted risk of catastrophic vehicular launch and entry/landing failures. However, the radiation risk from an SPE (solar particle event) can be greatly reduced within a heavily shielded location of the vehicle (on the order of 20 g/cm2) and by avoiding close passage to the sun.

The report did not give special consideration to the galactic cosmic ray hazard. These cosmic rays consist of 85% protons, 14% helium nuclei, and 1% heavier nuclei. They have extremely high energies and penetrate many cm of tissue or other materials. Particles heavier than helium (known as HZE particles because they are highly charged) are very densely ionizing so that, even though the flux of such particles is small, they have a significant biological impact.

The Design Reference Mission was set aside and updated at intervals. The near-term focus of manned spaceflight returned to low Earth orbit.

---

[66] https://www.nasa.gov/pdf/373665main_NASA-SP-2009-566.pdf

International Space Station

Space Station Freedom's development in the early 1990s, now including European and Japanese modules, became a series of frustrations given an unenthusiastic Congress. As noted above, in an encouraging move – much to the credit of Administrator Goldin – NASA and the Russian Space Agency in October 1993 agreed to a merger of their space station projects – Freedom and Mir-2 – thereby leading to the creation of the International Space Station (ISS).

The ISS has become a remarkable joint project among the US, Russia, Europe, Japan and Canada. Its first component was launched into orbit in 1998 and the station has been continuously occupied since November 2000. Development and assembly of the station continues, with several major new Russian elements scheduled for launch starting in 2020. From the point of view of furthering the exploration of the Solar System by astronauts, the essential contribution of the ISS and Russia's Mir has been in understanding the health issues involved in long-duration space flight.

Lacking the artificial gravity of the giant rotating space station in Arthur C Clarke's *2001 A Space Odyssey*, the ISS crew must set aside 2½ hours each day for exercise to cope with the adverse effects of their microgravity environment. The ISS, operating in low Earth orbit, is partially shielded from solar and cosmic ray radiation by the underlying planet so the full health consequence of a human mission to Mars has yet to be determined. Compared to operating in deep space outside the Earth's protective magnetosphere, ISS crew receives about one-third the radiation dose. So, it is not feasible to have ISS astronauts fully simulate a 1,000-day Mars mission just using the ISS. The planned lunar station *Gateway* could, however, carry out such a simulation of travel to and from Mars. The heavy, highly charged galactic cosmic rays account for more than 50% of the radiation dose equivalent that the ISS crew experience. The Station is in a 51.6-degree inclination orbit so that, because the geomagnetic field diminishes with latitude, HZE particles that are deflected at lower latitudes now can penetrate the station and its crew when the ISS is overflying higher latitudes.

Learning how to assemble and service large structures in Earth orbit and how to deal with the political and logistical issues of an international project has been another essential, positive lesson for undertaking another such project (specifically an international lunar base) in the future. If experience with extremely long-duration spaceflight in the absence of gravity in the Russian station Mir and the International Space Station can convince NASA's medical specialists that manned exploration missions to the surface of Mars are, in fact, acceptably safe (there will doubtless be no shortage of astronaut volunteers), then NASA's Design Reference Mission 5 is an attractive way to proceed – and probably as affordable as any other approach.

As of May 2020, 240 crew from 18 countries had visited the International Space Station, many of them multiple times: from the US 151 individuals, from Russia 48, from Japan 9, from Canada 8, from Italy 5, from France 4, from Germany 3, and one each from Belgium, Netherlands, Sweden Brazil, Denmark, Great Britain, Kazakhstan, Malaysia, South Africa, South Korea, Spain, Sweden and the United Arab Emirates. Quite a record of international collaboration! [67]

As of 2020 the longest time spent continuously in space was 438 days – accumulated by Russian cosmonaut Valeri Polyakov in the Mir station in the mid-1990s. On an earlier flight he had logged an additional 240 days.

---

[67] https://en.wikipedia.org/wiki/List_of_human_spaceflights_to_the_International_Space_Station

2000 US Presidential Election

The 2000 election was contested between George W Bush and Al Gore. Gore won the popular vote but America's Electoral College vote was decided for Bush following a highly controversial recount of the ballots in the state of Florida and an appeal decided by the US Supreme Court (a 5 to 4 decision). Gore would remain prominent as an environmental activist focused on the looming climate catastrophe. In 2007 he would be awarded the Nobel Peace Prize jointly with the UN Intergovernmental Panel on Climate Change. Looking back from 2020, the 537 Florida votes in question (out of almost 6 million) have quite possibly sealed the climate fate of increasingly hot Planet Earth.

The early years of the new century were potentially exciting times for NASA as President George W Bush, like his father George HW Bush, took an early interest in space exploration and wished to leave an Apollo-like mark in history. President Bush II on 14 January 2004 announced a New Vision for Space Exploration intended to "extend human presence across our Solar System."

Administrator O'Keefe was a key architect of the new policy. There were to be several specific goals:

- Complete the International Space Station by 2010.
- Develop and test a new spacecraft, the Crew Exploration Vehicle, by 2008, and conduct the first manned mission no later than 2014. This spacecraft was to be capable of ferrying astronauts and scientists to the Space Station after the Shuttle is retired and beyond our orbit to other worlds
- Return to the Moon by 2020, as the launching point for missions beyond.

The President announced that this initiative would:

Send a series of robotic missions to the lunar surface to research and prepare for future human exploration. Using the Crew Exploration Vehicle, extended human missions would be undertaken to the Moon as early as 2015, and no later than 2020 and use it as a stepping-stone for more ambitious missions. A series of robotic missions to the Moon, similar to the *Spirit* rover that was (at that time) sending remarkable images back to Earth from Mars, would explore the lunar surface beginning no later than 2008 to research and prepare for future human exploration. Using the Crew Exploration Vehicle, humans would conduct extended lunar missions as early as 2015, with the goal of living and working there for increasingly extended periods. The experience and knowledge gained on the Moon would serve as a foundation for human missions beyond the Moon, beginning with Mars. NASA would increase the use of robotic exploration to maximize our understanding of the solar system and pave the way for more ambitious manned missions. Probes, landers, and similar unmanned vehicles would serve as trailblazers and send vast amounts of knowledge back to scientists on Earth.[68]

He concluded his address with the inspirational thought: "Mankind is drawn to the heavens for the same reason we were once drawn into unknown lands and across the open sea. We choose to explore space because doing so improves our lives, and lifts our national spirit. So let us continue the journey." To meet this ambitious plan, Administrator O'Keefe announced a transformation of organization to change NASA into "a leaner, more focused agency by developing an organizational structure that recognizes the need for a more integrated approach to science requirements, management, and implementation of systems development and exploration missions." The goal was to "eliminate the 'stove pipes,' promote synergy across the agency, and support the long-term exploration vision in a way that is sustainable and affordable," through an organizational structure "that affixes clear authority and accountability".[69]

---

[68] http://history.nasa.gov/SEP%20Press%20Release.htm
[69] http://www.nasa.gov/home/hqnews/2004/jun/HQ_04205_Transformation_prt.htm

The deck chairs were appropriately rearranged. 2015 has, of course, come and gone while plans for astronaut missions into deep space are still tentative. Presidents have the authority to direct NASA to adopt new priorities but Congress holds the critical purse strings. The 535 members of Congress decide levels of taxation that apply across the nation and, also, the authorization and appropriation of funds. In the case of NASA these funds directly benefit only some of the States. So, since (with some notable exceptions) the principal concern of members of Congress is getting re-elected, in the absence of some compelling reason (e.g., a heated Cold War between the US and China), there is not broad motivation to greatly increase spending in just some of the States. Following the Apollo years and the end of the US-USSR Cold War, Congress generally finds it easiest to just kick the can further down the road when grand ideas for astronautic deep space exploration are proposed by the White House.

In June of 2004 the 'President's Commission on Implementation of US Space Exploration Policy' (the "Aldridge Commission" chaired by Edward C. "Pete" Aldridge, Jr.), was charged with making recommendations on implementing the President George W Bush's "Vision for Space Exploration to the Moon, Mars, and Beyond". The report called for a transformation of NASA, building a robust international space industry, a discovery-based science agenda, and educational initiatives to support youth and teachers inspired by the vision. President George W Bush said in a statement:

> I am confident that the Commission's report will help Congress, NASA, other government agencies, the private sector, the international community, and the American public to work together to undertake the next steps in our journey into space for the benefit of generations to come.

Constellation Program

A major policy step was the scheduled mandatory retirement of the Space Shuttle in 2010. In fact the final Space Shuttle flight was that of Atlantis on July 8, 2011. With this guidance and support from the White House, NASA under Administrator Sean O'Keefe pushed forward with the *Constellation Program* with the goal of completing the International Space Station, returning to the Moon by 2020, and with a manned flight to Mars as the ultimate goal. Congress passed NASA's Authorization Act in 2005 that charged NASA "to manage human space flight programs to strive to achieve: (1) returning Americans to the Moon by 2020; (2) launching the Crew Exploration Vehicle close to 2010; (3) increasing knowledge of the impacts of long stays in space on the human body; and (4) enabling humans to land on and return from Mars and other destinations".[70]

Development of progressively more capable heavy lift launchers – Ares I to Ares V – was to be the first step of the Constellation Program in the period from 2005 to 2009 under Administrators O'Keefe and Michael Griffin. In 2020 we still await the first flight of the Saturn V replacement; the development of a new expendable Saturn V-class launcher has, in the absence of funding for its intended application (a return to the Moon) become something of an albatross around NASA's neck. Moreover, the development of a reusable super-heavy lift launcher called *Starship* by the SpaceX company – if successful – may well limit the future of the NASA development of its Space Launch System.

2009 Obama Administration Space Policy

Barack Obama was inaugurated as the 44th US President in January 2009 and his Administration chose to take a fresh look at the direction of the manned spaceflight program. Meanwhile, there was interest in the immediate development of the Crew Exploration Vehicle and the Ares I rocket that together would enable the US to fly crew to and from the International Space Station. Plans for the more powerful Ares V rocket and associated missions beyond low Earth orbit were not addressed and, effectively, indefinitely postponed. In May 2009, the President announced his nomination of Charles Bolden, retired Marine Corps Major General and Space Shuttle astronaut, as NASA Administrator and soon a new review of

---

[70] https://www.congress.gov/bill/109th-congress/senate-bill/1281

manned spaceflight direction was carried out. In September 2009 recommendations were made to NASA by yet another distinguished committee *Review of US Human Space Flight Plans* Committee of experts, looking at options for the post Space Shuttle era. Mars was not proposed as a near term goal but, rather, the Committee, chaired once more by Norman Augustine and including Sally Ride among its members, concluded that:

> A human landing followed by an extended human presence on Mars stands prominently above all other opportunities for exploration. Mars is unquestionably the most scientifically interesting destination in the inner Solar System, with a history much like Earth's. It possesses resources, which can be used for life support and propellants. If humans are ever to live for long periods on another planetary surface, it is likely to be on Mars. But Mars is not an easy place to visit with existing technology and without a substantial investment of resources. The Committee finds that Mars is the ultimate destination for human exploration; but it is not the best first destination.[71]

In summary, the Committee concluded that (then) current plans (the Constellation Program) to return astronauts to the surface of the Moon by 2020 were unsustainable given plausible NASA budgets and that a plan to proceed directly to Mars is even less feasible. It therefore recommended a 'flexible' approach, one that gradually built up human exploration capability toward the eventual goal of exploring the Martian surface – by means of journeys to lunar orbit, to the Earth's Lagrange Points (locations in deep space where a body is in gravitational equilibrium between Earth and Sun), to near-Earth objects (asteroids), and to the vicinity of Mars (flybys and orbit).

In the report the Mars surface remained the 'real' destination even if the timeframe was distant. The report spelled out the clear implications for the development of a giant new expendable launch vehicle for crew and cargo, for the greater involvement of international partners and of the commercial sector. Other recommendations addressed the phasing-out of Space Shuttle operations and the greater exploitation of the International Space Station.

By, in effect, indefinitely postponing a crewed mission to the surface of Mars the Augustine Report broke through the political/financial unrealism of previous post-Apollo plans for human missions into deep space that had landings on Mars as their goal. The report cogently argues the case for its recommendation:

> The Flexible Path is a viable strategy for the first human exploration of space beyond low-Earth orbit. Humans could learn how to live and work in space, gaining confidence and experience traveling progressively farther from the Earth on longer voyages. This would prepare for future exploration of Mars by allowing us to understand the long-term physical and emotional stress of human travel far from the Earth. It would also validate in-space propulsion and habitat concepts that would be used in going to Mars. The missions would go to places humans have never been to, escaping from the Earth/Moon system, visiting near-Earth objects, flying by Mars thereby continuously engaging public interest. Explorers would initially avoid traveling to the bottom of the relatively deep gravity wells of the surface of the Moon and Mars, but would learn to work with robotic probes on the planetary surface. This would allow us to develop new capabilities and technologies for exploring space, but ones that have Earth-focused applications as well. It would also allow us to defer the costs of more expensive landing and surface systems. From the perspective of science, it would demonstrate the ability to service observatories in space beyond low-Earth orbit, as well as return samples from near-Earth objects and (potentially) from Mars. This flexibility would enable us to choose different destinations, or to proceed with the exploration of the surface of the Moon or Mars. This allows us to react to discoveries that robots or explorers make (such as indications of life on Mars) or eventualities that are thrust upon us (such as a threat from a near-Earth object).[72]

These recommendations for the future were much more realistic because the US is unlikely to ever totally abandon the human element of its space exploration program but, also, may never bite the bullet and actually allocate the resources

---

[71] https://www.nasa.gov/pdf/396093main_HSF_Cmte_FinalReport.pdf
[72] ibid.

needed to send a crew to the surface of Mars on what would, necessarily, be as expensive as Apollo and more risky by a large margin but without a compelling justification.

The Augustine Report described above provided the basis of President Obama's space policy, announced in a speech at the Kennedy Space Center in April 2010 where he committed to the completion of the design of a new heavy lift launch vehicle by 2015, predicted a manned spaceflight to Mars orbit by the mid-2030s with an earlier manned mission to an asteroid by 2025. What was notably missing in this was any proposal to return astronauts to the Moon. Also notable was direction that NASA would rely on launch vehicles operated by private aerospace companies, with NASA paying for flights for government astronauts.

The Constellation program including the Ares launchers was begun in 2007 and cancelled in October 2010. A year later, NASA identified the similar Space Launch System (SLS) as its planned new vehicle for astronaut exploration missions beyond low Earth orbit. As of 2020 the first SLS launch was expected to take place in 2021, a mighty long development.

It is evident that, in the several decades since Apollo, the US has lacked sufficiently powerful motivation (of the kind provided by the Cold War) to do more than gradually build up NASA's manned spaceflight infrastructure – the Space Shuttle (now phased out after 135 missions), the International Space Station (funding planned through 2024) and, presently, the Space Transportation System – until the pieces are in place to once again support astronauts in journeys into deep space. Other domestic funding priorities (and there are many especially after the Covid-19 collapse of economies around the world) preclude doing more. A more-or-less-level budget appropriation by Congress is what NASA can reasonably hope to maintain unless and until some compelling new motivation arises. As a result, NASA's plans to launch astronauts to explore Mars look, very optimistically, to the 2030s for execution. Astronaut missions to the surface of Mars would surely require NASA's budget to be greatly augmented and to remain at that level. The fundamental problem facing advocates of human Mars missions well into this 21st Century of ever-more-capable robotic missions is WHY given that robotic missions continue to make increasingly comprehensive progress. Given the intense annual competition for Federal funding it is not surprising that successive Congresses choose to kick this particular can down the road.

The history of manned spaceflight since the end of the Apollo program demonstrates that NASA can sustain a commitment to astronaut travel into deep space for half a century by imaginative *planning* alone. The return of astronauts (not just from the US) to the Moon is a relatively near-term prospect while, in the case of astronaut missions to Mars, the author can see little reason other than the advent of a China-initiated Space Race why *planning* by NASA cannot continue into the indefinite future or until a revolutionary new propulsion system – superluminal warp drive?? – comes into being. Alternatively, given the enthusiastic commitment by the owner of the SpaceX company, the dream of human exploration/colonization of Mars may be an effort that can be left to the private sector.

End of President Obama's second term in office January 2017

The reader will be well acquainted with the nature of the four-year term (2018 to 2020) of the new Republican president, one Donald J Trump, a psychopath who would do his best to end the nation's two-century experiment with constitutional democracy. He would set a record as only one of three US Presidents to be impeached and, in his case, impeached *twice* during his term in office.   His second trial by the Congress was for no less than treason: *incitement to insurrection* as he contested the result of the 2020 election won convincingly by Joseph Biden. Trump urged his cult-like supporters (who had come to Washington from all over the country) to storm the Capitol where Congress was in session carrying out the required confirmation of the results of the 2020 election in which Trump was the loser (232 to 306 in the Electoral College). There were 81 million votes for Joe Biden and, remarkably, no less than 74 million for Trump. Obviously the US is in much trouble - continuation of the societal polarization that led to the Civil War (1861 to 1865) and its aftermath.

The mob of Trump supporters failed in their evident intent to assassinate Vice President Pence (who was overseeing the electoral count) along with the Leader of the House Democrats. Five people were killed in the riot and several badly wounded by the mob. Although all this was captured in detail by video cameras, Trump was acquitted by most of the Republican Senators in his subsequent trial. (The US Senate in 2021 is equally divided between Democrats and Republicans so that the two-thirds majority required to convict was, as in Trump's first trial, practically unobtainable.) Congress therefore will establish an independent commission - modeled on the inquiry into the 11 September 2001 attacks on New York and the Pentagon - to investigate the 6 January attack on the US Capitol by the Trump klan. In addition, Trump faces numerous lawsuits. All this is ongoing at the time of writing and may well have implications for US space policy.

General Health of NASA's Deep Space Exploration Programs in 2020

Operations of NASA's several robotic exploration missions have continued with success and the launch of a highly ambitious mission to Mars (sample collector *Perseverance* with a small rotorcraft) has taken place toward collecting samples from an ancient lakebed for eventual return to Earth. Collaboration with ESA will enable the return, in 2031, of the Perseverance samples. NASA's ongoing robotic mission operations – Mars Reconnaissance Orbiter and Curiosity Rover in particular – continue to advance our knowledge of Mars – making the potential science contribution of any crewed missions increasingly questionable. Missions to carry out exobiology investigations in the outer Solar System are in planning as are missions to asteroid targets. In brief, in 2020 the scientific US space exploration program, with European cooperation, appears to be in robust health (as was also true when President Reagan proposed to terminate it in 1980!).

CHAPTER 7

EXPLORATION PROGRAM CANCELATION AND RECOVERY

Spanish Inquisition, Solar System Exploration Committee, Observers & Mariner Mk2, Flagship Augmentation Missions, Exoplanets, Comet Halley Armada

The Spanish Inquisition

Turning the clock back four decades to the November 1980 Presidential election of movie-star-turned-Republican governor-of-California, Ronald Reagan, it is necessary to recall how NASA's Solar System Exploration Program almost went out of business. The Program, out of the blue, ran into what amounted to a perfect storm. As Monty Python has famously noted, nobody expects the Spanish Inquisition – least of all did the author, now at NASA HQ. Unlikely as it may seem, the problem, political in nature, had its roots in recent disconnected events in distant countries: Iran and Kashmir.

Firstly: the Iranian revolution of early 1979 led to the seizing of the US embassy in Tehran in that November and the holding hostage of 52 diplomats. The attempt in April 1980 by President Jimmy Carter to rescue the hostages was a disastrous failure with major consequences for his re-election bid in November. Ronald Reagan (1911-2004) would be the President of a new Republican Administration.

Secondly: NASA's charismatic Associate Administrator for Space Science, Tim Mutch died in October 1980 in a climbing accident in the Himalayas and his successor, Burton Edelson (1926-2002), was not named until 1982. Tim was much liked and admired, had been the leader of the Viking lander imaging team, was a professor on leave from Brown University and was author of two books about the geology of the Moon and Mars; his leadership would be sorely missed in the following months.

With the incoming Republican President, support for NASA's Solar System Exploration overnight disappeared. Ronald Reagan (1911-2004) in his first inaugural address had famously declared "Government is not the solution to our problem; government *is* the problem" (he maybe was looking far ahead to a US government's response to a future pandemic catastrophe) and, given his neoliberal supply-side 'Reaganonomics' philosophy, some might say he spent the next 8 years doing his best to prove that to be the case; others have canonized the President. Reagan's Director of the Office of Management and Budget (OMB) David Stockman (much later a Wall Street financier and in 2007 the subject of a criminal investigation, later dropped) immediately set out to make significant cuts in government spending and, for whatever reason, included NASA's exploration program in his sights.

Reactions to this baffling change in policy ranged from shock and anger (that of the author and many others) to grief as reported by M. Darby Dyer, Professor of Astronomy at Mount Holyoke College: "Graduate students wept openly in the hallways, and veteran faculty shook their heads".[73]

NASA meanwhile was without an Administrator as Robert Frosch had retired from that position in January 1981. His successor, James Beggs (1926-2020), did not take office until July along with Hans Mark who assumed the position of Deputy Administrator. Dr Mark had been Secretary of the Air Force and had an extensive background in space science for

---

[73] "The Exoplanet Next Door", p54-59, *Scientific American,* February 2019

he had been Director of NASA's Ames Research Center and might ordinarily have been expected to be a strong advocate for the Space Sciences, including Solar System Exploration. It emerged, on the contrary, that Dr Mark was not enthused about exploration of the planets. In 1975 he had noted the substantial cost of a program from which he believed "no fundamental or unexpected discovery" had emerged. And the program itself, he observed two years later, was running out of steam: "we have reached the point in the planetary exploration where, for missions planned between now and the early 1980s, we will have done just about everything we can given our current technology. In other words, we soon will have 'saturated' our capabilities."[74]

Dr Mark's thinking about a long-range plan for NASA included a hiatus in planetary exploration until the construction of a space station had been completed as a base for deep space launches and sample return missions. The influence of Von Braun was evident and it is obvious, now that the Shuttle era is behind us, that Dr Mark's thought process was exactly back-to-front. The author also suspects lingering bad blood between NASA Ames and JPL – the Pioneers versus the Mariners. As for JPL, it would have to seek other sponsors, which to Dr Mark evidently meant the military; it would seem probable that the Reagan Administration would have liked to have JPL play a major role in the expensive (and potentially destabilizing) *Strategic Defense Initiative* that was taking shape in the President's thinking. Probably not by coincidence, in 1982, General Lew Allen (1925-2010), an exceptionally capable and impressive individual who had previously been Chief of Staff of the Air Force, became the Director of JPL; he served there until 1990.

In the convenient absence for many months of the NASA Administrator and the Associate Administrator for Space Science, David Stockman's axe was aimed at the Space Sciences Program. The Planetary Program was judged to be of lesser priority than the Astrophysics Program (an arguable point but not really relevant) where the development of the Hubble Space Telescope was underway. Dr Mark, as noted, accepted that the Jet Propulsion Laboratory would be surplus to the Agency's needs but Administrator Jim Beggs, once he was in office, fortunately did not share these views as detailed by George Washington University Professor John Logsdon in a blow-by-blow reconstruction of this unfortunate episode.

The dire outcome intended by the OMB was averted by a combination of public and planetary science community protests and direct political intervention with the White House, including the President himself, by JPL/Caltech allies; after some months of intensive lobbying by supporters of the exploration program only (!) the VOIR mission and the US spacecraft of the ESA-NASA Ulysses heliophysics mission (a 2-spacecraft joint ESANASA mission to study the polar regions of the Sun) also under development at JPL were cancelled. Such mission cancellations are, fortunately, rare but this was, in fact, considered a significant reprieve because the Voyager mission beyond Saturn (i.e., the Uranus and Neptune encounters) and the Galileo mission under development were also threatened with cancellation along with JPL itself. Later, Phoenix-like, the Magellan mission was to emerge from the ashes of VOIR. President Reagan was not, in fact, uninterested in space technologies, quite the contrary. He was fundamentally antagonistic to the Mutual Assured Destruction doctrine – a strategic offense doctrine – that maintained the peace during the Cold War and, instead, looked to a system of defense that came to be known as the Strategic Defense Initiative (SDI) and, later, Star Wars. The US State Department notes: "There were several reasons why the Reagan Administration was interested in pursuing the technology in the early 1980s. One was to silence domestic critics concerned about the level of defense spending. Reagan described the SDI system as a way to eliminate the threat of nuclear attack; once the system was developed, its existence would benefit everyone. In this way, it could also be portrayed as a peace initiative that warranted the sacrifice of funds from other programs. Privately, Reagan was quite adamant that the goal of U.S. defense research should be to eliminate the need for nuclear weapons, which he thought were fundamentally immoral". [75]

Defensive weapons in orbit were combined with ground-based units; the initiative was publicly announced in March 1983. As the US State Department notes: "critics both in the United States and around the world called the SDI initiative a clear violation of the 1972 Antiballistic Missile Treaty. That treaty had committed the United States and the Soviet Union to refrain from developing missile defense systems in order to prevent a new and costly arms race".[76]

---

[74] *Into the Black: JPL and the American Space Program, 1976-2004* by Peter J. Westwick, Yale University Press, 1 Oct 2008
[75] https://2001-2009.state.gov/r/pa/ho/time/rd/104253.htm
[76] ibid.

The weapons in question included 'directed energy' weapons – various kinds of lasers and particle beams (hence Star Wars), and a hypervelocity rail-gun – along with various kinds of space-based interceptors. One of these was called 'brilliant pebbles' – high velocity projectiles. As will be described later (Chapter 15) the sensors and cameras that were developed and manufactured for this SDI application were to become components of the 1994 Clementine lunar orbiter mission – an unusual, in fact unique, collaboration between the Department of Defense and NASA. Much research and testing was carried out but the system was never deployed; under President Bill Clinton the name of the program changed to the Ballistic Missile Defense Organization with regional rather than global scope.

NASA's program of deep space exploration had, to this point (1980), been a major success for the Agency – and, indeed, the nation – in that there had been much public interest and support for the exploration of the Moon, Mars, Venus and the Voyagers launched in 1977 to the outermost Solar System together with the Galileo orbiter & probe mission that was now in development. In the early days of deep space exploration the formulation of mission priorities had generally followed the recommendations of the National Academy's Space Sciences Board. The less expensive Pioneer missions had been very successful but, being spin-stabilized, had not returned the high-quality images that provided both much of the science return and, also, the attention of the tax-paying public. Caltech Professor Bruce Murray, JPL's Director from 1976 to 1982, termed the 'flagship' missions (Viking, Voyager and Galileo) *Purple Pigeons*, as opposed to what he called *Grey Mice* meaning the Pioneer missions managed at NASA Ames Research Center). The VOIR mission proposed for a new start in President Carter's last budget submission would have used synthetic aperture radar from orbit to image the surface of Venus at high resolution and had 'purple' potential for breakthrough science and strong public interest. With the apparition of Comet 1P/Halley about to take place in the near future (1986), JPL had been pushing for an even more 'purple' mission that would have used new, advanced propulsion to achieve an extremely challenging rendezvous with Comet Halley (the comet is in a retrograde orbit about the Sun).

As a backup to such a risky development, a study had also been undertaken of a joint Comet Halley mission with ESA where NASA would have supplied a Mariner/Voyager class flyby spacecraft and ESA a probe. The National Academy's Space Science Board, however, gave the thumbs down to this collaborative project so NASA abandoned all plans for a Comet Halley mission. ESA, however, wisely continued with its Giotto probe – its very first deep space mission. It was launched on an Ariane rocket out of Korou in French Guiana.

NASA's priorities were focused on bringing the Space Shuttle into operation. This then was the situation when the Administration passed from Jimmy Carter to Ronald Reagan. According to space policy historian John Logsdon of the George Washington University: NASA Administrator James Beggs, in a September 1981 letter to Stockman, proposed elimination of the planetary program as a way to meet NASA's budget targets, a move that would make JPL "surplus" to NASA. John Logsdon further observed "You can question whether Beggs was serious or playing the 'Washington Monument' game, referring to a classic ploy by the National Park Service to respond to budget cuts by closing some of its most visible landmarks, like the Washington Monument, in a bid to restore funds." Logsdon noted: "The problem was, it was acceptable" to the Administration, which proposed in November 1981 to not just cut future planetary programs but also cancel Galileo.[77]

For two years or more NASA's Solar System Exploration Program was left dangling while the Agency and the community of planetary scientists sought to find a solution that would satisfy the neoliberal cost cutters of the Reagan Administration (assuming that cost was indeed the issue rather than Star Wars). Fortunately, the two Voyager spacecraft had already been launched and were on their way to produce spectacular results that even the new Administration could not ignore.

President Reagan was elected to a second term in 1984 at a time when US-Soviet relations were near their lowest ebb in large part because of the destabilizing effect of the Space Defense Initiative. A new arms race in space seemed underway that would be more affordable to the US than to the USSR. Consequently the Doomsday clock was reset to 3 minutes to midnight. Reagan's second term in office (to January 1989) overlapped with that of Mikhail Gorbachev in the USSR who was General Secretary of the Communist Party between 1985 and 1991 (when the Soviet Union was dissolved). The combined leadership of

---

[77] http://www.thespacereview.com/article/2191/1

these two men was of great consequence, the first evidence of which was the December 1987 signing of the Intermediate-Range Nuclear Forces Treaty. This eliminated all short- and intermediate-range nuclear and conventional missiles, as well as their land-based launchers. More than 2,600 missiles had been eliminated by 1991; this was followed by 10 years of on-site inspections. The world began to breathe a little easier – the Doomsday clock moved back to six minutes to midnight.

The dissolution of the USSR led to the independence of the former Soviet republics, fifteen in all including Georgia, Ukraine and the Baltic States – Estonia, Latvia and Lithuania. This was a time of great optimism about the future. In the US, George HW Bush succeeded Ronald Reagan and served in the White House from January 1989 to January 1993 during which time he waged the successful (first) Gulf War to remove the forces of Iraq's Saddam Hussein from Kuwait. In 1989 the US political scientist Francis Fukuyama had published a famous essay entitled *The End of History?* in the journal *The National Interest* arguing that Western liberal democracy was set to become the final form of government.

Well, it hasn't turned out that way. In Russia Mikhail Gorbachev resigned and handed over his powers to Boris Yeltsin (in office 1991-1999) who proposed to transform the economy from socialist to capitalist principles. He proved woefully unable to stabilize a government notable for corruption during a period of low oil prices, inflation and economic collapse. On the last day of 1999 Yeltsin resigned and passed the presidency to his Prime Minister, Vladimir Putin. History had started again *with a vengeance*.

Program Recovery: Solar System Exploration Committee 1981-86

Among the first to come to grips with the existential nature of the problem posed by the Reagan Administration was John Naugle (1924-2013) who had been the Agency's Chief Scientist and, now retired from NASA, was a member of the NASA Advisory Council. The Council agreed to the creation of a strategic planning effort by a new Solar System Exploration Committee (SSEC) whose makeup was selected by Dr Naugle. The working assumption was that *cost* was, in fact, the basic concern of the 'small government' Republican Administration. Comprised of two dozen senior scientists and key experienced managers (John Naugle, Noel Hinners, David Morrison, Arden Albee, Kinsey Anderson, James Arnold, Charles Barth, Thomas Donahue, Michael Duke, Lennard Fisk, Lawrence Haskin, Donald Hunten, Harold Klein, Eugene Levy, James Martin, Harold Masursky, John Niehoff, Toby Owen, Donald Rea, Larry Soderblom, Edward Stone, Joseph Veverka, Laurel Wilkening) the SSEC in a series of detailed reports developed recommendations for a cost-constrained program of high-quality science missions that it recommended be carried out over the next two decades ("Through Year 2000"). The author was the executive director of the multi-year effort.

The program was designed to provide a balance between the numerous planetary science interests – inner planet missions, outer planet missions, missions to comets and asteroids and discovery of planets about other stars. Importantly, it was also designed so that new mission starts would take place at intervals such that the overall budgetary requirement would remain essentially level at an amount close to its 1983 appropriation. At this level of expense these missions would represent the Core program. The key here was to maximize the inheritance of technology and not, like the Galileo Jupiter Orbiter and Probe – under development at that time – approach, seek maximum satisfaction of every potential science goal.

Inner Solar System Missions: Observers

For the inner Solar System it was concluded that commercial Earth-orbiting satellite hardware has most of the capability needed to satisfy orbital remote sensing science missions. For missions beyond Mars the Mariner/Voyager technology had proved itself. So, a three-part strategy emerged that included an ongoing series of relatively inexpensive missions based on commercial Earth orbiters (to be called *Observers*) – beginning with Mars and followed by a similar mission to the Moon. Five Observer missions were identified by the SSEC including three to Mars and one to the Moon. It was expected that NASA Ames, because of their creative participation in the SSEC study of low cost inner planet missions, would compete to manage these missions but, curiously, the Center did not. So the Observers were assigned to JPL.

Outer Solar System Missions: Mariner Mk II

A second series of missions to the outer planets and comets would be more ambitious and would be proposed for new starts at intervals consistent with the level annual budget. The latter were classed as *Mariner Mk 2* missions, similar in scope to past Mariner and Voyager missions (Mariners 11 & 12 had been renamed Voyager 1 & 2); these did not require new technologies. The dual-spin Galileo Jupiter Orbiter approach was unnecessarily complicated. The SSEC identified nine missions to bodies beyond the inner Solar System beginning with a rendezvous with a short period comet. These were an obvious fit for JPL management. A few years later, as told in Chapter 10, a new NASA Administrator, Daniel Goldin, would famously require planetary missions to be *faster, cheaper and better*. The SSEC Observers and Mariners would indeed be somewhat faster in their implementation than their predecessors because no new technology would be required. They would not be better than before – but would fully maintain the science standards of the program. They would be financially constrained (cheaper) in scope and cost – and only one launch per mission was now considered necessary to stay within acceptable levels of risk. Since risk could not be entirely eliminated, the Program Director (the author) later directed that critical spares should be built to allow a rapid re-flight if it should ever be necessary.

Augmentation Missions: Mars and Comet Sample Return, Exoplanets

The third class of missions was called *Augmentation* missions (or now *Flagship* missions) that would be proposed only at intervals in competition with other flagship Space Science missions such as e.g., the Hubble Space Telescope. Mars Sample Return and Comet Sample Return missions were identified as the highest priority augmentations – missions that would allow the full force of laboratory instrument analysis to be brought to bear. The analyses of meteorites and lunar samples had demonstrated just how deep were the insights that could be achieved.

For many years JPL had been carrying out studies of how to implement a Mars sample return mission – a complex and expensive undertaking that, nevertheless, did not pose challenges much beyond what had been met by previous deep space missions like Viking in particular. There are many reasons why it would be more challenging than returning samples from the Moon, something that had long-since been accomplished by Russia. The SSEC report *An Augmented Program* provides an in-depth scientific rationale for Mars Sample Return and describes in detail how such a mission would be carried out by the Agency. At that time, an international collaboration with another space agency would not have been an option for NASA to seriously consider as it is now (with ESA) decades later (Chapter 16).

The SSEC core program recommendations, published in a series of four detailed reports titled *Planetary Exploration Through Year 2000* were presented to, and accepted by, the NASA Advisory Council in 1983. The effort continued for several years and the chairmanship passed after one year from John Naugle (1923-2013) to Noel Hinners (1935-2014) and, later, to David Morrison of the University of Hawaii. The recommendations for an Augmented Program were published in another in-depth report in 1986. The author, as Deputy Director and later Director of the Program Office was Executive Director of the SSEC and provided continuity of leadership throughout. Conveniently, while Noel Hinners was SSEC chairman (1981-1982) he became Director of the National Air and Space Museum sited on the other side of Independence Avenue from the author's office at NASA HQ thereby making weekly meetings possible.

John Niehoff, Alan Friedlander and their colleagues at Science Applications Inc (SAIC) provided essential independent cost analyses and systems engineering support while Gene Giberson (1923-1999) Assistant Director for Flight Programs at JPL, Jim Pollack (1938-1994) of NASA Ames and Mike Duke of NASA Johnson provided the needed Center coordination to complete the effort. Gene's long experience at JPL and his engaging personality made him especially valuable to, and appreciated by, the author. Five subcommittees supported the SSEC: Outer Planets: Toby Owen, chair, Terrestrial Planets (solid bodies): Arden Albee, chair, Terrestrial Planets (atmospheres): Tom Donahue (1921-2004), chair, Small Bodies: Laurel Wilkening (1944-2019), chair, and Mission Operations: Jim Martin (1920-2002), chair. (Laurel incidentally, would later be Chancellor of the University of California, Irvine.) Together, these subcommittees involved another two-dozen senior planetary scientists from various fields along with several senior managers. As a result the recommendations – science justification and implementation detailed in a series of reports, represented a balanced consensus across the complete range of planetary sciences.

Notable by their absence on the SSEC were Carl Sagan and Bruce Murray, both highly motivated and articulate advocates for the exploration of the planets – one with a focus on Mars. They, together with Lou Friedman, had in 1980 founded *The Planetary Society*, an organization dedicated to the exploration of the Solar System and the search for extra-terrestrial life. They took on lobbying of the Administration and Congress to that end and, as a result had their own agenda and would not have been appropriate for membership of the SSEC. They did, however, meet with the SSEC and argued for a focus on Mars exploration and the search for evidence of life, past or present. The return of samples from Mars and a clear path to manned exploration – ideally as an international effort – were what The Planetary Society leaders envisioned was needed to reinvigorate space exploration. The SSEC was impressed but, in the present programmatic circumstances, unconvinced and continued to pursue a balanced Core program without a particular emphasis on Mars. Mars did, however, receive much attention in the 1986 recommendations for an Augmented exploration program. The small community of planetary scientists whose research was focused on the Moon was largely absent from the debate that led to the selection of mission priorities.

Space policy historian John Logsdon has argued that the tide for a continuing program of Solar System exploration was turned by "outreach by insiders, like the president and trustees of Caltech (which runs JPL for NASA), to key members of Congress. Those efforts got the attention of Senate Majority Leader Howard Baker, who wrote President Reagan recommending that the program be continued. "In the end game, they were successful," Dr Logsdon said, in restoring funding for Galileo and opening the door for future missions. "Nothing that the community did on its own would have produced the result they got."[78]

Dr Logsdon's point is well taken but, without the SSEC's rational, balanced and funding- capped new direction for the program, approved by the NASA Advisory Council, it is equally unclear that lobbying alone would have sufficed to save the program.

As a near-term way of getting the program back on track the SSEC had proposed that a modestly de-scoped version of VOIR be undertaken. Major cost savings were made by the use of spares from Voyager, Galileo, Ulysses and even Mariner 9. With essential support from the NASA Advisory Council this mission, named for the great Portuguese explorer Ferdinand Magellan, was approved, funded and soon was in development – Chapter 8. Similarly the first Observer mission, Mars Observer, was started, to renew Mars exploration after a gap of many years. However, this Observer mission was not approved, as had been hoped, as the first in a *continuing line item* but just as a single mission. Therefore, in 1989, the author initiated study of a somewhat different approach: a program of planetary missions – *Discovery* missions – that were to be similar in scope to the Physics & Astronomy Program's *Explorer* missions i.e., a modestly priced continuing line item in NASA's budget open to proposals from the planetary science community that would not have to be "sold" every year. Years later this would prove to be the backbone of the Program for a long time to come.

---

[78] ibid.

The proposed first Mariner Mk 2 mission was to be a rendezvous with Comet Kopff that was to be preceded by a flyby of an asteroid: *Comet Rendezvous Asteroid Flyby* (CRAF). A second MM2 mission, *Cassini*, was also planned, one that would orbit Saturn. This latter mission attracted the interest of ESA scientists who proposed to build a Titan entry probe that would hitch a ride to Saturn on the orbiter. The combined mission proposal, named *Cassini-Huygens*, had much support from both sides of the Atlantic and along with CRAF was approved to go forward. All of the newly planned missions were to be launched – indeed, like all NASA missions – using the Space Shuttle.

Exoplanets

Besides the recommendations for Augmentation Missions that would return samples from Mars and from a comet, the SSEC also recommended that NASA should make an increasing effort to search for planets orbiting other stars – *exoplanets* – because such evidence was mounting regarding the feasibility of such discoveries. George Gatewood of the Allegheny Observatory provided the SSEC with the necessary background to fully appreciate the potential of such an effort. Using the 76 cm Thaw refractor at the Observatory together with a new Multichannel Astrometric Photometer he had been a pioneer in developing the astrometric approach to such a hunt by making precise, long-term measurements of a star's position – to detect slight back-and-forth "wobbles" produced by the gravity of unseen planets moving around it. A second, similar approach to the hunt was described in the SSEC report: one that measures the velocity of the star relative to Earth by measuring the Doppler shift in the star's spectral lines. Given the then-recent detection by the *Infrared Astronomical Satellite* (IRAS) of disks made up of fine solid particles orbiting nearby stars (Vega, Formalhaut and Beta Pictoris) the prospects for success were considered very real and, as such, would open a new dimension to Solar System exploration. Accordingly, the SSEC Augmented Program report included a recommendation that "Activities involving the search for, and the study of, other planetary systems, should be carried out by NASA's Solar System Exploration Division. These activities are consistent with the Division's established goals and especially with the goal of understanding the origin and evolution of our solar system". This search, carried out over the following decades using both ground-based telescopes and the *Kepler Space Telescope* has, as is well known, proved to be a triumph for NASA and for astronomers in Europe. The discoveries as of 2019 are reported in the second part of this history: The Outer Solar System and Beyond.

In making the case for these several augmentation missions the SSEC explicitly recognized that new technological capabilities would be required – the reason why they would be much more expensive than the core program missions. The new technologies included: low thrust propulsion, aero-capture, automated autonomous rendezvous and docking, artificial intelligence for semi-autonomous roving vehicles and remote sampling, and advances in communication technology. Although the Mars and comet sample return missions have not yet been carried out, all of these advances have, in fact, been made over the thirty years since the SSEC report was delivered. So, undertaking Mars and comet sample return missions in the coming decades will be sufficiently challenging but should encounter no major technology barriers. At the time of writing – in early 2020 – NASA and ESA are together moving toward carrying out a first Mars sample return with launches in 2026 (Chapter 15). The Coronavirus pandemic emergency will likely have an impact.

The successful re-emergence of the Solar System Exploration Program from its travails was to be met, like all of NASA programs, by a stunning and tragic setback when in 1986 the Challenger exploded soon after launch and its crew perished. By now the author had become the Director of the Solar System Exploration Program. His predecessor, Jesse Moore, had the great misfortune to have been promoted at this time to become the Associate Administrator of the manned spaceflight program.

A delay of several years ensued for all programs – this included Galileo, Ulysses, Magellan, Mars Observer, CRAF and Cassini-Huygens – while the Space Shuttle problem was analyzed, understood and (expensively) fixed. The delay inevitably increased the costs of the approved planetary missions as their 'marching armies' had to be supported for additional years. CRAF eventually suffered cancellation (being resurrected by ESA's *Rosetta* mission) and the Mars Observer was de-scoped (removing one instrument) and redirected toward launch on an expendable Titan. The Galileo and Magellan missions remained as payloads on the Space Shuttle but now they were to be mated – in a move judged to be prudent by the

Johnson Space Center – to a solid upper stage (the *Inertial Upper Stage*) instead of the liquid-fueled *Centaur*. The less powerful IUS meant that the JPL mission designers had to exercise considerable ingenuity in coming up with new trajectory designs. Rescheduling the launches of all these missions also required some give-and-take. Because the ESA *Ulysses* spacecraft needed to receive a gravity-assist at Jupiter in order to place it into a polar orbit about the Sun, the Galileo and Ulysses launches were now in competition for the same launch window. The problem was resolved by the (much appreciated) willingness of ESA to absorb the costs involved in having Galileo take the first opportunity with Ulysses following a year later.

With all this drama, the 1980s were proving to be a time that most would like to put behind us. However, the Comet Halley apparition in 1986 was the sole (very) bright spot as described below. NASA's top leadership underwent a number of changes in the 1980s. James Beggs served as Administrator from 1981 to 1985. Deputy Administrator Hans Mark left NASA in September 1984. William Graham who was briefly Acting Administrator until mid-1986 replaced Mr. Beggs. James Fletcher (1919-1999) who had been Administrator in the 1970s returned to serve to April 1989. Dale Meyers was acting Administrator for a month, and Richard Truly was then Administrator till April 1992.

Five Administrators in the 1980s was clearly a sign that NASA was not in the greatest health. The Voyager encounters with Jupiter in 1979, Saturn in 1980 were, however, outstanding bright spots that served to maintain morale in the planetary science community, at JPL and, indeed, in the author's Program Office at NASA HQ. The international effort in 1986 – reported below – that coordinated Soviet, ESA and Japanese spacecraft encounters with Comet Halley, supported by NASA's Deep Space Network, also served to help weather an extraordinarily difficult decade. Eventually Magellan was launched in May 1989 and Galileo in October 1989 – at which point the author concluded he badly needed a break and handed over the reins to his successor, Wesley Huntress.

Mars Observer did not get launched until September 1992 (about which more later, Chapter 10) and, alas, was lost as it neared its destination. Meanwhile, in 1989, as previously mentioned, workshops were initiated to identify relatively low cost Discovery missions that would have strong science justification and underpin both the planetary science community and the Agency's implementing Centers – not necessarily just JPL. In the mid-1980s, while the SSEC was working to save NASA's Solar System Exploration Program, there were a number of events and activities, described below, that served to lighten the spirits of the planetary science community, the staff of the Jet Propulsion Laboratory, the author and his NASA program management team.

Halley Armada

With this one exception, space missions to comets and asteroids in orbits that spend most of their lives in the outer Solar System are treated in a second volume of this history. The 1986 encounters with Comet Halley are discussed here because they were important in the recovery of NASA's Solar System Exploration Program after it was threatened by cancellation. The eagerly awaited Comet Halley apparition led the space agencies of the US, the Soviet Union, Europe and Japan all to develop plans to encounter the comet as it passed through the Inner Solar System. Initially, a major competition seemed in store. However, given the way events unfolded, the Comet Halley missions provided the most successful international cooperation ever in space science exploration – in no small part because NASA (on the advice of the National Academy of Sciences) chose to forego the opportunity to carry out its own mission and, instead, chose to provide support for the other agencies.

Given the nature of Comet Halley's retrograde orbit about the Sun, its orbital inclination and its extreme velocity as it nears the Sun, a rendezvous mission would have required a propulsive capability far beyond that provided by upper stages like the Centaur. There were two candidate technologies, both of which needed to operate over months rather than minutes: a giant solar sail (propulsion by radiation pressure of sunlight) and an ion drive engine (propulsion by means of plasma accelerated by electric fields) that would use very large solar panels for power. Neither technology existed and would be very challenging, expensive and risky given the available time for development. Such a Comet Halley rendezvous would surely have been a defining 'purple pigeon' for JPL and Bruce Murray, its Director and advocate.

Unsurprisingly, it became clear that such a mission was not affordable and that the development was generally too risky – so NASA settled for a flyby mission that it would carry out in collaboration with ESA who would supply a small close-encounter probe to be called *Giotto* (named after artist Giotto di Bondone, who had used the appearance of Comet Halley in 1301 as the model for the Star of Bethlehem in his 1304 painting titled "Adoration of the Magi" in the Scrovegni Chapel in Padua, Veneto, Italy.)

Collaborative plans were quite advanced when the US National Academy's Space Science Board concluded that a flyby mission would not achieve sufficient science return – so NASA dropped the plan, no doubt much to the annoyance of ESA as well as many in JPL and NASA. ESA, however, had a greater commitment to the enterprise and decided to continue with Giotto and to launch it on an Ariane rocket from its launch site near the equator in Kourou, French Guiana. This took place successfully on 2 July 1985. Incidentally, a NASA guest at the Giotto launch, the author, was very impressed that the ESA team had arrived there in style on an Air France *Concorde*.

The Soviets pursued an imaginative plan of their own (Chapter 4) that involved passing by Venus on the way to Comet Halley with two spacecraft, VEGA 1 and 2, dropping-off balloons into the upper Venus atmosphere on the way. The Halley-bound spacecraft were developments of the earlier Venera probes and were designed by the Babakin Space Center. The Japanese missions were more modest and consisted of two small flyby probes named Sakigake and Suisei. Sakigake was Japan's first-ever deep space probe; it and Suisei were launched on Mu-3SII rockets out of the Kagoshima Space Center on the southernmost tip of Japan. The spacecraft had modest instrumentation that served to measure the solar wind and magnetic field as they flew by Comet Halley at considerable distances. Sakigake, unlike Suisei, carried no imaging instruments. Somewhat embarrassed, NASA still had an important role to play because it had (and has) the world's most capable Deep Space Network of communication facilities located around the Earth whose support could magnify the return of each of the spacecraft. Also, NASA led an international ground-based program of telescopic observations of Comet Halley throughout its apparition – the *International Halley Watch* – that was a valuable science complement to the brief encounters of the spacecraft.

The InterAgency Consultative Group (IACG)

Even during the Cold War period, space scientists and program managers in the US, Europe and the Soviet Union maintained good relations so it was not difficult to set up an organization to coordinate the activities of the four space agencies. The creation was straightforward because no exchange of funds nor technology transfer were involved and because the function was simply advisory. The *InterAgency Consultative Group* (IACG) was formed and met annually during the years leading up to the 1986 encounters. Each Agency hosted one of the annual meetings that took place in Padua Italy (1981), Kagoshima, Japan (1982), Tallinn, USSR (1983), Budapest, Hungary (1984), and Cocoa Beach, USA (1985). Roger Bonnet, Director of ESA's science programs, Rudiger Reinhard, Hugo Fechtig and colleagues represented ESA; Roald Sagdeev, Karoli Szego and Sasha Zakarov represented Intercosmos; JAXA was equally represented; the author, JPL's Gene Giberson, Ray Newburn & colleagues, and other scientists represented NASA. After Comet Halley had passed by in 1986 the IACG held a science meeting in Heidelberg after which, on 7 November, Pope Paul II invited all the attendees to the Vatican Palace in Rome in celebration of a unique and successful international collaboration in science.

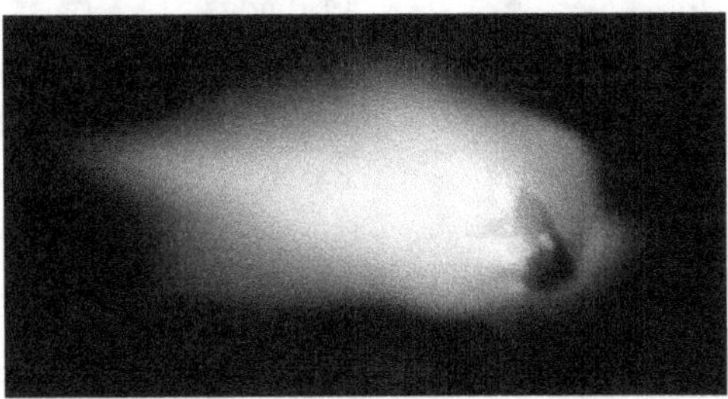

Spacecraft observations of Halley, a visitor from the coldest depths of our Solar System, supported the 'dirty snowball' comet model that Fred Whipple had long espoused: a nucleus composed of a mixture of ices – water, carbon dioxide, methane and ammonia – and of dust. Whipple had argued that, warmed by the Sun as they fall inwards, ices at or near the surfaces of comets change directly from a solid to a gas causing jets of volatiles to burst out to create the comet's coma. The Giotto probe had come within 600 km of the nucleus and returned its best image from a distance of 6,500 km 95 seconds before its closest approach. The rather crude composite image includes details on the nucleus and the dust jets emanating from the sunlit side.

The nucleus turned out to be peanut-shaped with a size of 16x8x8 km. Contrary to what might have been expected, the nucleus is among the darkest objects in the Solar System having the reflectivity of coal. The surface of Comet Halley is largely composed of dusty, non-volatile materials and only a small portion of it is ice. This indicates, as the science teams noted, that comets are more like icy dirt-bags than dirty snowballs. Tracking data allowed an estimate of mass of about 2.2×10^{14} kg and of density: ~0.6 gm/cc. This low density implies a loose structure akin to a rubble pile. Comet Halley's surface was only partially mapped; it is very rugged and includes at least one crater. The rotational period of 177 hours estimated by Earth-based telescopic observations of the coma contrasts with spacecraft imaging observations of gas jets that implied a period of 53 hours. It is judged to be the most active of all the periodic comets. Comet Halley's next perihelion will be 28 July 2061.

The Magellan launch took place successfully on 4 May 1989 and the launch of the Galileo mission to Jupiter took place in late 1989 so the Program was apparently – but deceptively – in a stable state once more. The author therefore concluded that this would be a good time to move on from HQ back to a NASA Center far away from Washington DC. In doing so, he missed the opportunity to work in the NASA HQ of a new Administrator, Daniel Goldin who, by spring 1992 had been appointed by President George HW Bush. Mr. Goldin brought with him welcome enthusiasm for planetary exploration but something of a messianic style and an emphatic commitment to carrying out missions differently – "faster, cheaper, better". This slogan could, in fact, have been a summary for the SSEC's proposed Core program, one that avoided the need for technology advances but the author had not been smart enough to have done so.

Decades of experience by JPL and other Centers in developing a rigorous project management approach for deep space missions had led to fewer failures (and, as a major cost saving, had eliminated the need to launch two planetary spacecraft to provide backup). Unimpressed, Mr. Goldin concluded that the Agency suffered from "bloated bureaucracy" and needed trimming down to size.[79]

Mr. Goldin had worked at NASA's Lewis Research Center on electric propulsion systems and, for 25 years worked at the TRW Space and Technology Group in California. He must have had some bad experiences. Bloated bureaucracy was certainly not obvious at JPL where the author had worked for six years and, later, as the Director for the Solar System Exploration Program had responsibility for the planetary missions led by JPL and NASA Ames. On the contrary, to the author (who has only limited experience of the manned spaceflight Centers) the Centers carrying out the science programs seemed embodiments of the very best that a government program run by fallible beings could reasonably hope to be – especially when there is inherent great risk and potential for national embarrassment – something with which both NASA and the Russians had experience.

During Mr. Goldin's tenure near-term crewed exploration beyond Earth orbit was abandoned and, as a result, the Clinton Administration's 1996 National Space Policy officially removed human exploration from the national agenda.

All NASA science missions must necessarily seek (and hope to find) an optimal balance between performance, cost and risk. This is what the SSEC had proposed, mainly by avoiding the need for technology advances and by limiting the Core Program science ambitions relative to e.g. Viking and Galileo. The Administrator's "faster, cheaper, better" philosophy simply seemed to point to less-ambitious science, less balance among the science goals and higher-risk missions with a narrow focus on Mars. In the latter regard he likely was encouraged by Carl Sagan.

The SSEC recommendations largely fell by the wayside though Magellan, Mars Observer and Saturn-bound Cassini-Huygens continued, as did Jupiter-bound Galileo. The spin-stabilized Pioneer missions that had been managed by NASA Ames were perhaps the kind of missions that Mr. Goldin had in mind: they were certainly somewhat faster and cheaper – and, certainly, successful. However, few could argue that they were *better* than the 3-axis-stable Mariner/Voyager missions that had revolutionized our knowledge of the Solar System all the way out to Neptune. Now, fortunately, the Discovery approach already being formulated was, in fact, directly responsive to the Administrator's cheaper command. Relatively inexpensive Discovery missions would provide much of the scientific balance that otherwise would have been lost. Also, the search for planetary systems about other stars would take off to huge success thanks to a highly innovative Discovery mission.

---

[79] https://history.nasa.gov/dan_goldin.html

Happily, the creativity of the science community ensured that the science return from these otherwise grey mice (as Bruce Murray would no doubt have termed them) was still substantial. In fact, taken altogether, the Discovery missions have proved to be both very fruitful and a highly desirable stabilizer for the Solar System Exploration Program; Director Wes Huntress deserves a lot of credit. Coupled with the ongoing Magellan, Galileo and Cassini-Huygens missions, the overall program assumed a welcome balance of scope and risk. In addition, because NASA followed the Astrophysics Explorer approach in soliciting *mission* proposals directly from Principal Investigators (rather than *instrument* proposals through a traditional Announcement of Opportunity (AO) for a NASA-preselected mission the program opened itself to new ideas and to implementing organizations besides JPL including e.g. the Applied Physics Laboratory of Johns Hopkins University. Healthy competition!

CHAPTER 8

VENUS FINALLY REVEALED

Magellan, Venus Express, Akatsuki, Phosphine

1989 Magellan

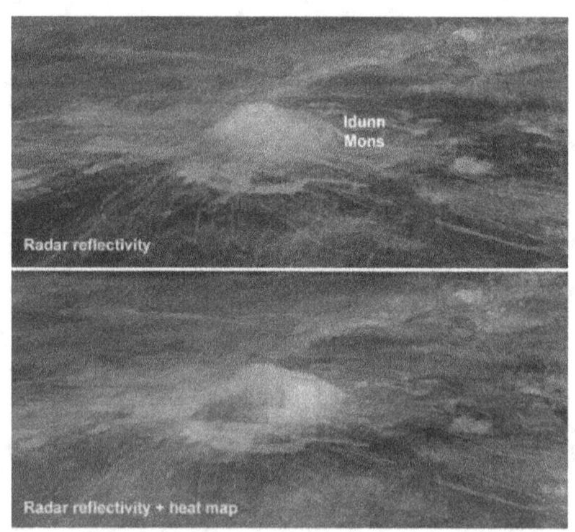

The tragic Challenger disaster occurred on Tuesday, January 28, 1986, when the Space Shuttle *Challenger* exploded 73 seconds into its 10th flight killing all seven crew members – five NASA astronauts, one payload specialist, and a civilian schoolteacher. After much programmatic trauma – across-the-board in NASA in both the return to flight of the Shuttle and in its Solar System Exploration Program – the Magellan Venus radar-mapping mission (son of VOIR) was built, tested and launched on Space Shuttle Atlantis on 4 May 1989. The goal was to complete the high-resolution (km or better) mapping of Venus already begun (from 30°N to the N pole) by the Soviet Union in the early 1980s. With such high-resolution mapping the true nature of our sister planet would be revealed for the first time.

In contrast to the way in which a radar altimeter such as that on the Pioneer Venus Orbiter was used to map the cloud-hidden surface of Venus, synthetic aperture radar imaging is more complex and not intuitive to understand – so the reader is referred to online sources for proper explanation. This radar imaging approach is extensively used from aircraft and has a lengthy history. It is sufficient to know that the Magellan's antenna, oriented off-nadir, transmitted pulses (several thousand per second) of microwaves through the cloud cover as Magellan moved in its orbit. Each pulse produced an echo as it was reflected off the side surfaces of rocks, cliffs and geologic features; these echoes were received by the same 3.7meter parabolic antenna and recorded for transmission back to giant antennas of the Deep Space Network. Heavy-duty signal processing software at JPL took all of these data and constructed an image of the illuminated Venus surface.

JPL already had essential experience with radar mapping from Earth orbit as they had previously managed the SEASAT mission. This had used synthetic aperture radar (and other instruments) to monitor oceanographic phenomena such as sea-surface winds, temperatures, and wave heights. Several Federal Agencies were involved in setting the mission requirements.

SEASAT was launched in June 1978 on an Atlas out of the Vandenberg Air Force Base in California into a near-polar orbit where it operated for 3 months. One of the scientists at JPL who was heavily involved in the development of this first radar mapping from space was Charles Elachi who, much later, would become JPL's eighth Director in 2001. Besides its importance for Magellan, SEASAT also provided the experience that underlay a series of Space Shuttle imaging radar experiments that were carried out through the turn of the century.

Martin Marietta designed and built the Magellan spacecraft. JPL provided project management. Gordon Pettengill of MIT was the Principal Investigator of the synthetic aperture radar experiment. Gravimetry (PI Georges Balmiro of Centre National d'Etudes

Spatiale) and radio science occultation measurements were the other science experiments of the mission. Two-dozen Co-investigators and 17 Guest Investigators supported Professor Pettengill. Among the latter were three Russian scientists: Efraim Akim, Alexander Basilevsky and Alexander Zakharov.

Magellan had a succession of Project Managers at JPL: Douglas Griffith, James Scott, Anthony Spear and John Gerpheide (1925-2016). Steve Saunders was, throughout, the Project Scientist with Ellen Stofan his deputy. The spacecraft and its upper stage rocket (a two-stage solid fuel Inertial Upper Stage [IUS]) were carried to Earth orbit in the payload bay of STS Atlantis, and deployed six hours into the flight after 5 orbits. From there, after a 150-second burn, Magellan began a 15-month journey to Venus – much longer than normal because the launch of Galileo later that year created a conflict for the optimal launch date. Magellan's launch ended a fraught eleven-year gap in US space exploration launches since the Voyagers. The Galileo spacecraft was launched months later (also on the Shuttle) on 18 October 1989 and the ESA Ulysses spacecraft was launched on the Shuttle a year later. After this, planetary exploration spacecraft returned, with some relief, to using expendable rockets.

The three-axis-stable, solar-powered Magellan spacecraft made significant cost savings by taking advantage of residual hardware and designs left over from previous missions: Galileo contributed the computer system, the 3.7-meter-high gain antenna had been left over from Voyager and the medium-gain antenna was a spare from the Mariner 9 mission. Multi-layered thermal blankets and a special white paint served to protect the spacecraft during its travels in toward the Sun. The high-gain antenna was not ideal for the radar function and the elliptical orbit also complicated the task so that consequently it took 4,000 commands each orbit to operate the synthetic aperture radar. The X-band downlink communication was to be at the very high rate of 269 kbps. On occasion during the mission high temperatures required pointing the large dish antenna toward the Sun so that the rest of the spacecraft could cool off in the antenna's shade.

The Magellan spacecraft was inserted into a 3-hour elliptical polar orbit – periapsis 295 km and apoapsis 7,760 km – about Venus on 7 August 1990. Closest approach was near the equator at 10°N. Range at the North Pole was ~2,000 km. The radar mapping was carried out from the low altitude portion of each orbit with playback to the DSN during the high-altitude portion. In order to point the fixed antenna to the side for mapping and then to point at Earth for playback, this required the appropriate adjustment of the spacecraft's attitude each orbit. Magellan completed six mapping cycles beginning with two that together covered 96% of the planet. The third cycle served to provide stereoscopic coverage of about 20% of Venus.

The width of the imaged swath on each orbit varied from 17 to 28 km depending on the altitude of the spacecraft in the lower part of its elliptical orbit. The slow rotation of Venus provided a slight overlap in the coverage from orbit-to-orbit. By acquiring at least four independent observations of the surface over the course of the mission the speckle-like noise in the images was reduced. Because brightness in such an image is a function of surface roughness, angle of incidence, and electrical properties of the surface the interpretation of the radar images required the learning of some new skills by the science team.

The fourth mapping cycle was devoted to mapping the gravitational field of Venus, a process that involved pointing the antenna continuously at Earth to measure changes in the signal's Doppler shift caused by the small variations in spacecraft velocity in passing over areas of greater and lesser gravitational attraction.

At this point, with much already accomplished, the Magellan team successfully carried out an experimental energy-saving approach – aero-braking – to approximately circularize the elliptical orbit. The onboard propulsion unit was used to lower the periapsis altitude of the spacecraft to about 180 km so that it skimmed through the uppermost atmosphere where frictional drag had the effect of gradually lowering the apoapsis altitude from 7,760 km down to 540 km. The benefit of this was in allowing the acquisition of much higher resolution gravimetric data (cycle 6). By 15 May 1991, after 1,792 orbits, Magellan had mapped over 80% of the surface with a resolution between 200 and 500 meters. By the end of the radar mapping 98% of the surface had been imaged at resolutions better than 100 meters.

Now the Pioneer Venus altimetry and the limited coverage of the Venera 15 & 16 radar maps could be properly interpreted; Venus was revealed to be a body quite unlike any other in our Solar System, one that has few impact craters on a surface that is mostly (85%) covered by volcanic materials – the remainder by highly deformed mountain belts. There are no clear signs of plate tectonics but a suggestion that such dynamic crustal activity may yet commence.

In this instance – with few craters to count – determining the areal density of craters provides only limited help in establishing relative ages of different areas. The team concluded that, in fact, the surface is all of roughly the same age and is relatively young. In spite of the thick hot atmosphere, the lack of water and the light winds at the surface mean that erosion is minimal. The mean surface age is estimated to be about 500 million years suggestive of a global resurfacing event at that time – perhaps the way in which a terrestrial planet lacking plate tectonics relieves the build-up of internal heat. Lava channels thousands of kilometers long imply that the very low-viscosity lava erupted at a very high rate. The Magellan and Venera radar maps have provided the planetary science community with the same break-through information that Mariner 9 had achieved at Mars twenty years earlier – albeit Venus does not have the same exciting variety of processes and exobiology promise that does Mars.

The Venus surface observed by Magellan is characterized by a mixture of volcanism, fracturing processes and impacts with a notable absence of degradation processes and of plate tectonics. Volcanoes are everywhere and at least 100,000 small (less than 15 km across) volcanoes shaped like inverted shields have been counted. In smooth volcanic plains there are more than 100 volcanoes greater than 100 km in diameter. Although most appear extinct and none were erupting, Magellan did observe ash-flows near the summit of Maat Mons, Venus's highest volcano.

Maat Mons is a huge shield volcano, 395 kilometers in diameter whose three-dimensional cone-like shape (vertical dimension greatly exaggerated in the image) has been reconstructed by combining the side-looking radar data with radar altimetry. Its 30-km diameter summit caldera contains five smaller collapse craters. Ash flows near the summit and on the northern flank detected by the radar sounding imply relatively recent activity – a possible explanation for the variability of sulphur dioxide and methane in the middle atmosphere that was noted in the Pioneer Venus data in the early 1980s.

There is variety in the form the volcanoes take, for example one that the science team describes as a 'petal' type (40 by 60 km in size) is located in eastern Aphrodite Terra. There are radial, star-like fracture systems called novae; and circular rings of fracture called *coronae* – oval-shaped features that may be formed by upwelling of warm material. Some of the innumerable small volcanoes have the appearance of flattened domes or 'pancakes' and form 'shield fields'. The Magellan science team suggests that the pancakes are formed by highly viscous lava eruptions taking place under the extreme conditions of Venus's atmospheric pressure.

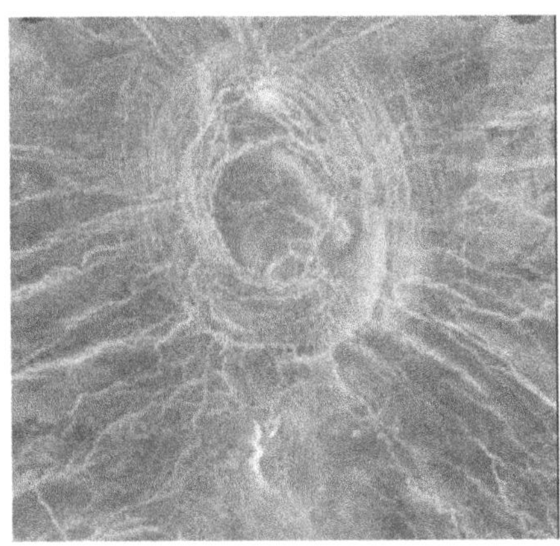

Venus sports what may be the largest circular structure – 2,600 km in diameter – in the Solar System: *Artemis*, named after the twin sister of Apollo and Greek goddess of the hunt. The feature lies between the rugged highlands of Aphrodite Terra to the north and the relatively smooth lowlands to the south. It is centered on 30°S, 135°E. Its circular shape and size make Artemis the largest corona identified to date – one that could encompass most of the US from near Denver to the West Coast. It includes an interior topographic high surrounded by the 2,100-km-diameter, 25- to 200-km-wide, 1- to 2-km-deep circular trough, called Artemis Chasma thought to be related to the upwelling of hot material in the form of plumes or diapirs (an intrusive structure resulting from mobile material – in this case hot, less dense magma – that has been forced upward into brittle surrounding rocks.)

The analysis continues of the geologic and atmospheric history of Earth's nearest neighbor and of what might almost have been our twin. It seemed likely that Venus remains volcanically active and so this would be a major question for ESA's Venus Express mission to address when it arrived at Venus in April 2006.

2005 Venus Express

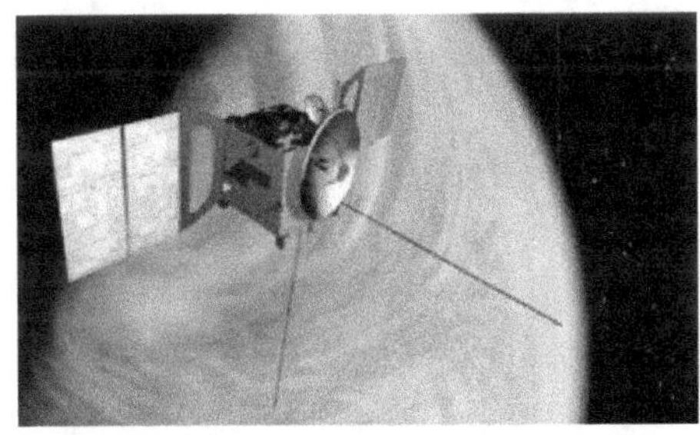

The European Space Agency carried out its first mission to Venus – the Venus Express Orbiter that was launched on a Soyuz-Fregat from the Baikonur Cosmodrome in Kazakhstan on 9 November 2005. The Venus Express prime contractor was EADS Astrium, Toulouse, France. EADS led a team of 25 subcontractors from no less than 14 European countries. The Project Manager was Don McCoy, the Project Scientist Hakan Svedhem and the Spacecraft Operations Manager Andrea Accomazzo. Just as for NASA's early Mariner missions, the Venus Express spacecraft and its instrumentation were similar to that of its predecessor Mars Express (Chapter 13). The science team Principal Investigators were Stanislav Barabash (Space Plasma and Energetic Atoms), Tielong Zhang (Magnetometer), Vittorio Formisano Planetary Fourier Spectrometer, Jean-Loup Bertaux (UV and IR Spectrometer), Bernd Häusler (Venus Radio Science), Pierre Drossart and Giuseppe Piccioni (UV/Visible/Near-IR Imaging Spectrometer) and Wojciech Markiewicz (Venus Monitoring Camera).

Necessarily, the thermal control system required more efficient radiators and reflective gold insulation instead of black. The solar arrays were much smaller. Three of the six science instruments were spares from Mars Express while two were modified instruments that were designed for ESA's 2004 Rosetta mission to Comet Churyumov-Gerasimenko.

Venus Express entered orbit in April 2006 and began monitoring the clouds and atmosphere using UV and IR instrumentation. Observations continued for the following nine years aiming successfully at gaining an improved understanding of the planet's atmospheric dynamics. Venus Express also had instrumentation to monitor surface temperatures in order to look for evidence of contemporary volcanism. Both of these goals were well satisfied.

The evolution of the thick, hot Venus atmosphere is of particular interest to Earthlings – many of us painfully contemplating continuing climate change – because the extreme surface temperature of Venus is the result of an intense greenhouse effect due to the infrared properties of its constituent carbon dioxide atmosphere plus the contribution of some water vapor and sulfuric acid aerosols. The Venus Express team concluded that the clouds reflect 80% of the insolation, the atmosphere absorbs another 10% so that only 10% actually makes it to the surface. The very high surface temperature arises because the atmosphere serves as a powerful greenhouse trapping the thermal radiation and creating "an amazing 500°C difference" between surface and cloud tops.[80]

---

[80] http://www.esa.int/Our_Activities/Space_Science/Venus_Express/Greenhouse_effect_clouds_and_winds

The Venus atmospheric circulation is extreme as the team reports:

At the level of the cloud tops, the atmosphere rotates at a formidable velocity, with wind speeds up to 360 km per hour. The speed of the winds then progressively decreases to almost zero at the planet surface, where it becomes a gentle breeze, only able to raise dust. Two enormous vortices, with very complex shapes and behaviours, rotate vertically over the poles, recycling the atmosphere downwards. The internal vortex structure is highly variable: its centre of rotation is offset from the South Pole and the feature changes shape about every 24 hours; it drifts around the Pole periodically every few days. The vortex at the North Pole, the only one previously observed in some detail, has a peculiar double 'eye' shape, surrounded by a collar of cool air. It completes a full rotation in only three Earth days.[81]

Venus Express provided the first view of the south polar region and found that, just like in the north previously observed by the Pioneer Venus orbiter in 1979, the atmosphere there is also characterized by a raging vortex. The science team reports "heated air from equatorial latitudes rises and spirals towards the poles, carried by the fast winds. As the air converges on the pole and then sinks, it creates a vortex much like that found above the plughole of a bath".

Venus Express made a number of other surprise discoveries: a remarkably cold region high in the planet's atmosphere that may be frigid enough for carbon dioxide to freeze out as ice or snow; a high-altitude ozone layer; and a mysterious layer of sulphur dioxide far above the main cloud layer. Although Venus has no internally generated magnetic field the spacecraft carried a magnetometer to study the interaction between the solar wind and the uppermost atmosphere. This experiment has detected the loss of atmosphere from both the day and the night-side of Venus.

Venus most probably once had oceans like its sister planet Earth. In the 1980s James Kasting of Penn State University described how hydrodynamic atmospheric loss (molecular breakdown and heating of water vapor in the stratosphere by solar UV leading to the hydrogen expanding and accelerating to escape velocity) could have carried away an ocean's worth of hydrogen in a matter of a few tens of millions of years. Kasting and Kevin Zahnle of NASA Ames further concluded that the escaping hydrogen would have 'dragged' much of the oxygen with it but would have left behind heavier carbon dioxide. Lacking surface water, the carbon dioxide would have remained in the atmosphere unable to form carbonate rocks – leading instead to an ever-increasing greenhouse build-up in temperature.[82]

In addition to the insolation trapped at the surface by the atmosphere there is also the trapped heat from the interior (probably comparable to Earth's heat flow) to consider in interpreting the surface that Magellan revealed. Active volcanism is a clear possibility addressed by the Venus Express's long-term monitoring of surface temperature. The Venus Express team has identified infrared radiation coming from three volcanic regions mapped by Magellan – "perhaps relatively fresh lava flows that had not yet experienced significant surface weathering."[83]

The lava flows are estimated to be young – less than 2.5 million years old. Atmospheric monitoring by Venus Express determined that there had been "a sharp rise in the sulphur dioxide content of the upper atmosphere in 2006–2007 followed by a gradual fall over the following five years". Moreover, the spacecraft's Venus Monitoring Camera has mapped thermal emission from the surface through a transparent spectral window in the planet's atmosphere leading to observations of localized changes in surface brightness between images taken only a few days apart. Therefore still-volcanically-active Venus does seem likely.

After 8 years in orbit Venus Express had used up all of its propellant – fuel that, as the end of the mission came closer, had allowed it to 'surf' in and out of the Venus atmosphere. The spacecraft had carried out "a series of low passes during

---

[81] ibid.

[82] http://www.atmos.washington.edu/%7Edavidc/papers_mine/Catling2009_SciAm_AtmEscape_Preprint.pdf

[83] http://m.esa.int/Our_Activities/Space_Science/Venus_Express/Hot_lava_flows_discovered_on_Venus

2008 to 2013, diving into the planet's upper atmosphere to altitudes only 165 km above the surface in order to measure the density of the upper polar atmosphere which proved a surprising 60% thinner than predicted".[84]

By July 2014, by means of a series of small thruster burns, Venus Express was placed in an orbit with 460 km periapsis and 22-hour period. Now the spacecraft's orbit began to decay, "dipping further and further into Venus' atmosphere, before the mission lost contact with Earth (November 2014) and officially ended in December 2014". The atmospheric surfing provided a number of insights that would have been hard to gain in any other way. The Venus atmosphere was found "to be rippling with atmospheric waves – and, at an average temperature of minus 157°C, colder than anywhere on Earth". The team explains:

Atmospheric gravity waves are similar to waves we see in the ocean, or when throwing stones in a pond, only they travel vertically rather than horizontally. They are essentially a ripple in the density of a planetary atmosphere – they travel from lower to higher altitudes and, as density decreases with altitude, become stronger as they rise. The second type, planetary waves, are associated with a planet's spin as it turns on its axis; these are larger-scale waves with periods of several days.[85]

2010 Japanese Space Agency's Akatsuki

The Japanese Space Agency has also included the exploration of Venus in its plans and on 20 May 2010 it launched its Akatsuki spacecraft (the name means 'Dawn' or 'Daybreak') to Venus from its Tanegashima Space Center on an H-IIA 202 rocket to study the atmosphere of Venus from orbit. The science payload comprised five imaging systems that covered wavelengths all the way from the UV to the mid IR. Together these were planned to record the distribution of sulphur dioxide in the clouds, the structure of high-altitude clouds, the location of surface hot spots indicative of active volcanism, and heat radiation emitted from the lower reaches of the atmosphere. The spacecraft also carried an Ultra-Stable Oscillator "for high precision measurement of distance and communication".

The spacecraft reached Venus on 6 December 2010 as planned but the engine burn required to insert it into orbit failed to operate for the required 12 minutes and Akatsuki continued in its orbit about the Sun, returning to Venus in November 2011 and December 2015. The main engine was judged to be no longer operable so the team in a heroic effort developed a plan to use the attitude control thrusters to slow down the spacecraft sufficiently to be captured by Venus. Because these thrusters do not need oxidizer, the 65 kg of oxidizer that remained in the main engine tank was vented overboard in October 2011 to lighten the spacecraft. On 7 December 2015 Akatsuki successfully entered Venus orbit after a 20-minute burn of its 4 attitude control thrusters. Inevitably, instead of achieving the original 30-hour orbital period Akatsuki was now in a 9-day orbit. An adjustment of the orbit in March 2016 has placed it in a 13-day 14-hour orbit with a 400 km periapsis and a 440,000 km apoapsis. After a potentially catastrophic event five years earlier this was a remarkable recovery. The latest information about the status of the orbiter is as follows: "A follow-up thruster burn on 26 March 2016 lowered Akatsuki's apoapsis to about 330,000 km (210,000 mi) and shortened its orbital period from 13 to 9 days. The orbiter started its 2-year period of "regular" science operations in mid-May 2016".[86]

For the science team there has inevitably been enormous frustration but the team operating the spacecraft has had a unique learning experience that should serve them well in the future. Akatsuki's long wave IR camera captured what appears to be another new cytherean phenomenon namely bow-shaped clouds that encompass all the way from the southern to the northern hemisphere. These are presumably related to the V-shaped clouds that have been observed at UV wavelengths,

---

[84] http://sci.esa.int/venus-express/33010-summary/
[85] http://sci.esa.int/venus-express/55141-venus-express-goes-gently-into-the-night/
[86] https://en.wikipedia.org/wiki/Akatsuki_(spacecraft)

Future Missions to Venus

At the present time (2020) and with the world turned upside down by Covid-19, there is uncertainty in planning to return to Venus. The Indian Space Agency has studied a mission combining an orbiter and an atmospheric probe. The Russian Space Agency has considered Venera-D that would include a radar-mapping orbiter and a lander to be launched in 2026. In the US planetary science community there is strong interest. Two-dozen Venus mission proposals have been submitted to NASA since the completion of the Magellan mission; none have made it to a new start. NASA's most recent solicitation for Discovery-class missions has received a number focused on Venus. The February 2019 issue of *Scientific American* includes an article by M. Darby Dyer of Mount Holyoke College, Suzanne E. Smrekar of JPL and Stephen R. Kane of UC Riverside that summarizes the current understanding of the divergent 4.6 billion year evolutions of the innermost planets: Mercury, Venus, Earth and Mars. They make the appealing case that there is "reason to believe Venus became a habitable world before Earth and spent more than a billion years with conditions needed for life". Their evolution models for each of the terrestrial planets indicate that Mercury never stood a chance of being habitable in the absence of surface liquid water (note: water in ice form has been detected at the poles of Mercury as is also true for the Moon). Their modelling of the evolution of Venus, on the other hand, indicates that Venus would have had liquid water on its surface – released from it interior by volcanism and with a contribution by comet impacts – to almost 1 billion years ago.

The goal of a proposed new NASA orbiter mission – *VERITAS* (Venus Emissivity, Radio Science, InSAR, Topography, and Spectroscopy)– submitted at the most recent Discovery mission proposal opportunity – is to acquire data to better understand what makes a terrestrial planet habitable so that we can more reliably speculate about the nature of the ever-growing number of exoplanets in our Milky Way galaxy. "VERITAS would map Venus' surface to determine the planet's geologic history and understand why Venus developed so differently than the Earth. Orbiting Venus with a synthetic aperture radar, VERITAS charts surface elevations over nearly the entire planet to create three-dimensional reconstructions of topography and confirm whether processes, such as plate tectonics and volcanism, are still active on Venus. VERITAS would also map infrared emissions from the surface to map Venus' geology, which is largely unknown. Suzanne Smrekar of NASA's Jet Propulsion Laboratory in Pasadena, California, is the principal investigator. JPL would provide project management."[87]

The role of plate tectonics has been essential in the release of water and other volatiles from Earth's interior and in causing long-term (~100 million years) climate cycles. On Earth, therefore, plate tectonics has been of fundamental importance in how our planet has evolved; the apparent absence of this dynamic phenomenon on Venus is, therefore, of fundamental significance. Dyer, Smrekar and Kane believe the violent processes that led to the initiation of plate tectonics on Earth (around 3 billion years ago) may be happening on present-day Venus with its warm and thin lithosphere where Magellan radar images "show compelling similarities between features on Venus and terrestrial subduction zones." Artemis Corona is cited as an example: this is a circular formation about 2,600 km in diameter on the Venus equator "that is similar in scale and shape to the Aleutian trench that lies under the ocean along the coast of Alaska".

It is argued that although Venus has no magnetic field at present it would have had a field comparable to that of Earth's up until about a billion years ago – providing the means for protecting the planet's atmosphere from loss to space. Given this perspective, it is possible that life might have made a start on Venus as on Earth only to become extinct when the runaway greenhouse transformed Venus. The authors take note of the apparent absence of plate tectonics on Venus as an important factor in the story of how life could originate on a planet.

A second Discovery proposal focused on Venus is named DAVINCI+ (Deep Atmosphere Venus Investigation of Noble gases, Chemistry, and Imaging Plus). "DAVINCI+ would analyse Venus' atmosphere to understand how it formed and evolved and determine whether Venus ever had an ocean. DAVINCI+ would plunge through Venus' inhospitable atmosphere to precisely measure its composition down to the surface. The instruments are encapsulated within a purpose-built descent sphere to protect them from the intense environment of Venus. The "+" in DAVINCI+ refers to the imaging component of the mission, which includes cameras on the descent sphere and orbiter designed to map surface rock-type.

---

[87] https://www.jpl.nasa.gov/news/news.php?feature=7597

The last U.S.-led, *in-situ* mission to Venus was in 1978. The results from DAVINCI+ have the potential to reshape our understanding of terrestrial planet formation in our solar system and beyond. James Garvin of NASA's Goddard Space Flight Center in Greenbelt, Maryland, is the principal investigator. Goddard would provide project management."[88]

US-NZ Rocket Lab Venus Mission in 2023

To the author's considerable surprise, the 24 September 2020 edition of the International NY Times has an article that reports the plans of a private aerospace manufacturer, Rocket Lab, to launch a small spacecraft to Venus in just a year's time. The company operates a lightweight two-stage rocket known as Electron, which provides dedicated launches for smallsats and CubeSats. The company was founded in New Zealand in 2006 by engineer Peter Beck and has its headquarters in California. Launch would take place out of the Mahia Peninsula on the North Island of New Zealand. The mission was conceived well before the phosphine discovery announcement (see below) and details are few. Among the scientists working on this project is Sara Seager of MIT who was part of the team that announced their radio astronomy discovery of phosphine in the atmosphere of Venus. The Rocket Lab spacecraft is planned to fly by Venus and drop off a small probe instrumented with an infrared spectrometer.

Possible Life on Venus

With the recent evidence for phosphine (PH_3) in the clouds of Venus at about 50 km altitude (reported in the 14 September 2020 issue of *Nature Astronomy* by Jane Greaves of Cardiff University, Clara Sousa-Silva of MIT and colleagues) the possible discovery of microbial life on Venus has become front-page news.[89]

The University of Cardiff-MIT team identified an electromagnetic wavelength that is a rare signature of life and then went out to look for it on Venus using the most powerful radio telescopes available. Their identification of an absorption line of phosphine (PH_3) in the spectrum of Venus was first made using the James Clerk Maxwell Telescope in Hawaii, and then confirmed using the Atacama Large Millimeter/submillimeter Array (ALMA) in Chile. "This was an experiment made out of pure curiosity, really – taking advantage of JCMT's powerful technology, and thinking about future instruments," said Greaves, who is based at Cardiff University. "I thought we'd just be able to rule out extreme scenarios, like the clouds being stuffed full of organisms. When we got the first hints of phosphine in Venus' spectrum, it was a shock!"[90]

"Luckily, conditions were good at ALMA for follow-up observations while Venus was at a suitable angle to Earth. Processing the data was challenging, however, as ALMA isn't usually looking for subtle effects in bright objects like Venus." "In the end, we found that both observatories had seen the same thing – faint absorption at the right wavelength to be phosphine gas, where the molecules are backlit by the warmer clouds below," said Greaves.[91]

The signal is small and indicative of ~20 ppb abundance but, if confirmed, is significant because phosphine is extremely hard to make. In nature it is made by some species of anaerobic bacteria that live in landfills and marshes. "Many alternative ways of producing phosphine can be ruled out by this new study, but the team say that confirmation of the presence of life on Venus is still a long way away".[92] And planetary spectroscopy experts caution that observing a single spectral line is insufficient for identification.

Phosphine has long since (1979) been reliably observed in the atmosphere of *Jupiter* including by Reinhard Beer and Fredric Taylor (two fellow Brits at JPL) who made observations near the 5 micron IR window of Jupiter and determined that

---

[88] Ibid.

[89] https://www.nature.com/articles/s41550-020-1174-4

[90] http://www.spaceref.com/news/viewpr.html?pid=56261

[91] https://www.cambridgenetwork.co.uk/news/hints-life-discovered-venus

[92] https://www.sciencedaily.com/releases/2020/09/200914112219.htm

PH$_3$ "is a significant contributor to the continuum opacity in the window and in fact defines its short-wavelength limit".[93] Given that Jupiter has a hot, predominantly hydrogen atmosphere the issue of life did not arise.

Regarding the question of why the infrared is not the spectral region where phosphine has been detected on Venus, Fred Taylor comments "the phosphine is within the clouds, which are opaque at the shorter wavelengths, making IR detection more difficult and less sensitive. Jupiter has gaps in the clouds, which is why we could observe phosphine there."

Fred notes that the amounts of PH$_3$ claimed for Venus are very small; "even averaging many observations from a large telescope they only got a signal to noise ratio of about 10; Caltech wisdom was that at least 100:1 is needed for a discovery. The microwave lines are well spaced and I suspect that the next strongest one is either too weak or obscured by water features or both".

So there is clearly much progress to be made before life in the clouds of Venus becomes the science discovery of the 21st Century.

---

[93] *Icarus* 40, 189-192

CHAPTER 9

RETURN TO MERCURY

MESSENGER, Bepi Colombo

Mercury is a challenging planet to reach and, especially, to study from orbit given its relative proximity to the Sun. As a result, as the new century dawned, the most recent NASA mission to Mercury had been the Mariner 10 mission that launched back in 1973 and that flew by Mercury three times. A return was overdue. Mercury's unique end-membership of our Solar System also ensures that its nature will intrigue and that its exploration will continue.

2004 NASA MESSENGER

Presently (2020), the most recent mission to Mercury was NASA's aptly named MESSENGER, whose science team was evidently seeking world leadership in acronym creation: MErcury Surface, Space ENvironment, GEochemistry and Ranging. This Discovery project was designed built and managed by the Johns Hopkins Applied Physics Laboratory with David Grant, Project Manager and Sean Solomon of the Carnegie Institute of Washington, the Principal Investigator.

A reminder: Mercury, the smallest and innermost of the four terrestrial planets orbits the Sun every 88 days. Its rotation period is 59 days creating a 3:2 spin-orbit resonance, rotating on its axis three times for every two revolutions. Its orbital eccentricity, 0.205, is the largest of all the planets so that at its greatest distance from the Sun (0.47 AU) it is about 1.5x further from the Sun than at its closest distance (0.39 AU). The obliquity (axial tilt) of Mercury is close to 0 degrees so that, even as close to the Sun as the planet is, the poles receive almost no insolation. This is a situation much like that on the Moon – the floors of craters near the poles are in perpetual darkness; observations by the 70-meter Goldstone radar in the 1990s had identified high radar reflectivity near the poles indicating the likelihood that ice would be found there.

As noted, Mercury had been the target of only one previous mission – Mariner 10 launched in November 1973. The spacecraft had made three flyby encounters that had provided some basic information about Mercury: its density, its magnetic field strength and its battered lunar-like surface with the giant Caloris basin. These three flybys had all been forced to observe the same illuminated side of Mercury so that more than half of the surface still remained to be mapped. Much fundamental characterization of this innermost planet was needed requiring an orbiter. This seemingly called for a heavy lift launch vehicle and hefty in-space propulsion to achieve orbit. Expensive. The problem is that traveling inward to the Sun, a spacecraft picks up so much velocity that orbit capture becomes a real challenge. MESSENGER's secret – that allowed it to compete in the Discovery program – was to use a complex trajectory that had been designed by Chen-wan Yen of JPL a dozen years earlier. This trajectory required a series of Venus and Mercury gravity assists to gradually slow the spacecraft down to a velocity that would allow insertion into orbit about Mercury with only the propellant that could be launched on a Delta II. Assuming a long-lived spacecraft and Voyager-class navigation, this complicated trajectory solved the performance problem!

In July 1999, NASA selected MESSENGER as the seventh Discovery mission. Six years later the mission managed by Johns Hopkins Applied Physics Laboratory was ready for launch from the Cape. Following a gravity-assist flyby of Earth one year after launch in August 2005, two flybys of Venus in October 2006 and again in June 2007, and a January 2008 flyby of Mercury, MESSENGER had reduced its velocity enough to enable use of its rocket engine to enter orbit about Mercury on 18 March 2011. After fully checking out the spacecraft data gathering began on 4 April 2011.

The mission's goal was to understand better the origin, evolution and present state of Mercury by producing a high-resolution geological map of the surface, by measuring its surface chemical composition and its magnetic field. The comprehensive instrument payload comprised a multispectral imaging system (PI Scott Murchie of Johns Hopkins Applied Physics Lab APL), gamma ray and neutron spectrometers (PI William Boynton of the University of Arizona), X-ray spectrometer (PI George Ho of APL); magnetometer (PI Mario Acuna of NASA Goddard SFC), laser altimeter (PI David Smith of NASA GSFC), Radio Science (PI also David Smith), Atmospheric & Surface Composition Spectrometer (PI William McClintock of U. of Colorado) and an energetic particle and plasma spectrometer (Barry Mauk of APL). Having only a tenuous atmosphere, Mercury was a "clean" target for all these measurements.

The earlier Mariner 10 triple flyby mission had determined the density of Mercury to be 5.427gm/cc so that it was already known from modeling that Mercury has a large iron core, likely molten, one that occupies more than half its volume. As a result that mission had established that Mercury has a magnetic field – albeit, on account of the slow rotation, much weaker than Earth's (about 1% of the strength). This is strong enough to deflect the solar wind and create a magnetosphere. The mantle of silicates is modeled to be 500 to 700 km thick while the crust is estimated to be 35 km deep.

Mariner 10 had provided a general characterization of Mercury as, crudely, a bigger, hotter Moon – with an ancient heavily cratered surface, a large magma-filled basin (appropriately named Caloris) and extensive plains comparable to the lunar maria. The craters are much like those of the Moon but, because of Mercury's greater gravity, the rays created by ejecta do not extend as far. The formation of Mercury's largest basin, Caloris, 1,550 km diameter, was an exceptionally violent event that led to disruption of the crust at the antipode – "unusual, Hilly terrain known as 'Weird Terrain'".[94]

Especially in the north Mariner 10 had observed smooth plains that appeared to be the result of flood volcanism. A half-dozen basins had been observed including the giant Caloris Basin and 400 km wide Tolstoy Basin that is multi-ring in form and has an ejecta blanket that extends for 500 km. The basins contain smooth plains much like the lunar maria and accordingly these were also interpreted to be lava flows. The Mariner 10 science team also noted collapse structures in some craters that appeared to be of volcanic origin. They also identified eleven volcanic domes that included a 1.4 km high dome in the center of Odin Planitia, a large basin in the Tolstoy quadrangle.

All this was known before MESSENGER arrived at Mercury with the task of observing half of the previously unseen surface toward completing the global mapping of geology and mineralogy. With these data it would be possible to establish the history of this extreme end-member of the Solar System's planets. A first mapping of the polar-regions was a particular mission priority – as was achieving an improved understanding of Mercury's internal structure, magnetic field and magnetosphere.

MESSENGER's 12-hour orbit was highly elliptical and of near-polar inclination (80deg) so that the spacecraft spent only limited time close to the hot surface of Mercury: Closest approach was 200 km while the furthest was 10,400 km. Just as when Mariner 9 mapped Mars from a highly inclined elliptical orbit, the spatial resolution of the mapping was not uniform with latitude requiring oblique viewing of the southern hemisphere. In a year MESSENGER had mapped the planet from pole to pole completing its primary mission. Being still in excellent shape, MESSENGER then entered a series of extended missions. During these many months orbiting in the gravity field of Mercury it was necessary to use the rocket engine periodically to maintain the periapsis altitude and avoid gradually losing height and crashing. Inevitably, in time MESSENGER used up all of its propellant and the mission ended when the spacecraft crashed on 30 April 2015.

The complete mapping of Mercury has not led to any revelations regarding the general nature and history of Mercury but has provided further evidence that Mercury experienced limited volcanic activity until perhaps a billion years ago. The MESSENGER data provide much more detail for theorists to reconstruct its origin and subsequent history. Like the other terrestrial planets Mercury would have accreted in the first few million years of the solar nebula's collapse 4.6 billion years ago. The gravitational energy of the accreting planetesimals and of the subsequent differentiation into core, mantle and

---

[94] https://en.wikipedia.org/wiki/Mercury_(planet)

crust would led to a completely molten body. As the crust solidified the planet began to preserve and accumulate impact craters, some of which were large enough to be classified as basins. The end of the late terminal bombardment about 3.8 Byrs ago is as far back as we can observe direct evidence of Mercury's evolution. The oldest terrain on Mercury is that of the inter-crater plains that the science team describes as "gently rolling hilly plains".[95]

There are other plains of a different kind that fill craters much like the lunar maria – their origin is evidently flood volcanism. MESSENGER also identified pyroclastic deposits that suggest that perhaps Mercury is not quite the long-dead planet that it had seemed. Geologist Jim Head of the science team has reported on the discovery of volcanic vents: "These huge vents, measuring up to 25 km in length, appear to be the source of tremendous volumes of very hot lava that have rushed out, carving valleys and creating the teardrop-shaped ridges in the underlying terrain".[96]

The volcanic vents are not easy for the non-specialist to identify as they often take the form of sunken pits surrounded by bright material. From orbit the vents may be mistaken for impact craters but are irregular in shape. MESSENGER has identified 51 pyroclastic deposits and, within the southwest rim of the Caloris Basin, nine overlapping volcanic vents have been mapped. The age of this volcanic complex is uncertain but "it could be of the order of a billion years".[97]

If this age is supported by further analysis then it would seem that Mercury managed to retain a breath of life much later than had been previously supposed for, based on the areal density of the impact craters, it is evident that most of Mercury's surface has been inactive for billions of years. The decay of radioactive elements in the huge core would seem to be a long-lasting source of the needed heat.

The catastrophic formation of the giant Caloris basin had consequences on the opposite side of Mercury: there is a large area of unusual hilly "chaotic" terrain at the exact antipode of Caloris and this likely was created by intense seismic waves resulting from the basin formation.

Previous missions to Mercury had discovered very large scarps – evidence of tectonic activity. MESSENGER's imaging coverage has also led to the discovery of much smaller fault scarps – "cliff-like landforms that resemble stair steps."[98]

Tom Watters, Smithsonian senior scientist at the National Air and Space Museum in Washington, D.C. has reported:

The young age of the small scarps means that Mercury joins Earth as a tectonically active planet, with new faults likely forming today as Mercury's interior continues to cool and the planet contracts. Large fault scarps on Mercury were first discovered in the flybys of Mariner 10 in the mid-1970s and confirmed by MESSENGER, which found the planet closest to the sun was shrinking. The large scarps were formed as Mercury's interior cooled, causing the planet to contract and the crust to break and thrust upward along faults making cliffs up to hundreds of miles long and some more than a mile (over one-and-a-half kilometers) high.[99]

In the last 18 months of the MESSENGER mission, the spacecraft's altitude was lowered, which allowed the surface of Mercury to be seen at much higher resolution. These low-altitude images revealed small fault scarps that are orders of magnitude smaller than the larger scarps. The small scarps had to be very young, investigators say, to survive the steady bombardment of meteoroids and comets. They are comparable in scale to small, young lunar scarps that are evidence Earth's Moon is also shrinking.

This active faulting is consistent with the recent finding that Mercury's global magnetic field has existed for billions of years and with the slow cooling of Mercury's still-hot outer core. MESSENGER from its near polar orbit has made observations of radar-bright areas using its neutron spectrometer to measure average hydrogen concentrations. Participating Scientist David Lawrence and colleagues conclude that the data demonstrate the presence of water ice more than tens of centimeters thick but they note that it is not located at the immediate surface. Evidently the ice is buried

---

[95] https://en.wikipedia.org/wiki/Inter-crater_plains_on_Mercury
[96] http://sciencenetlinks.com/science-news/science-updates/mercurys-volcanoes/
[97] https://en.wikipedia.org/wiki/Mercury_(planet)#Surface geology
[98] https://airandspace.si.edu/exhibitions/exploring-the-planets/online/solar-system/mercury/surface.cfm
[99] https://www.nasa.gov/feature/the-incredible-shrinking-mercury-is-active-after-all

beneath a dark layer of regolith 10 to 20 cm thick.[100]

The MESSENGER mission came to a planned and dramatic end at 3:26 pm EDT on Thursday April 2015 when it crashed into Mercury's surface at about 14,000 km/hour and in doing so created yet another crater on the planet's surface.

2018 ESA/JAXA BepiColombo

The analysis of the MESSENGER data will continue for many years – long after the spacecraft ran out of propellant to sustain its orbit and impacted Mercury on 30 April 2015. By the time of MESSENGER's demise, plans were already well advanced for the next mission to Mercury: *BepiColombo*, the name of a joint mission of the European Space Agency (ESA) and the Japanese Aerospace Exploration Agency (JAXA). The mission is composed of two satellites: the Mercury Planet Orbiter (MPO) and the Mercury Magnetospheric Orbiter (MMO). The mission will perform a comprehensive study of Mercury's magnetic field, interior structure, and surface. It set off on an Ariane 5 from Europe's Spaceport in Kourou on 20 October 2018. ESA's Mercury Transfer Module (MTM) is carrying the orbiters to Mercury using a highly innovative combination of solar electric propulsion and gravity assist flybys. There will be with one flyby of Earth, two at Venus, and no less than six at Mercury! When it arrives at Mercury in late 2025, it will suffer temperatures in excess of 350°C and will gather data during its 1-year nominal mission and possible 1-year extension.

The mission is named after Giuseppe Colombo, a scientist, engineer, and mathematician who designed the first interplanetary gravity-assist maneuver during the 1974 Mariner 10 mission, a technique now commonly used by planetary probes. Both spacecraft are to be inserted into polar orbits – one of 2.3 hour period and 400 x 1500 km altitude the other (the Magnetospheric Orbiter) an orbit of 9.2 hour period and 400 x 12,000 km altitude.

As reported in SpaceRef, "BepiColombo took a final glimpse of Earth on 11 April 2020, a day after its closest approach to the planet to perform a gravity-assist flyby. An image, showing Earth as a bright crescent against the blackness of the Universe, was captured at 14:24 UTC by one of the monitoring 'selfie' cameras mounted on Mercury Transfer Module (MTM), one of the three components of the BepiColombo mission."[101]

---

[100] ibid.

[101] http://spaceref.com/mercury/bepicolombo-bids-farewell-to-earth-and-the-moon.html

CHAPTER 10

RETURN TO MARS 1:

FAILURES, ALH84001 & FIRST SURFACE MOBILITY

Mars Observer, ALH84001, Faster, Cheaper, Better, Mars Pathfinder, Mars Global Surveyor, Mars 96, Mars Climate Orbiter, Mars Polar Lander, Nozomi

Back to 1983 with the Magellan mission development underway and Deputy Administrator Hans Mark now supportive – "a terrific job" – of the new program proposed by the Solar System Exploration Committee.[102] At the Presidential level, Ronald Reagan's interest in NASA focused on the Agency's planned development of Space Station Freedom that he had announced in his January 1984 State of the Union Address to Congress: SS Freedom was to be used by astronomers, would be a microgravity laboratory for scientists and commercial researchers and, unsaid, be a means of proving long-duration human space travel – one day to Mars. The Mars Observer proposed by the SSEC would be the first link in the chain of robotic and crewed missions that one day would see astronauts land on Mars. Mars Observer was given its go-ahead in February 1984 for a 1990 Space Shuttle launch.

There was a good deal of grass-roots support for human exploration of Mars with the expectation that it would be preceded in an orderly way by robotic missions. A passionate interest in human exploration of Mars was (and still is) shared by a number of vocal individuals and organizations. In the 1980s the *Mars Underground* emerged out of the University of Colorado, founded by a number of enthusiasts including Penelope Boston, Chris McKay and Carol Stoker, now all working at NASA Ames Research Center where Boston is the Director of NASA's Astrobiology Institute and where McKay and Stoker have been carrying out astrobiology fieldwork and Mars exploration technology development for many years.

The Underground held an annual series of workshops entitled 'The Case for Mars' and documented detailed architectures by which the first crewed missions to Mars could most effectively be undertaken. Among the supporters of the Mars Underground was former NASA Administrator Thomas Paine. Another organization, *The Planetary Society*, founded in 1980 by Carl Sagan, Bruce Murray and Louis Friedman, gave voice to strong support for manned Mars exploration. The Society today has over 40,000 members from more than one hundred countries.

So: happy days again! In Pasadena the Planetary Society led by Carl Sagan and supported by Roald Sagdeev of Moscow's Space Research Institute (IKI) was hopeful that, though modest in scope, this new Mars mission initiation (Mars Observer) after more than a decade would be a turning point for NASA. They envisioned, with reason, that Mars exploration conducted as an international program by the USA and the Soviet Union could, and likely would, bring enormous benefits to both sides.

---

[102] *Why Mars: NASA and the Politics of Space Exploration* by W. Henry Lambright, page 88

Eventually, major international collaboration in space has indeed flourished: Space Station Freedom has evolved to become the International Space Station that flies overhead today. Mars exploration collaboration with the Soviets, however, never did happen because both nations' space programs had setbacks – including NASA's Challenger disaster in January 1986. The Soviets had experienced their traumatic loss of a Saturn V-class launch vehicle, the N1 Moon rocket, in 1969 but had not had a disaster comparable to the loss of the Challenger and its crew. The Soviets had, in fact, developed a re-useable space plane much like the Space Shuttle but it made just one flight, in November 1988, and did so without a crew. There was one big difference between the two vehicles: the main engines of the Soviet Shuttle were at the base of the external tank rather than on the orbiter and, so, would not be recovered.

The Russian Space Shuttle was called *Buran* ("Blizzard") and had been launched successfully on 15 November 1988 on the back of a big *Energia* booster. Following the dissolution of the Soviet Union the Buran program was cancelled along with the Energia after only this flight and now nothing remains – in 2002 the Russian Space Shuttle was destroyed as a result of the collapse of its hangar roof at the Baikonur Cosmodrome.

Inspired by The Case for Mars conferences yet another organization – *The Mars Society* – was formed in 1988 and is entirely focused on promoting the human exploration and settlement of Mars. The founder is a very energetic aerospace engineer, Bob Zubrin, who had worked for Martin Marietta. There are over 10,000 members and associates in more than 50 countries around the world. They have created research centers in the Arctic, in Utah, Europe and Australia "where scientists and engineers can live and work as if they were on Mars, to develop the protocols and procedures that will be required for human operations on Mars, and to test equipment that may be carried and used by human missions to the Red Planet".

As already discussed in Chapter 6, some of Zubrin's ideas about using *in situ* Martian resources (the CO_2 atmosphere) to carry out manned Mars exploration have in fact found their way into NASA's planning for eventual human missions to Mars.

A similarly direct and practical effort to bring forward the date for the human exploration of Mars has been that of Pascal Lee who is co-founder and chairman of the *Mars Institute* – "dedicated to advancing the scientific study, exploration, and public understanding of Mars. The Institute is "headquartered at NASA Ames Research Center (USA), with offices in Toronto (Canada) and Stavanger (Norway)". [103]

An astronomer and space scientist at the SETI Institute near NASA Ames, Lee established and leads the *Haughton Mars Project* (HMP) – an international, multidisciplinary field research project centered on science and exploration studies at the Haughton impact crater and surrounding terrain on Devon Island in Arctic Canada. Haughton Crater contains a large variety of Mars-like geological features and, located, as it is, on an isolated, uninhabited island with no infrastructure, this site makes an ideal analogue for Mars exploration research including advances in artificially intelligent mobile landers.

Lee has led over 20 HMP field expeditions to date. Among them, in May 2003, Lee led an Arctic winter expedition to drive the Mars Institute's Mars-1 Humvee Rover from Resolute Bay on Cornwallis Island across the Wellington Channel's 40 km of sea ice to Cape McBain on Devon Island. The crossing was a success and the Mars-1 has since been serving as a mobile field lab and concept vehicle for future pressurized rovers to be used on the Moon or Mars.

---

[103] https://en.wikipedia.org/wiki/Mars_Analogue_Research_Station_Program

Another force pushing the US toward carrying out the human exploration of Mars sooner rather than later is Apollo 11's Buzz Aldrin who has been a long-time proponent of using cycling spacecraft that would swing endlessly between the orbits of Earth and Mars to facilitate the travel of astronauts to and from the planet (a lunar cycler spacecraft could likewise be envisioned). His vision of how NASA might best proceed begins with the establishment of bases on both the near side and far side of the Moon and includes landings on Phobos, the inner moon of Mars. The Aldrin scenario involves creating an infrastructure of multiple spacecraft whose orbits cycle between Earth and Mars – a neat idea but probably not one for the near term.

John Niehoff of SAIC proposed to NASA a related approach for a series of robotic sample return missions. "His Interplanetary Platform (IP) would transport smaller vehicles between Earth and Mars. It would provide them with 'keep-alive' solar cell-generated electrical power, thermal control, course-correction propulsion, and other requirements typically provided by a throwaway spacecraft bus. The IP would cut costs over the course of the program because it would need to be launched onto its interplanetary path only once. As the IP flew without stopping past Mars or Earth, the smaller vehicles would separate to land on or go into orbit around the planet or would leave the planet to rendezvous and dock with the IP".[104]

Quite independently from NASA's science programs, US Department of Energy research carried out in the 1980s proved to be very important for the search for life on Mars because this research raised the serious possibility that Mars might be the home of an extreme form of extant microbial life – kilometers deep underground. As Richard Monastersky describes in a 1997 paper in *Science News* titled "Deep Dwellers: Microbes thrive far below ground":

In the 1980s DOE came under pressure to address its toxic waste problems at various nuclear facilities. Uranium, heavy metals, and organic compounds were contaminating groundwater deposits at these sites because the department had improperly discarded such materials during the Cold War drive to produce plutonium for bombs. Deep microbes, if they existed, could affect the waste problem by either hindering or aiding the spread of toxic materials through the ground. So in 1985 Frank J. Wobber of DOE launched a subsurface science program, part of which aimed to assess the existence of life in rock. The researchers developed special anti-contamination procedures for drilling into the ground and obtaining rock samples.

The DOE-sponsored team tested these techniques by drilling several boreholes at the Savannah River nuclear processing facility in South Carolina. By 1989, Wobber, Tommy J. Phelps of the Oak Ridge (Tenn.) National Laboratory, and their colleagues documented the existence of a diverse community of bacteria and archaea at a depth of 500m. The Savannah River discoveries seemed remarkable until Phelps, Tullis Onstott, and others studied rocks from the Taylorsville Basin in eastern Virginia. Joining forces with Texaco Oil Co., which was drilling exploratory wells in the basin, the scientists tested for microbial residents in much deeper rocks.

In the late 1980s, researchers sponsored by the Department of Energy found microbes living in rock 500 meters below the surface in South Carolina. In the last 4 years, Onstott and other researchers have pushed the envelope of life much deeper, extending it to nearly 3 km below ground. In some cases, these microbes have apparently remained prisoners of the deep for millions of years, making such colonies veritable living fossils. "This is a revolution that's expanding the limits under which we know life to exist," says Onstott. "I think we're recognizing that life is a bit more tenacious than we had given it credit for, and there's a much broader realm of possibilities where life can survive. The discoveries not only redefine how scientists view the modern Earth, they also raise intriguing questions about the origin of life and the possibility that microbes survive today beneath the surface of Mars and other planets".[105]

In the early 1990s modelling by Stephen Clifford then of the Lunar and Planetary Institute in Houston, as discussed later in this Chapter, demonstrated that Mars probably has an extensive hydrosphere. The implications for Mars were not lost on world famous entomologist E.O. Wilson who in his 2006 book *The Creation* writes:

---

[104] https://www.wired.com/2013/12/linking-space-station-mars-the-imuse-strategy-1985/
[105] https://geoweb.princeton.edu/research/geomicrobio/scinews.html

Deep beneath our feet, and extending for at least two miles down, is another and in some respects far greater world: vast unexplored populations of bacteria and microscopic fungi, collectively called the SLIMES (subterranean litho-autotrophic microbial ecosystems). The inhabitants may collectively outweigh all of the living matter on the surface of the planet. They depend not on surface energy or organic matter drawn from Earth's surface but on independently ("autotrophically") derived sources of chemical energy in the dissolved minerals that surround them (hence lithos, stone). If somehow Earth's surface were burned to a crisp, the life below would likely persist. Then someday, perhaps a billion years into the future, it might evolve new forms of life that could repopulate the surface. The discovery of the SLIMES has given added hope to scientists that life will be found on the bitter cold and powder dry planet of Mars – not at the surface but far beneath it, at the level of liquid water.[106]

The Challenger disaster of 1986 inevitably led to some serious re-thinking about the direction of the US space program, discussed in Chapter 6: Astronaut Spaceflight After Apollo. In 1988 Burton Edelson (1926-2002), NASA's Associate Administrator for Space Science and John L McLucas (1920-2002), previously Secretary of the Air Force, reported in a contribution to the journal *Space Policy* that "Soviet General Secretary Gorbachev has proposed a joint US-Soviet program to explore the planet Mars" and described the benefits that would be gained by both sides. They pointed out that combined robotic Mars missions "could take place in the 1990s, and a first step towards manned exploration could be the writing of a development and flight plan aiming for the first decade of the 21st century". Alas, it didn't happen.[107]

The possibility floundered in part as a result of the failure of the Soviet's own near-term robotic Mars exploration effort. This was to have been carried out by two spacecraft – *Phobos 1* and *2* – that were to rendezvous with that small Martian moon to learn more about its composition and origin. Launched in 1988, Phobos 1 was lost in Earth orbit and Phobos 2, though it reached its target, did not complete its mission, one that had included the landing of two probes on the moon.

And so a real opportunity for international collaboration was lost – besides the planned contribution by NASA of its Deep Space Network (DSN), a number of other nations, including Sweden, Switzerland, Austria, France, West Germany, were to have been participants. Thus ended the visionary proposal of joining the hands of Cold War enemies in exploring Mars. It should be noted that, at the level of scientists and NASA program managers, relations with the Soviets were always very cordial and have remained so: in the 1980s there were biannual exchanges of information about deep space exploration plans that were held in Washington and Moscow. Deputy Associate Administrator Sam Keller (1930-2014) – who, later, was appointed NASA Associate Administrator for Russian Programs in 1992 – and the author headed the US team and on the Soviet side Valeri Barsukov (1928-1992) of the Vernadski Institute led their side. Raold Sagdeev of IKI was also linked in. The four space agency (Soviet Union, ESA, JAXA and NASA) effort to study Comet Halley at its coming apparition was (Chapter 7) also being carried out during the 1980s with close coordination between the agencies – so an international collaboration to explore Mars was by no means fanciful. One valuable result of the agency-level meetings has been the annual Brown-Vernadsky Microsymposium, organized by Jim Head and Alexander Basilevsky, that has now been taking place for nearly three decades – a partnership between Brown University and Russia's Vernadsky Institute for Geochemistry and Analytical Chemistry. NASA's recovery from the January 1986 Challenger disaster (and the critical Rogers Report into its cause) was not helped by the fact that in December 1985 Administrator Beggs had to take leave to deal with a legal problem. Meanwhile, two years earlier, in 1984, Hans Mark had left NASA to become Chancellor of the University of Texas system. James Fletcher (1919-1991) who had been Administrator from 1971 to 1977 stepped into the breach, served again from May 1986 to April 1989 and oversaw the return of the Shuttle to flight in September 1988. The two-year hiatus in launch capability impacted all of NASA's programs causing massive cost overruns as well as the launch delays. The

---

[106] *The Creation: An Appeal To Save Life On Earth,* E. O. Wilson, Norton 2006, p 119 Permissions WW Norton Inc, 500 Fifth Avenue, NY 10110
[107] http://www.sciencedirect.com/science/article/pii/0265964688900100

backlog of deep space mission launches began to be cleared when Magellan lifted off in May 1989, Galileo in October 1989, and Ulysses in October 1990. The Hubble Space Telescope launched in April 1990.

The Mars community of scientists had begun to look beyond the anticipated Mars Observer mission with some surface missions of limited scope – *Mars Environmental Survey* (MESUR) mission under consideration including a network of landers with seismometers and meteorology instruments. In late 1989, after seeing Magellan and Galileo on their way, the author had retired as Director of the Solar System Exploration Program to a sabbatical year at the Smithsonian Air & Space Museum and then to a position at NASA Ames; his successor was Wesley Huntress who very quickly became an advocate of the Discovery mission approach that was being studied by a science advisory committee as a potential program similar to the Astrophysics *Explorers*. These Explorer missions were (and still are) a continuing series of small, focused missions that respond to proposals from the astrophysics community and provided a backbone to the Astrophysics Program while the flagship *Great Observatory* missions (Hubble, Compton Gamma Ray, Chandra X-Ray, and Spitzer Space Telescope) were gradually undertaken. The Congress had been unwilling to support a line item for the planetary Observers but, by opening up the program to project management by laboratories other than JPL (specifically the Applied Physics Laboratory, APL, of Johns Hopkins University), the Congress had a change of mind. No doubt Senator Barbara Mikulski of Maryland helped in this.

In April 1992 Richard Truly had moved on as NASA Administrator and President George HW Bush had appointed to NASA leadership Daniel Goldin – whose background at TRW Inc. was in military space applications. Lennard Fisk (on leave from the University of Michigan) remained as Associate Administrator for Space Science and Wesley Huntress the Director of the Solar System Exploration Program. Mr. Goldin (who would be reappointed by Presidents Bill Clinton and George W Bush) brought with him great enthusiasm for Mars exploration and, encouraged by Carl Sagan, was eager to establish a path to the first manned exploration of Mars. He was not a fan of the way NASA was managed, seeing it as overly bureaucratic and, as a result, sclerotic. His perspective appeared to apply to the science programs as much as any other part of NASA. This was a puzzle to many as the science programs were, if anything, overly ambitious, competitive and challenging – the opposite of sclerotic. It was never clear exactly what troubled Mr. Goldin – perhaps too much reliance on the advice of external bodies like the National Academy of Sciences, perhaps too many project progress reviews, too little technological innovation?? Certainly, in the author's experience, there was plenty of bureaucracy at NASA Headquarters given the need to tread carefully through the various Washington DC minefields. Another problem of the Washington environment is an affliction commonly known as *Potomac Fever*, unfortunately not fatal to those who catch it.

Although Mr. Goldin was eager to see astronauts set foot on Mars he chose to de-emphasize the ongoing development of Space Station Freedom and, instead, to get behind robotic Mars missions as precursors to the first crewed Mars mission. But the robotic Mars missions would have to be carried out differently – smaller, more frequently and individually much less expensively. *Faster, better, cheaper!* was his mantra – the opposite to the *purple pigeon/grey mouse* paradigm favored by Bruce Murray when he was Director of JPL. More risky too – in a speech at JPL on May 28, 1992, Mr. Goldin declared: "Be bold – take risks. A project that's 20 for 20 isn't successful. It's proof that we're playing it too safe. If the gain is great, risk is warranted. Failure is OK, as long as it's on a project that's pushing the frontiers of technology".[108]

Perhaps this somewhat relaxed philosophy with respect to risk can be understood by Mr. Goldin's previous management experience of military space endeavors; their 'black' nature means that failures are not front-page embarrassments – moreover, the financial constraints on the Department of Defense are less severe than NASA's. Although in many commercial contexts such a bold philosophy no doubt makes good sense, Mr. Goldin overlooked the fact that NASA and JPL were not 20:20 successful – the Apollo fire at the Cape, the Apollo 13 fuel cell explosion, the Challenger disaster (the Columbia disaster would be on the later watch of Administrator Sean O'Keefe), the Galileo antenna near-disaster and the Hubble mirror near-disaster were always in the back of the minds of the Agency's engineers, managers and scientists for whom failure was definitely not a conscious option, quite the opposite. These experiences, not to mention the continuing failures in the Soviet program, might have led other NASA Administrators to question whether deliberately adding risk was such a good idea. But Galileo and Hubble were flagship missions with much new technology so, Mr. Goldin evidently supposed that future simple missions ('grey mice') on smaller launch vehicles (Deltas instead of Titans) would be intrinsically

---

[108] http://www.jpl.nasa.gov/jplhistory/the90/

less risky and would have less taxpayer money at stake. That was the belief and initially it seemed to hold. To fit within the launch capability of a Delta, Mars orbiters would need to carry less fuel for Mars orbit insertion and this meant that aerobraking would need to be applied – a months-long process in orbit, not without risk, requiring perfect navigation (fortunately, one of the things for which JPL is renowned). As it happened, however, soon enough JPL would again not be batting 20 for 20.

A fully successful mission that would exemplify and justify Mr. Goldin's philosophy has been the *New Horizons* mission encounter with Pluto and its moon Charon that took place in 2015 and is now traveling deeper into the Kuiper Belt. Charles Boldin was by then NASA Administrator. This mission and its complex birthing pains are reported in detail in a book written by PI Alan Stern some years before the Project actually got underway.

In October 1992, Mr. Goldin appointed Associate Administrator Lennard Fisk to the position of Chief Scientist, a position without funding authority. The two had not found common ground as is revealed in an interview for NASA's Oral History Project. 8 Dr Fisk subsequently returned to academia. Wesley Huntress became Associate Administrator of the Office of Space Science and Applications (OSSA). The Mars Observer loss a year later (in August 1993) had the counter-intuitive effect of leading NASA to re-emphasize exploring Mars – but with a different approach.

1992 NASA Mars Observer

As discussed earlier – Chapter 7 – the planetary science community and the NASA Program Office were very conscious of the need to carry out a cost-constrained deep space exploration program in order to avoid the Reagan Administration's termination of the NASA-JPL program of deep space missions. To this end the Solar System Exploration Committee had recommended that maximum use be made of already-developed technology. In the case of inner Solar System orbiter missions – Mars and the Moon in particular – it had been determined that most of the technology of Earth orbital platforms was now sufficient. So a competition had been held to select a contractor to work with JPL to carry out the first such mission, Mars Observer, after a two-decade hiatus in Mars exploration. The Mars Observer spacecraft was based on the RCA Satcom's spacecraft bus, propulsion, thermal protection, and solar array. RCA meteorological satellite designs provided the Attitude and Articulation Control System, the command and data handling subsystem, and the power subsystem. The propulsion system and high-gain antenna were new elements of the spacecraft.

The Mars Observer science objectives included both geoscience (elemental and mineralogical surface composition, topography, gravity and magnetic field) and climatology (atmospheric structure and circulation, seasonal sources and sinks of volatiles and dust). An eight-instrument payload (along, of course, with a radio science team) had been selected. In the recent wake of the Challenger disaster the Mars Observer launch was now to be carried out using an expendable rocket (the Titan III with a TOS upper stage). The delay and new plan began to drive up the mission cost and it was necessary to deselect an instrument – the visual and IR mapping spectrometer (VIMS). This de-selection was the painful last responsibility of the author who, soon after, left the Program Office believing the mission was again in good shape and in good hands.

The original MO Project Manager was Bill Purdy who later left JPL to take up a position in industry. His successor was David Evans. Arden Albee, Professor of Geology at Caltech was Project Scientist. The Mars Observer science instruments were: Camera System (PI Mike Malin of Malin Space Systems), Laser Altimeter (PI David Smith NASA Goddard), Thermal Emission Spectrometer (PI Phil Christensen of Arizona State U.), Pressure Modulated IR Radiometer (PI Dan McCleese of JPL), Gamma Ray Spectrometer (PI Bill Boynton of U. Arizona), Magnetometer (PI Mario Acuna NASA GSFC). Mars Observer was launched from the Cape on 25 September 1992. This was just a month after Hurricane Andrew had struck the Florida coast (and probably accounts for the need to clean the spacecraft of particulate contamination). The spacecraft was closing in on Mars for its planned orbit insertion on 24 August 1993 when – disaster! – all communication was lost on 21 August. Following an investigation it was concluded that the mission failure had most likely been caused by leakage of fuel and oxidizer vapor through an improperly designed valve. During the cruise the vapor mix would have accumulated in feed and pressurant lines resulting in an explosion when the engine was restarted for a routine course correction.

The unrealized science goals of Mars Observer would, of course, remain valid. Consequently the Mars Observer science instruments would be re-flown on four subsequent Mars orbiters: Mars Global Surveyor launched in 1996, Mars Climate Orbiter launched in 1998, Mars Odyssey launched in 2001 and Mars Reconnaissance Orbiter launched in 2005. Of these, the Mars Climate Orbiter failed after reaching Mars. Arguably, a redesign of the faulty valve and a re-flight of the Mars Observer would have been much faster, cheaper and better.

This mission failure served to confirm Mr. Goldin in his view that NASA and JPL's approach to carrying out space exploration was badly flawed. In this he evidently overlooked the fact that Mars Observer was precisely NASA-SSEC's proposed approach to carry out inner planet missions both faster and cheaper than before by building on proven commercial technology – in this case RCA's meteorology satellites. The basic concept was to reduce costs by buying the spacecraft on a fixed-price contract and selecting only scientific instruments that were already mature. The approach had been critically reviewed by highly regarded managers and considered sound. The Challenger accident had been the biggest impediment to success and the expanded time in development had increased the Mars Observer's cost. The Challenger accident had also added to earlier Program management concerns (specifically, concerns of the author) now that dual launches were no longer the standard approach as they had been up to Voyager, about how to recover from seemingly inevitable occasional failures of launch vehicles and spacecraft. At the author's direction key spares, available for subsequent missions, for a re-flight of the Mars Observer had been acquired (an additional cost) and would in fact be used for the several missions needed to recover the science. The Mars Global Surveyor mission was the first of these and was effectively Mars Observer-light. Ironically the Mars Observer came to be regarded by Mr. Goldin as one more *Battlestar Galactica* because its advertised cost had grown by a factor of three compared with the original SSEC plan – an advertised cost, however, that now included not only the cost of extended development and critical spares but also the cost of the Titan III, the Transfer Orbit Stage and operations. The author is also inclined to think that there may also have been some creative bookkeeping at JPL during the time it took to recover from the Challenger disaster.[109]

The Presidential election of November 1992 saw Bill Clinton's victory with Albert Gore as his Vice-President. Space was not a high priority during their eight years in office (1993-2001). The limping Space Exploration Initiative of President George HW Bush (to send humans back to the Moon and ultimately to send astronauts to Mars, Chapter 6) died at this point. Space Station Freedom came close to cancellation and its planned capability was downsized. However, Dan Goldin remained as NASA Administrator and proposed that Space Station Freedom be redesigned to combine US- and Russian-built elements. Benefits would include budget savings and, just as important, political considerations. In Russia the space program was falling apart in the wake of the dissolution of the Soviet Union in December 1991 so, on both sides, there were compelling considerations for cooperation; by the end of 1993 the renamed International Space Station included partnerships with Russia, Europe, Japan and Canada. This was a major achievement by the Administrator.

The possibility of a *Mars Together* program looked like it might follow but a series of unfortunate events in the US and in the Russian Mars programs submerged that attractive idea. On the US side the key event was the failure of the Mars Observer.

Now Dan Goldin and Wes Huntress intended to establish a new era in Mars exploration: relatively inexpensive missions launched every 26 months, encouragement of collaboration with international partners, leadership of a Mars Sample Return mission to pursue the question of life on Mars, and robotic missions as precursors to enable manned missions to Mars. There would be four broad goals in the new Mars Exploration Program: determine if life had ever arisen on Mars; characterize the Martian climate; characterize the geology of Mars; and prepare for human exploration. Locating sources of Martian water – ice and liquid, past and present – would be the link between all of these goals. The science goals of the Mars Observer remained as a high a priority as before but now the instruments that had flown on that mission would be separated and flown on a series of faster/cheaper/better missions. The new Mars Surveyor Program with its several orbiters would regain the Mars Observer science from orbit and, also, would begin a new series of low cost surface science missions. Thus NASA "would send

---

[109] *Why Mars: NASA and the Politics of Space Exploration* p119 By W. Henry Lambright, Johns Hopkins University Press

two low-cost spacecraft, an orbiter and a lander, to Mars every 26 months over the course of 10 years". This was named the Mars Surveyor Program and was successfully sold as a line item to the Clinton OMB. Missions would be cost-capped at $175M and limited to a three-year development. For the orbiter science "faster" had, however, somehow evolved to become much slower![110]

The broad planetary science balance that had been the intention of the Solar System Exploration Committee would, nevertheless, still be maintained by the ongoing missions – Magellan, Galileo & Cassini-Huygens – and by relatively inexpensive Discovery missions proposed competitively by the planetary science community.

In June 1993, an important paper by Stephen Clifford of the Lunar and Planetary Institute in Houston had been published in the *Journal of Geophysical Research* and titled 'A Model for the Hydrologic and Climatic Behavior of Water on Mars'. Clifford argued that if the planetary inventory of out-gassed H_2O exceeded the pore volume of the Martian cryosphere (the part of the crust always below freezing: perhaps 2 km deep at the equator and perhaps 6 km deep at the south pole) by more than a small amount then a sub-permafrost ground water system of global extent would be the consequence. Below the ice-saturated cryosphere at kilometers depth would be a ground-water table. In this circumstance, low-temperature hydrothermal circulation would take place – upward vapor flux and returning liquid flux – driven by the Martian geothermal gradient.[111]

Given the previously described discovery of extremophile microbes at kilometer depths on Earth, the possibility that microbial life might flourish at depth on Mars seems like a real possibility. However, exploration of this deep subsurface realm whether by astronauts or by means of robotic spacecraft is a challenge that no one is seriously contemplating at present. At the request of the author (then Director of the Center for Mars Exploration at NASA Ames) in 1993 the Sandia National Laboratory examined coiled tube drill technology as a means of drilling to kilometer depth – a significant challenge but probably achievable. Subsequent thinking has considered penetration to depth by using an electrically heated probe to melt down through the rock.

NASA moved forward with the Mars Surveyor Program – to launch two spacecraft, an orbiter and a lander, to Mars at each 26-month opportunity. In February 1994 President Clinton's budget proposal included a new start for the *Mars Surveyor Program* and Congress approved. JPL would manage the first mission: the *Mars Global Surveyor*. Lockheed-Martin was awarded the contract to build the spacecraft in 28 months toward the 1996 launch window. The mission would take advantage of Mars Observer spares: 5 instruments, electronics and software but would use aero-braking to minimize the fuel needed for to reach the desired orbit.

A second Mars mission was also selected to launch at the end of 1996: *Mars Pathfinder*, a small prototype rover. This was selected as the second mission in the Discovery Program – it was not really in the spirit of that program but was a convenient source of funding to prepare two missions for launch in the late 1996 launch opportunity. But before either of these missions launched, on 6 August 1996 there was a bombshell announcement by researchers at NASA Johnson Space Center who had been using an electron microscope to study a meteorite that was known on the basis of Viking isotopic studies to have originated on Mars.

SNC Meteorites

There is a rare group of meteorites – the *SNC* meteorites – that has gained much attention because we now know that these igneous rocks were blasted off the surface of Mars. Because of their unusual elemental and isotopic compositions, they long ago stood out to researchers as anomalies compared to all other meteorites. It is only since the 1970s that their importance has been clarified.

The first of these unusual meteorites was recovered two hundred years ago (1815) near the village of Chassigny in the Haute Marne province of France, the next was found in India near the town of Shergahti in 1865 and the third in 1911 in Egypt near the village of El-Nakhla. Remarkably, today more than a hundred SNCs have been recovered from all over the

---

[110] *Why Mars: NASA and the Politics of Space Exploration* p121, By W. Henry Lambright, Johns Hopkins University Press
[111] http://www.lpi.usra.edu/meetings/earlymars2012/reprintLibrary/Clifford_1993.pdf

world including North Africa and the Antarctic. About three quarters of all SNCs are shergottites. Nakhlites make up the rest for only one other chassignite has been found. From their distinctive mineralogy and chemistry, including oxygen isotope abundances and aqueous weathering products, these rocks were recognized to have originated from a single planetary body; intensive study of the nakhlites in the 1970s determined that, because of their youthful age and thermal history, these meteorites could not have come from an asteroid – Mars was the favored candidate source. In 1983 the definitive link to Mars was established through an analysis of gases (argon, krypton, and xenon) within shock-melted glass found in shergottites: they were determined to match the unique composition of the Martian atmosphere as measured by the Viking landers just a few years earlier. All the SNCs record evidence in the form of shock alteration of the impact that blasted them into space.

Although the shergottites divide into two compositional classes with differing ages – 180 million years and 165-475 million years – they are most easily understood if they come from a single impact event. This is because they evidently must derive from some of the very youngest terrains – the central Tharsis volcanic plains or the Olympus Mons volcano – that cover only a small fraction of the Martian surface. All of the nakhlites are also young having formed from basaltic magma about 1.3 billion years ago. Similarly, the formation age of Chassigny is dated at 1.35 billion years. Nakhlite ejection from Mars took place about 11 million years ago as determined by measuring the effect of exposure to cosmic rays while the rocks languished in orbit about the Sun until arriving on Earth within the last 10,000 years. Because all nakhlites have essentially identical cosmic ray exposure ages and seemingly identical formation ages it is probable that a single impact on Mars led to their formation. The cosmic ray exposure ages of the shergottites vary from 0.5 to 19 million years – presumably because some were shielded inside large rocks for part of the time they spent in space.

Even without knowing exactly where on Mars the SNCs originated, investigators have gained important insights. JC Bridges and PH Warren of Leicester University report:

"SNC chemistry in general reflects the Fe-rich mantle of Mars (which contains twice as much FeO as the Earth's mantle), the late accretion of chondritic material into the mantle, and possibly the presence of a plagioclase-rich magma ocean, which acted to variably deplete the mantle in Al. The high FeO contents of the SNC melts are associated with high melt densities (allowing the ponding of large magma bodies) and low viscosities, both of which are consistent with the large scale of many observed Martian lava flows."

The Mars meteorite that has received enormous attention is the famous *Allan Hills 84001*. It created a major stir when an electron microscope image was interpreted to show a fossilized bacterium. ALH84001 is classified along with the SNCs because of its mineral composition and its traces of gas but it is different in one key way – its age is much, much older, about 4.1 billion years. Its Martian origin was only recognized in 1994. ALH 84001 is an igneous rock mainly composed of orthopyroxene – $(Mg,Fe)SiO_3$ – and is the first and, so far, only member of the so-called "OPX group" of Martian meteorites. The history of ALH 84001 is described by the Lunar and Planetary Institute in Houston as follows:

"Long after ALH 84001 crystallized from molten lava and cooled, about 4.0 billion years ago, it was heated again and deformed by a strong shock. This heating and shock were probably from the nearby impact of an asteroid or meteorite. Sometime after this impact, possibly about 3.6 billion years ago some kind of liquid flowed through ALH 84001 and deposited rounded globules of carbonate minerals. The only more recent event that we can see in ALH 84001 is another shock event. This shock may have come from the meteorite impact that lofted ALH 84001 off Mars. ALH 84001 was probably blasted off Mars about 16 million years ago; it could have been longer ago, but not much more recently. This age is actually how ALH 84001 was in space, exposed to cosmic rays, as it travelled between Mars and the Earth."[112]

---

[112] https://www.nasa.gov/centers/johnson/news/releases/2004/J04-025.html

This famous Martian meteorite was found by a team of meteorite hunters (as illustrated) from the ANtarctic Search for METeorites Program – ANSMET – which is sponsored by the Polar Programs Office of the U.S. National Science Foundation (NSF) and by grants from NASA's Solar System Exploration Office. The NSF has been funding the program since 1977 so that more than 21,000 meteorites have been collected, dominating the number found from outside Antarctica.

The ANSMET program has had two Principal Investigators. From 1976 to 1995, Professor William Cassidy of the University of Pittsburgh served that role. Since 1996, Dr Ralph Harvey of Case Western Reserve University has been the PI. Curation and characterization of the samples is supported by a partnership between NASA and the Smithsonian Institution. Currently ANSMET support comes from NASA's Near Earth Object program, with funding through the 2021-2022 season. The field workers include both senior researchers and graduate students; close to 200 individuals have participated over the years. After each field season the newly recovered specimens are shipped (still frozen) to the Antarctic Meteorite Laboratory at NASA's Johnson Space Center in Houston Texas where they are thawed, dried and carefully examined; small chips are broken off for initial description by the curatorial staff at NASA JSC and at the Smithsonian.

The Antarctic might seem a strange place to look for rocks falling out of the sky given that they are quickly covered by snowfall after they land on the extensive high-altitude ice fields throughout the continent. The East Antarctic ice-sheet flows toward the margins of the continent where its progress is occasionally blocked by the TransAntarctic Mountains. Old deep ice is pushed to the surface where it is exposed to strong winds and massive deflation that removes large volumes of ice and prevents the accumulation of snow. This leads to a concentration of the rocks and, given the contrast between the dark rocks and the ice, they become relatively easy to find by the science team searching on their snowmobiles. Nevertheless, most ice-embedded meteorites eventually slide and are lost to the ocean. Perhaps more examples of the OPX and Chassigny classes of Martian meteorites will be discovered in the coming years for surely the devastating impact that blasted ALH 84001 off of the surface of Mars must have sent tons of rocks into space.

ALH 84001: Microbial fossil?

The almost 2 kg meteorite in question was the first meteorite collected from the Allan Hills of the TransAntarctic Mountains back in 1984 and had, accordingly, been identified as Alan Hills 84001.

The Martian origin of Allan Hills 84001 had not been recognized until 1993. It differed from the other SNC meteorites in being an orthopyroxenite, an igneous rock with very low silica content – but also containing iron-rich carbonate. The sample dating showed that it had crystallized from melt about 4.1 billion years ago; the carbonate suggested that it had been weathered by liquid water. That alone made the meteorite very interesting but what created a sensation was the electron microscope image that showed a tiny, tiny structure that was interpreted to be a fossil bacterium. Indeed it was only tens of nanometers in size – that of theoretical nanobacteria and smaller than any known cellular life. If confirmed (and the finding was controversial) this would have been the first proof of the existence of life elsewhere in our Solar System.

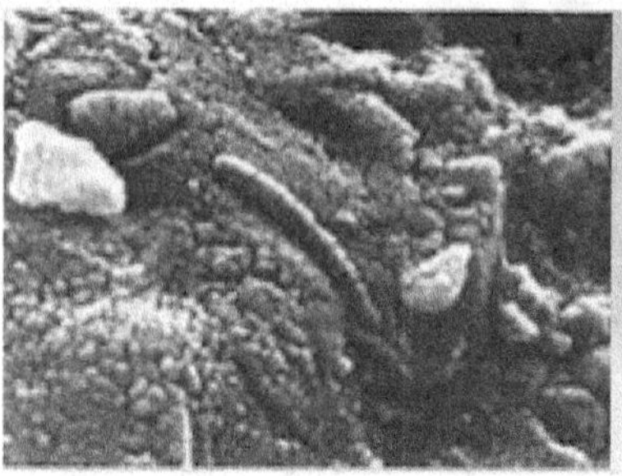

The biological attribution, unfortunately, would not stand the test of time. A NASA JSC paper published in 2004 reports that "magnetite, an iron-bearing mineral found in Martian meteorite ALH84001, was likely caused by inorganic processes, and that those same processes can be recreated in the laboratory, forming magnetite identical to that found in the Mars meteorite."[113]

As might be expected, the discovery received a great deal of attention including that of President Clinton who held an impromptu press conference and asked Vice-President Gore to convene a summit meeting to re-examine the US space program. This was indeed held in December 1996. There was renewed effort by NASA to soon carry out a Mars Sample Return mission perhaps by means of an international collaboration and for some years the goal of carrying out this mission would be the focus of NASA's long-range planetary exploration planning. It has returned.

In August 1996 President Clinton announced his new National Space Policy of which Space Science was, of course, but a part. A new NASA's programmatic thrust titled "Origins" was endorsed and NASA was directed to look for planetary bodies in orbit around other stars. An outgrowth of this new enthusiasm for searching for evidence of life elsewhere in our Solar System and beyond was the creation of the Astrobiology Institute at NASA Ames, led by Nobel Prize winning biochemist Baruch Blumberg.

1996 Mars Pathfinder: First Surface Mobility

For decades JPL had been developing and testing Mars surface mobility prototypes and had, in fact, proposed a very ambitious RTG-powered Mars Rover mission back in 1977 as follow-up to the Viking discoveries. To design a small-scale solar-powered rover was therefore not a big challenge. The bigger challenge was the terminal landing system. All of the world's space agencies that have tried to land a spacecraft on Mars have experienced failures of one kind or another though none because of rocks or slopes. Up until this time NASA had a solid record both on the Moon with the Lunar Surveyors, Apollo and the Vikings (though Viking 2 had landed with one leg on a small rock). The Viking lander panoramas showed that there were plenty of rocks on the Martian surface so landing safety was a continuing concern since the landing error ellipse at that time was about 200 x 100 km.

Given the limited precision of lander targeting and the ubiquity of rocks, the idea of using airbags to cushion the touchdown was a reasonable idea. (All new cars sold in the US since 1998 have been required to be fitted with airbags for the driver and passenger). Developing the airbag system for Pathfinder led to an approach of some complexity – the entire lander would be protected by a pyramid-shaped airbag system composed of 24 interconnected spheres measuring 5 meters tall and made of a material similar to that used in space-suits.

Rather than one big airbag that simply cushioned the vertical descent of the landing, this was a system that fully surrounded the lander and stayed inflated until any horizontal landing velocity was also eliminated – the plan was for the lander to bounce its way to a complete standstill before the airbags were deflated allowing the little Pathfinder rover to emerge. Although this innovative (and somewhat hair-raising) approach arguably introduced as much risk as it avoided, the

---

[113] ibid.

system was adopted and put through rigorous testing at the NASA Glenn Space Center in Ohio. Glenn has the world's largest vacuum chamber (this was pressurized with a simulated Martian atmosphere). Of the two missions scheduled to launch in late 1996, Mars Pathfinder launched second – on a Delta II on 4 December – but overtook the Mars Global Surveyor orbiter (which had launched in November) to arrive at Mars on July 4 1997. Both missions were managed by JPL. Tony Spear and Brian Muirhead provided successive Pathfinder Project Management. The Project Scientist was Matthew Golombek; Richard A. Cook was Mission Operations Manager; and John Wellman was Science and Instruments Manager. Jacob Matijevic was Rover Manager and Allan Sacks was Ground System Manager. Twenty planetary scientists formed the science team.

On 4 July 1997 the mission was successful in landing and deploying the tiny Sojourner rover on the surface in the Ares Vallis region of Chryse Planitia. This landing site is in an ancient flood plain and was chosen – as it had been for the Viking 1 lander – to be a relatively safe surface containing a wide variety of rocks deposited during that catastrophic flood billions of years earlier in the Hesperian era. Atmospheric entry took place inside an ablative entry capsule and, after separation from the capsule, was then slowed down by a supersonic parachute. At ~350 meters above the ground solid rocket gas generators inflated the airbags and finally 3 solid rockets fired when the lander was ~100 meters from impact. The lander first bounced about 15 meters into the air, bounced another 15 times and then rolled before coming to rest after ~2.5 minutes (and ~1 km) from the initial impact. Though neither elegant nor simple it was successful. The landing site was subsequently named in honor of Carl Sagan who had recently died.

After coming to rest the airbags were deflated and retracted, the solar panels deployed and daylight was awaited to survey the site and deploy the little rover. As it happened one of the airbags had not deflated fully and would be an impediment to the deployment of the rover. Happily, the problem was fixed and on the second day Sojourner was released, stood up and backed down a ramp. It was in effect a miniature solar-powered version of the Curiosity rover that would land on Mars in 2012: six-wheeled, 65 cm long, 48 cm wide, 30 cm tall and weighing just 10.5 kg. The solar panel powered a rechargeable battery.

The science payload of the immobile lander consisted of a stereo camera and an atmospheric structure/meteorology package. The little rover had acquired a striking name through an essay contest: *Sojourner Truth*, the adopted name of Isabella Baumfree (1797-1883), an African-American abolitionist and women's rights activist. Sojourner was equipped with cameras and an alpha/proton/X-ray spectrometer to determine the composition of the rocks and soil.

During its three months of operations, Sojourner traveled ~100 meters in total, always within just 12 meters of the mother craft. 550 images were returned. Within reach, as the panorama showed, there were plenty of rocks to analyze and compositional analyses were carried out on sixteen. They all proved to be volcanic in origin, some much like terrestrial andesites (igneous rocks made up of ~60% silicon dioxide and, on Earth, their composition is the average of the continental crust's). The largest meter-size rock about 5 meters northwest (named Yogi as it somewhat resembles the head of a bear looking away from the spacecraft) is a basalt whose shape and texture indicate that it was probably brought there by the flood.

With the successful operation of the Alpha Proton X-Ray Spectrometer (APXS) during the mission, geochemistry data were now available from three sites on Mars – albeit two not far apart. The Viking Landers had measured elemental abundances of nine soil samples at the Viking 1 and eight soils at the Viking 2 sites. These exhibited broad similarities – an iron-rich chemistry similar to that of palagonite (an alteration product of water and basaltic glass – certain areas of this mineral on Hawaii's Mauna Kea volcano provide a close spectral and magnetic analogue) – while the Pathfinder soils showed enrichment in silica and depletion in sulfur. As a group, Pathfinder rocks and soils are richer in SiO_2, TiO_2, and Al_2O_3 relative to Viking soils. They are depleted in SO_3. Pathfinder soils also have more MgO than Viking soils.

Sojourner operated successfully for almost three months. Communication failed after 7 October. It is thought that the battery, designed for one month of operations, may have failed after multiple charging and discharging. Overall Mars

Pathfinder was both an engineering and scientific success, something of a novelty that generated much public interest and demonstrated that JPL was capable of successfully carrying out faster/cheaper missions as well as purple pigeons.

The multiple local images of rocks taken by Pathfinder were used to develop a cunning way of increasing the effective spatial resolution – a technique that also could be applied to images acquired from orbit. Bob Kanefsky, Tim Parker, and Peter Cheeseman at NASA Ames developed an automatic algorithm that, by taking multiple images of an object (notably Yogi rock) and combined them to extract a significant increase in resolution. The technique would be applied later to orbital images acquired by the HiRISE camera of the Mars Reconnaissance Orbiter.

1996 Mars Global Surveyor (MGS)

The Mars Global Surveyor was launched on 7 November 1996 and was a major success, achieving many of the Mars Observer objectives that had been proposed by the SSEC a dozen years earlier: characterizing the geologic processes that have shaped Mars, determining the composition of surface materials, measuring the global topography, gravitational and magnetic fields, monitoring the weather on a global scale and monitoring the seasonal dynamics of surface and atmosphere: dust storms, polar caps and condensate clouds. The orbiter also provided high-resolution imaging and topographic information about landing sites for missions that would follow.

Lockheed Martin in Denver built the MGS spacecraft. JPL's Glenn Cunningham was the Project Manager and Arden Albee, Professor of Geology and Planetary Science at Caltech, was the Project Scientist (as he had been for Mars Observer). The science payload was a reduced version of the Mars Observer's – lacking Bill Boynton's Gamma Ray Spectrometer and Dan McCleese's Pressure Modulated IR Radiometer. These would be re-flown on later Mars orbiters (McCleese, however, would suffer a second disaster but would be lucky his third time). So, the MGS spacecraft carried six science instruments – Thermal Emission Spectrometer (Principal Investigator Phil Christensen of Arizona State U), laser altimeter (PI David Smith of NASA Goddard), Magnetometer/Electron Reflectometer (PI Mario Acuna, NASA Goddard), Radio Science PI Len Tyler Stanford U), and Imaging System (PI, Mike Malin of Malin Space Science Systems).

When the orbiter arrived in September 1997 it first fired its engine to enter a 45-hour highly elliptical orbit with 260 km periapsis and 54,000 km apoapsis. The intended science orbit was achieved by aero-braking: short propulsive burns served to lower the periapsis down to 110 km above the surface where atmospheric drag gradually – over 4 months – lowered the apoapsis so that eventually MGS achieved its desired orbit. A bent solar panel, slightly damaged at launch, was a source of concern during the aero-braking but proved to be a surmountable problem. The circular orbit from which MGS observed Mars for the following nine years (!) was of near-polar inclination at 378 km altitude and was sun-synchronous so that (ideally for the detection of changes on the surface) the lighting conditions of the images remained the same throughout the mission. This was particularly valuable for the study of the gullies that form on the walls of certain craters, most likely as a result of seasonal seepage of liquid water. Hundreds of gullies have been recorded on steep slopes by the CCD cameras: a narrow-angle framing camera capable of 1.4 meter/pixel resolution and two wide-angle line array 'push-broom' cameras. The images were of sufficiently high resolution to discover layered terrains that were clearly sediments, most probably laid down beneath water early in Martian history. A quarter of a million images were returned and have provided the science team and the broader planetary science community with the means to address the full range of questions about physical processes on the planet.

Another key instrument onboard MGS was Phil Christiensen's Thermal Emission Spectrometer that collected data characterizing surface mineralogy; spatial resolution was a few kilometers. During the aerobraking phase the TES team made an important discovery of hematite – iron oxide (Fe_2O_3) – along with evaporite deposits in the Terra Meridiani region. "Christensen and his team studied over four million TES spectra and, after removing the effects of the atmosphere, they were left with the distinctive spectral curves of the surface materials. These were then compared with spectra measured in the laboratory for a wide variety of minerals".[114] The depth and shape of the hematite fundamental bands reveal that this hematite is the crystalline and coarse-grained (greater than about 10 micrometers in diameter) gray variety of the iron

---

[114] http://www.psrd.hawaii.edu/Mar03/Meridiani.html

oxide". Because hematite forms on Earth in several ways, most involving water, this discovery would lead to the selection of Terra Meridiani as the landing site for the Mars Exploration Rover Opportunity and, much later, for ESA's ill-fated Schiaparelli pathfinder lander.

The TES instrument has also identified a large area of Mars – "eleven times larger than the five volcanoes on the Island of Hawaii"– characterized by the mineral olivine. This is a magnesium iron silicate that is common on Earth in igneous rocks and weathers quickly in the presence of water into minerals such as hydroxide mineral goethite, the phyllosilicate chlorite, the clay mineral smectite, the iron oxide maghemite and hematite. The regions where olivine is found are evidently those where the climate has been dry from the time when the olivine was exposed. The laser altimeter provided by the Maria Zuber & David E. Smith team from the NASA Goddard Space Flight Center has created a definitive high resolution topographic map of Mars that shows the North-South dichotomy of Mars more clearly than ever (this is still not well understood). Many highly eroded or buried craters were identified that had been too subtle for previous observation. Canyons within the polar ice caps were mapped.

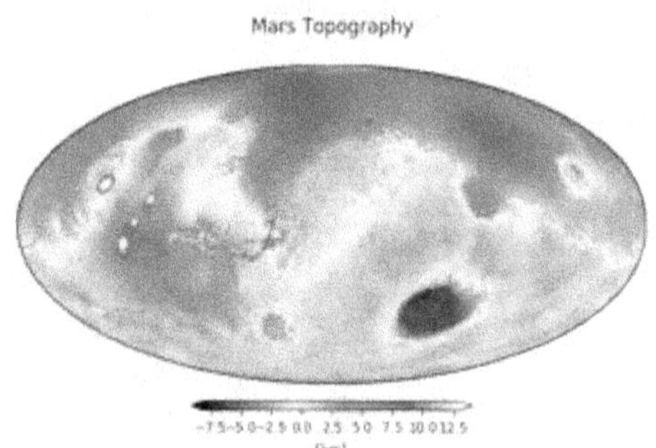

MGS magnetometer observations have confirmed that Mars does not presently have a global magnetic field. Evidently it did once, but the lack of magnetization in the large impact basins has allowed Mario Acuna's team to establish that the Martian dynamo had ceased to operate more than 3.5 billion years ago. Once Mars had ceased to have a global magnetic field it lost its protection against atmospheric erosion by the solar wind and any surface water would also have gradually been lost. Nevertheless, today in some regions "Mars does have very strong crustal magnetic fields, more than 30 times stronger than those of Earth".[115]

Given the geologic diversity of Mars, for Mike Malin's imaging team it was Christmas almost every day (as it has been for all Mars orbiter imaging teams). A remarkable image, 1.5 km across, shows detail in the gullies on the walls of Newton Crater in Sirenum Terra, a large region in the southern hemisphere centered at 40°S 150°W. Newton Crater is about 300 km across. There the temperature is far too cold and the pressure far too low for stable liquid water to be the erosive force. Repeated imaging of the gullies has led to the discovery of changes that the team consider may represent dry granular rather than liquid flows. Given how many years MGS operated, the science team has been able to make observations through several annual cycles so that it discovered that "for three Martian summers in a row, deposits of

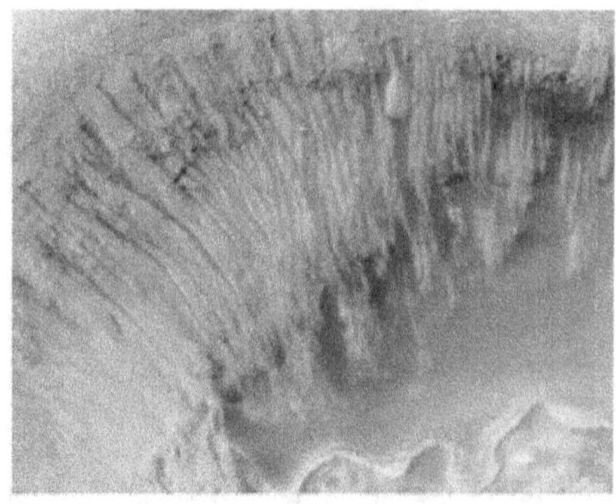

carbon-dioxide ice near Mars' South Pole shrunk from the previous year's size, suggesting a climate change in progress".[116]

The MGS imaging system also captured an intriguing phenomenon without any terrestrial analogy – the seasonal eruption

of geysers of gas and dust that takes place in the spring in the south polar region of Mars. This region differs from the north not only in the smaller size of the permanent ice cap but also in the surrounding terrain. In the north a vast dune field surrounds the cap while in the south the dunes are much more scattered. Here during the autumn and winter the seasonal CO_2 ice cap accumulates to a depth of about a meter. At the beginning of spring Mike Malin's MGS camera system recorded the formation of dark spots mainly on the ridges and slopes of the dunes.

Remarkably, these features initially were not on top of the ice but underneath. Spider web-like channels were observed that had evidently

---

[115] https://mgs-mager.gsfc.nasa.gov
[116] https://en.wikipedia.org/wiki/Newton_(Martian_crater)

been carved between the ground and the ice – the ice being effectively transparent to the sunlight. Then, as the surface under the ice warmed further, geysers of CO_2 gas and dark sand erupt. Amazing! The sand accumulates on the ice to form dark dune spots. Radial networks of these features are typically 150-300 meters across. This unearthly phenomenon was monitored by subsequent Mars orbiters, notably the Mars Reconnaissance Orbiter with its ultra-high resolution camera HiRISE. The image above on the right is, of course, an artist's interpretation: Ron Miller for NASA.

The Mars Global Surveyor eventually fell silent: on 22 November 2006 the Mars Exploration Rover Opportunity (see below) that had been using MGS as a relay failed to detect any signal from MGS during an attempt to get the orbiter to transmit to the rover. This was apparently the result of battery failure caused by a complex sequence of events involving the onboard computer memory and ground commands. Mars Observer-light had been an enormous success.

1996 Russia's Mars 96

The Russians also sought to take advantage of the 1996 Mars launch window with a flagship mission, Mars 96, that included an orbiter based on the earlier Phobos design, two surface stations and two penetrators (spear-like probes that would bury themselves in the surface on impact). France, Germany, other European countries and the United States provided instrumentation. The circumstances in Russia at the time were, however, grim for the teams working in the space program. This unfortunate episode is described at some length below to remind US planetary scientists (and the author) that, if we sometimes become frustrated by politics in Washington DC and by various setbacks, matters have been immeasurably worse for our colleagues in Russia. According to Russian documentation:

With the collapse of the Soviet Union in 1991, Mars-94 project lost its twin vehicle, leaving a single flight version of the spacecraft in the budget. France had also withdrawn from the project, due to overspending on its part of the project. Still, scientists from 21 countries participated in the Mars-96 project, contributing instruments and experiments, including two experiments from NASA. Mars-94 (M1) now aimed to deliver two small landers, MS, and two penetrators to the surface of Mars following its launch in October 1994. However financial problems in the post-Soviet Russia were continuously delaying the production of flight hardware, for example the Argus platform, which was critical for the independent pointing of science instruments toward their targets. In the end, the M1 program had to be delayed for two years to a less favorable window for launches to Mars in 1996. The follow-on M2 project, intended to carry a rover, had to move as well – to 1998. According to one source, a six-wheel rover for the mission was expected to have a mass of 35 kilograms and carry 15 kilograms of science instruments. Mission planners considered dropping the development of the Argus platform altogether and installing its scientific payloads directly onto the spacecraft, accepting all resulting limitations. However, in the end, Germany, the supplier of a TV complex for Argus, funded the development of the platform.

And

Despite the horrendous economic situation surrounding them, workers and engineers involved in the Mars-96 project made a seemingly impossible effort to prepare the spacecraft for launch, often working without pay and seeing many of their colleagues leave the industry. There were proposals to delay the mission to 1998, however mission officials had good reasons to fear the complete cancellation of the project by that time, due to the worsening financial situation in the country and a previous decision to scale-down all future planetary probes to fit into the Soyuz rockets rather than Proton. In the end, all critical tests on the spacecraft had been completed by the planned launch date in November 1996.

And

During final assembly in Baikonur, personnel discovered that the spacecraft did not fit into the payload fairing of the launch vehicle. As it transpired, the small landers were actually larger than their official specifications and prevented the complete folding of a boom holding the high-gain antenna. To resolve the problem, specialists cut a 30-centimeter segment from the structure. The Proton rocket (8K82K, Proton-K) carrying the Mars-96 spacecraft lifted off from Baikonur Cosmodrome on Nov. 16, 1996, at 23:48:52.75 Moscow Time (20:48:53 GMT). Its three booster stages and the first burn of the Block D2 upper stage went as scheduled, delivering the spacecraft to an initial orbit at an altitude of around 160 kilometers and prompting many observers and participants of the project at mission control to start celebrating a successful launch. The second firing of the Block D was scheduled to take place one hour, eight minutes and 52 seconds after liftoff and beyond the range of Russian ground control stations. The maneuver would deliver 3,150 meters per second in acceleration, followed by the firing of the ADU propulsion unit onboard Mars-96 to escape the Earth orbit and start a path to Mars. However, Block D misfired, apparently delivering 20 meters per second in the wrong direction".[117]

The end result was that the spacecraft crashed into the ocean somewhere between the Chilean coast and Easter Island. A tragedy for all involved.

1998 NASA Mars Climate Orbiter

The next Mars launch opportunity was in late 1998 and NASA prepared two spacecraft for that window: the Mars Climate Orbiter and the Mars Polar Lander. The Mars Climate Orbiter carried just a camera and Dan McCleese's spare Pressure Modulated IR Radiometer from Mars Observer. The spacecraft launched from the Cape on a Delta II on 11 December 1998 and arrived at Mars on 23 September 1999 at which time communications were lost. This loss was attributed to an embarrassing miscommunication among the software engineers who had confused English units (pound-seconds) with metric units (Newton-seconds); the spacecraft had approached Mars too closely, had entered the atmosphere and broken up. JPL's extraordinary record of precision navigation had been marred and the episode brought into question the faster/better/cheaper philosophy.

1999 NASA Mars Polar Lander

The Mars Polar Lander unfortunately fared no better. Its objectives were to analyze the soil and climate at very high southern latitudes: Planum Australe. The spacecraft was launched (once more on a Delta II) on 3 January 1999 toward a landing site near the South Pole, then in summer. On December 3 1999 the probe entered the atmosphere successfully but communications were lost and it was concluded that the descent engine had shut down early so that the vehicle had crash-landed heavily. The spacecraft had also released two instrumented penetrator probes that were based on military technology and intended to analyze the subsurface composition at a depth of about one meter. They were released but no communications followed. These multiple failures led to more egg on faces and more questions as to whether faster/better/cheaper had been the way to go.

The failures of both the Mars Climate Orbiter and the Mars Polar Lander were, as it happened, accompanied by other embarrassments elsewhere in NASA's programs and led to a special hearing on 22 March 2000 by the US Senate Subcommittee on Science, Technology and Space headed by Chairman Bill Frist of Tennessee on the subject of the current management challenges at NASA. Senator Frist noted in a lengthy opening statement that year 1999 proved to be a very difficult and challenging one for the agency:

---

[117] http://www.russianspaceweb.com/mars96.html

We have read the reports on workers searching for misplaced Space Station tanks in a landfill; loose pins in the Shuttle's main engine; failure to make English-metric conversions causing the failure of a $125 million mission to Mars; two-time use of rejected seals on a Shuttle's turbo-pumps; $1 billion of cost overruns on the prime contract for the Space Station, with calls from the Inspector General at NASA for improvement in NASA's oversight; workers damaging the main antenna on the Shuttle for communication between mission control and the orbiting Shuttle; urgent repair mission to the Hubble Telescope; approximately $1 billion invested in an experimental vehicle and currently no firm plans for its first flight, if it flies at all; and the lack of long-term planning for the Space Station, an issue on which the Subcommittee has repeatedly questioned NASA. This Subcommittee recognizes and appreciates the technical challenges and hurdles NASA must address to make its missions successful. However, based upon our initial review of the various investigation reports on these problems, the real culprit may be management. We cannot and should not dismiss good basic management as an essential component of success. It still gets back to the fundamentals of planning, of leading, of organizing and of controlling.[118]

Administrator Goldin responded, noting that:

Success cannot be prescribed only by returning to past techniques for conducting missions. Success in the past was often achieved at great expense, using large government contractor teams and massive documentation to verify the design and implementation of complex systems. This nation cannot afford to do business in that manner, nor do we need to. Revolutionary new technologies and approaches to engineering are emerging. Success in the future will be achieved by using technology to enable small teams geographically dispersed, operating in virtual environments, using new tools such as soft computing, neural nets and learning systems to enable more fault tolerant systems. NASA is a leader in developing collaborative engineering environments and design tools.

Administrator Goldin remained convinced that "These new directions will in the future enhance the quality and productivity of Faster, Better, Cheaper approaches".

Some, like the author, may question Mr. Goldin's assertion that "this nation cannot afford to do business in that manner, nor do we need to". Mr. Goldin had determined that "an in-depth review of the entire Mars Program should be undertaken by independent observers. The Mars Program Independent Assessment Team (MPIAT) was chaired by A. Thomas Young".[119] Tom Young was an excellent choice as he had headed the SSEC study that concluded that the cost-constrained approach of the proposed Observer approach was viable.

The Mars Program review in question determined that the direct cause of the MCO failure was the fact that the navigation data that had been provided to the JPL navigation team by Lockheed Martin were in English units rather than the specified metric units. Further:

However, it is important to recognize that space missions are a 'one strike and you are out' activity. Thousands of functions can be correctly performed and one mistake can be mission catastrophic. Mistakes are prevented by oversight, test, and independent analysis, which were deficient for MCO. Specifically, software testing was inadequate. Equally important, the navigation team was understaffed, did not understand the spacecraft, and was inadequately trained. Navigation anomalies (caused by the same units error) observed during cruise from Earth to Mars were not adequately pursued to determine the cause, and the opportunity to do a final trajectory correction maneuver was not utilized because of inadequate preparation.[120]

Ouch! In the case of the Mars Polar Lander:

---

[118] https://www.gpo.gov/fdsys/pkg/CHRG-106shrg78634/html/CHRG-106shrg78634.htm
[119] http://sunnyday.mit.edu/accidents/mpiat_summary.pdf
[120] http://sunnyday.mit.edu/accidents/mpiat_summary.pdf

The design of the lander precluded any communications from the period shortly before separation from the cruise stage until after Mars landing. The planned communications after landing did not occur, resulting in the determination that the MPL mission had failed. Extensive reviews, analyses, and tests have been conducted to determine the most probable cause of the MPL failure. This is documented in the "Report on the Loss of the Mars Polar Lander and Deep Space 2 Missions." Several possible failure causes are presented, which include loss of control due to spacecraft dynamic effects or fuel migration, local characteristics of the landing site beyond the capabilities of the lander, and the parachute covering the lander after touchdown. Extensive tests have demonstrated that the most probable cause of the failure is that spurious signals were generated when the lander legs were deployed during descent. The spurious signals gave a false indication that the lander had landed, resulting in a premature shutdown of the lander engines and the destruction of the lander when it crashed into the Mars surface. Without any entry, descent, and landing telemetry data, there is no way to know whether the lander reached the terminal descent propulsion phase. If it did successfully reach this phase, it is almost certain that premature engine shutdown occurred.[121]

The review had much else to observe including the "enormous management challenges" presented by these deep space missions and the relative inexperience of the Mars Climate Orbiter and Mars Polar Lander project managers. It was pointed out that "the number of projects at JPL has increased significantly" and that "there are not enough experienced managers for the large number of projects for which JPL is currently responsible. This situation requires significant involvement by senior management to compensate for the lack of experience."[122]

Not pointed out by the review was that the "faster, cheaper, better" approach was in fact the very reason that so many small projects were being undertaken. This was a problem that few had anticipated – that even relatively small projects, as complex as all deep space missions are, need experienced managers. Evidently, communications between the project managers at JPL and the program managers at NASA HQ had broken down under the pressure that was being felt from the top to carry out the new Mars program with strictly circumscribed resources.

The 'Young Report' was a major black eye to the Agency, the Administrator, the Science Directorate, and JPL. W. Henry Lambright in his book *Why Mars* describes the crisis in detail:

There was plenty of blame to go round including that of the Clinton White House that accepted that NASA had been asked to do too much with too little. NASA's new budget experienced its first increase (from $13.6B to $14.3B) in seven years. President Clinton in a letter of 27 March directed the Administrator to "conduct a more robust and successful program of sustained robotic exploration of Mars." Associate Administrator Weiler appointed Scott Hubbard of NASA Ames to a position at HQ that would be the single point of contact for all NASA Mars exploration. At JPL Director Ed Stone named Firouz Naderi as Hubbard's counterpart at JPL. The thrust towards Mars Sample Return that had been animating the Mars program would be abandoned. The search for past and/or current life on Mars remained the ultimate goal of the exploration but now the focus would be "follow the water".[123]

Administrator Goldin accepted his share of the blame for the Mars mission failures. He said, "In my effort to empower people I pushed too hard, and in doing so, stretched the system too thin. It wasn't malicious. I believed in the vision, but it may have made some failure inevitable."[124] He was not prepared to abandon "faster, better, cheaper" but would provide the resources needed to succeed. A new approach to Mars exploration would emerge as described in Chapter 12.

1998 JAXA Nozomi

---

[121] ibid.
[122] NASA's Mars Program After the Young Report, Parts I & II: Hearing Before the Committee on Science
[123] W. Henry Lambright, *Why Mars*, pp 157- 167, Johns Hopkins University Press, 2014
[124] https://www.nasa.gov/home/hqnews/Goldin/2000/jpl_remarks.pdf

There was yet another failure at the 1998 Mars opportunity. Japan's Institute of Space and Astronautical Science (ISAS) had launched the Nozomi probe on July 4, 1998; it was designed to study the upper Martian atmosphere and, given Mars' lack of an appreciable magnetic field, the atmosphere's interaction with the solar wind. The spacecraft used a lunar swing-by on September 24, 1998 and another on December 18, 1998 to increase the apogee of its orbit. It flew by Earth on December 20, 1998 at a perigee of about 1,000 km. Alas, a malfunctioning valve during the Earth encounter resulted in a loss of fuel and left the spacecraft with insufficient to reach Mars. Just like Mars Observer, for want of a functioning valve the mission was lost. A thousand items of hardware and software must function as intended for a space mission to be a success; only one thing has to go wrong for a mission to fail.

CHAPTER 11

RETURN TO THE MOON

Clementine, Lunar Prospector, SMART-1, SELENE, Chang'e 1, Chandrayaan, Lunar Reconnaissance Orbiter, LCROSS, Chang'e 2, GRAIL, LADEE, Chang'e 3, Chang'e 5-T1, Israel's Beresheet, India's Chandrayaan -2 Vikram

Following the end of the Apollo missions in 1972 it would not be until January 1994 that the US launched another mission – an orbiter – to the Moon: *Clementine*. Since then there have been more than a dozen, mostly successful lunar missions: Clementine, Lunar Prospector, SMART-1, SELENE, Chang'e 1, Chandrayaan, Lunar Reconnaissance Orbiter, LCROSS, Chang'e 2, GRAIL, LADEE, Chang'e 3, Chang'e 5-T1, Israel's Beresheet, India's Chandrayaan -2 Vikram. Of these, NASA has sent five orbiters and an impactor while most of the others have been orbiters that nations, relatively new to deep space exploration, use to develop their technology and to create their own mapping databases. The Moon is our near neighbor so mission costs are reduced: for launch vehicles and communications, light time and trip time. Just as in the 1960s the learning process can be rapid. China has the most complete robotic lunar exploration program, one that has already included a successful and, most recently, sample return. Both Israel and India have also launched landers to the Moon – unfortunately without success. And, although the Apollo samples had long since provided the 'big picture' of lunar evolution, there remained many science questions still to be settled. Among these is the question of whether water ice may be found in the permanently shadowed floors of polar craters – a question of great potential practical interest.

Caltech scientists Kenneth Watson, Bruce Murray (1931-2013), and Harrison Brown (1917-1986) first suggested this possibility of polar ice in 1961 inspired by the earlier thinking of Harold Urey (1893-1981). Because the lunar spin axis is nearly perpendicular to the ecliptic (the angle between the lunar spin axis and the ecliptic normal is presently just 1.54 degrees) there are today regions on the floors of polar craters that are always in shadow and have been for some billions of years. In the past the situation was not always so – a theoretical reconstruction of lunar evolution by Matthew Siegler and colleagues at the Planetary Science Institute in Tucson, Arizona in 2011 concluded that a large change in obliquity occurred about halfway through the Moon's outward orbital evolution – after which the Moon has been in the so-called Cassini state (deduced by Giovanni Cassini in 1693 as a 1:1 spin-orbit resonance and a rotational axis that maintains a constant angle of inclination from the ecliptic plane).

Impacts of comets over billions of years are a plausible source of lunar volatiles including water ice. Also, it is theorized that water molecules could be produced *in situ* by solar wind protons that combine with oxygen atoms in the silicate regolith. These would be trapped in the minerals' crystal lattices (as distinct from forming ice). Lunar water could be very valuable one day both by providing life support for astronauts on extended-duration missions and, especially, such water could be dissociated into hydrogen and oxygen to use as rocket propellants thereby creating a fuel station on the Moon. Further, the geometry of the lunar spin axis that creates the permanent shadowing of the floors of craters near the poles of the Moon also means that the Sun rarely sets on the rims of the craters. Yearlong sunlight means that at such locations reliable solar electric power would be available for a lunar base that might be established there.

1994 US BMDO-NASA Clementine Orbiter

The Clementine orbiter was a space exploration mission carried out by the US Department of Defense's Ballistic Missile Defense Organization (BMDO) and NASA in a unique, onetime collaboration. A lunar orbiter had been proposed by the Solar System Exploration Committee years earlier (Chapter 7) and was to have been carried out as the second of the Observer missions following the Mars Observer; these were to be based on proven Earth orbiter designs and to be a continuing program of low-cost inner planet orbiters. The escalating cost of the Mars Observer in the wake of the Challenger disaster and its explosion a few days before orbit insertion spelled the end of this Lunar Observer plan. The part of the US planetary science community that focused on lunar research was very discouraged and sought other ways to return to our Moon. One or two lunar scientists had a foot in the Ballistic Missile Defense Organization (BMDO) program and from their initiative the Clementine mission emerged. The design was based on technologies that BMDO was developing for the *'Brilliant Pebbles'* approach to missile interception — swarms of small three-axis-stabilized spacecraft carrying imaging sensors, intelligence and propulsion to effect a collision with an incoming missile. USAF Brigadier General Pete Worden who was Deputy for Technology in the BMDO played a major role in formulating the mission. He was later a professor of astronomy at the University of Arizona and then Director of NASA Ames Research Center. Gene Shoemaker headed the science team with Paul Spudis his deputy who tells the story in his book *The Value of the Moon*. The mission would serve to prove some of the technologies as well as return new information about the Moon and, it was intended, a small asteroid that was to be a follow-on target.

The Naval Research Laboratory in South East Washington DC designed and built the spacecraft. After a very short development, Clementine with seven remote sensing instruments was launched on 25 January 1994 on a Titan II rocket out of the Vandenburg Air Force Base. One of the important findings from the global mapping was the immense size of the South Pole-Aitken basin: 2,600 km across and 12 km deep.

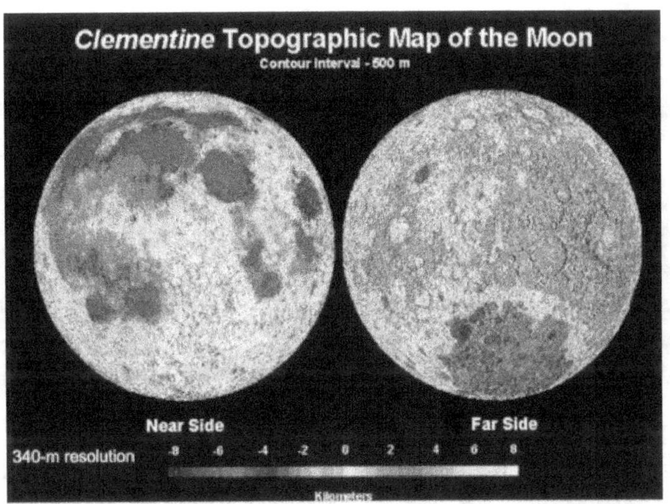

Most notably, the Clementine orbiter carried out an improvised bi-static radar experiment that targeted the permanently shadowed floors of polar craters — with the radar echoes being received by the Deep Space Network dishes. While the magnitude and polarization of the echoes were encouragingly consistent with icy crater floors, the results were not conclusive. Multi-spectral images were used to create mineralogical maps that have provided insights into the geologic evolution of the Moon. Laser altimetry by Clementine provided the first high-quality topographic maps of the Moon.

A malfunction unfortunately prevented the spacecraft leaving lunar orbit to travel, via an Earth flyby, to a planned encounter with the Near Earth asteroid 1620 Geographos after which, the reader will have surmised, the Clementine spacecraft was to have been lost and gone forever.

1998 NASA Lunar Prospector

The goals of the NASA missions that have returned to the Moon have been chosen to fill in key gaps in our knowledge. Besides the growing anticipation that water ice really is trapped in polar craters, lunar scientists had also been eager to make a definitive global map of the Moon's gravity field to better understand the nearside/farside dichotomy and the nature of the basin mascons. Lunar Prospector was the third mission in NASA's low-cost Discovery Program. It was managed by NASA Ames Research Center with the prime contractor Lockheed Martin. Thomas Dougherty was Project Manager. The Lead Investigator for the mission was Alan Binder of Lockheed Engineering and Science Co. There were six experiments onboard: Gamma Ray Spectrometer, Neutron Spectrometer, Magnetometer, Electron Reflectometer, Alpha Particle Spectrometer and Doppler Gravity Experiment.

The small spin-stabilized spacecraft was launched on a three-stage Athena rocket out of the Cape on 6 January 1998 toward a low-altitude polar orbit. Its purpose was to map the Moon's surface composition by means of gamma ray and neutron spectroscopy and, of course, look for deposits of polar water ice and to measure magnetic and gravity fields. Lunar Prospector's neutron spectrometer did indeed detect enhanced hydrogen concentrations at both the north and south polar regions. This was interpreted to indicate significant amounts of trapped water ice mixed in the regolith – possibly as much as 300 million metric tons. The South Pole Aitken Basin is 2,500 km in diameter (the blue area in the map) and is 12 km deep so that craters within the basin never see sunlight and are temperatures never rise above 100°K thereby acting as cold-traps of water from incoming comets and meteoroids So, water ice can accumulate there for billions of years.

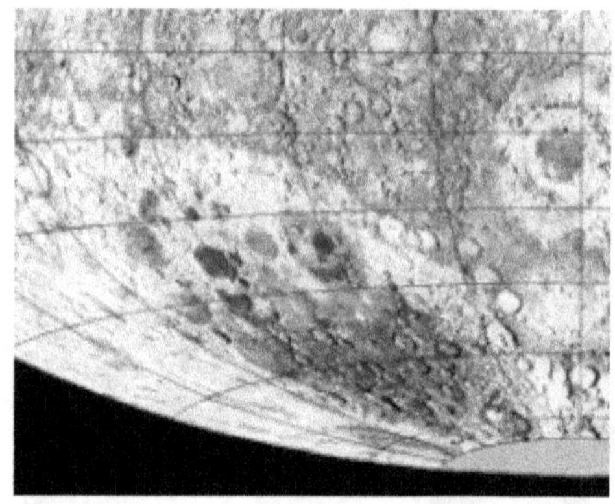

An alternative explanation was recognized to be the presence of the hydroxyl radical (OH) in minerals. The mission ended on July 31, 1999 when Lunar Prospector was crashed into shadowed Shoemaker crater near the south lunar Pole. It was hoped that the impact would release water vapor and that the plume would be detectable from Earth; disappointingly no plume was observed.

By means of Doppler tracking, Lunar Prospector did make an improved measurement of the near side lunar gravity field and identified three new large mass concentrations beneath the impact basins Mare Humboltianum, Mendel-Ryberg, and Schiller-Zucchius. Lunar Prospector also partially resolved four mascons in the large farside basins of Hertzsprung, Coulomb-Sarton, Freundlich-Sharonov, and Mare Moscoviense. The team used the data to conclude that the Moon has an iron core with a radius of 220 to 450 kilometers.

2003 ESA SMART-1

Following a gap of another five years ESA's SMART-1 (Small Missions for Advanced Research in Technology) was launched on an Ariane 5 on 27 September 2003 from Europe's Space Port in Kourou, French Guiana. A Swedish-designed spacecraft, it served mainly to demonstrate a highly efficient solar-electric ion drive propulsion system. The spacecraft traveled on a 14-month (!) transfer to the Moon and then entered a loose polar elliptical orbit for science operations. Although the spacecraft instrumentation included a small camera, X-ray and IR spectrometers the science return was, as planned, modest. On 3 September 2006 it was deliberately crashed into the Moon's surface.

2007 JAXA SELENE

The Japanese SELENE program of orbiters (Hiten flyby/orbiter, Kaguya, Okina and Ouna Orbiters) were extensively instrumented and have been notably successful, leading to improved global maps of lunar topography and far side gravity. The SELENE orbiters are named for the ancient Greek Titan goddess of the Moon. The Kaguya mission was launched on 14 September 2007 on a H-IIA rocket out of the Tanegashima Space Center and by 19 October had established a 100 km circular orbit about the Moon.

The spacecraft was equipped with more than a dozen remote sensing instruments including an altimeter, gamma ray spectrometer and imaging systems. Kaguya provided a novel approach to studying the inside of Shackleton Crater at the lunar South Pole where the crater floor is permanently shadowed and anticipated to hold water-ice deposits. Shackleton lies within the rim of the South Pole-Aitkin basin and is 12 km deep. The Kaguya's stereo camera succeeded in imaging the inside of the crater on 19 November 2007 by taking advantage of faint sunlight scattered from the upper inner wall near the rim. Somewhat disappointingly the derived surface reflectivity indicated that relatively pure water ice deposits are not to be found exposed on the crater floor. It was concluded that "water ice may be disseminated and mixed with soil over a small percentage of the area or may not exist at all".[125]

2007 CSA Chang'e 1

The Chinese Space Agency has begun a systematic program of missions to further explore the Moon with a series of spacecraft named for the Chinese Moon goddess. The Chang'e 1 orbiter was launched on 24 October 2007 from Xichang Satellite Launch Center and entered lunar orbit on 5 November. The science payload was made up of eight experiments: a stereo camera system, an interferometer spectrometer, a laser altimeter, a gamma ray and X-ray spectrometer, a microwave radiometer, and a high-energy particle detector and solar wind monitors. The project was successful in mapping lunar geology, surface composition and topography to serve in the planning of future probes. More generally, Chang'e 1 began the process of accumulating experience for the next stages in China's lunar exploration program. The mission was scheduled to continue for a year but was later extended; "It exercised a planned impact north of Mare Fecunditatis at 52.36 E, 1.50 S on 1 March 2009 at 08:13 UT".[126]

2008 ISA Chandrayaan Orbiter

The Indian Chandrayaan Orbiter and its Moon Impact Probe was a technology demonstration mission that had yet a different approach toward finding out whether the polar craters are traps for water ice. This was to be achieved by monitoring the ejector plume created by the impact of a probe that was to be released and de-orbited from the mother spacecraft – from which the impact would be monitored. Chandrayaan and its probe were launched on 22 October 2008 aboard a PSLV-XL rocket from a site north of Chennai; it was inserted into lunar orbit on 8 November 2008. Six days later on 14 November 2008 the Moon Impact Probe separated from the orbiter and, as planned, impacted near the Shackleton crater. The data acquired by the imaging spectrometer (Moon Mineralogy Mapper contributed by NASA and headed by Principal Investigator Carle Pieters of Brown University) did indeed provide confirmation of the presence of water vapor in the plume.

The same instrument subsequently served to create a mineralogical map of the lunar surface before Chandrayaan-1 ceased communication on 29 August 2009. Besides the imaging spectrometer data, Chandrayaan exercised an unusually large suite of experiments – not only from India but also a number of other countries: the Terrain Mapping Camera, Hyper Spectral Imager, Laser Ranging Instrument, Gamma Ray Spectrometer, X-Ray Fluorescence Spectrometer, Near IR

---

[125] https://ui.adsabs.harvard.edu/abs/2008Sci...322..938H/abstract
[126] https://nssdc.gsfc.nasa.gov/nmc/spacecraft/display.action?id=2007-051A

Spectrometer, Synthetic Aperture Radar, and Radiation Dose Monitor. As might be expected, the mission served to provide the Indian Space Agency with much valuable experience in operating spacecraft in deep space. After almost a year, the orbiter started suffering from several technical problems including failure of the star sensors and poor thermal shielding; it stopped communicating on 28 August 2009.

2009 NASA Lunar Reconnaissance Orbiter

NASA Goddard Space Flight Center's Lunar Reconnaissance Orbiter (LRO) was launched out of the Cape on an Atlas V rocket on 18 June 2009; it remains operational as of early 2020 in an eccentric polar mapping orbit. In an unusual arrangement, the Atlas V payload included a second lunar mission: the Lunar Crater Observation and Sensing Satellite (LCROSS), described below. Craig Tooley is the LRO Project Manager and Richard Vondrak the Project Scientist.

NASA's LRO and LCROSS were developed as an element of a new *Vision for Space Exploration* announced by the George W Bush on 14 January 2004: "a new plan to explore space and extend a human presence across our Solar System. We will begin the effort quickly, using existing programs and personnel. We'll make steady progress – one mission, one voyage, one landing at a time". These lunar missions were announced in the context of the STS-107 Columbia disaster that had occurred on 1 February 2003 with the loss of its international crew of seven: Rick Husband, William McCool, Michael Anderson, David Brown, Kalpana Chawla, Laurel Clark and Ilan Ramon. The new direction for NASA would encompass the completion of the International Space Station, the retirement of the Space Shuttle, the development of a new Crew Exploration Vehicle (*Orion*) with its intended first crewed flight in 2014 (still awaited in 2020), and the renewed exploration of the Moon by robotic spacecraft leading to crewed missions to the Moon by 2020! More generally, the lunar exploration was in anticipation of crewed missions to Mars. In practice, the new space policy subsequently evolved to the less ambitious policy of the following Obama Administration following the Presidential election of 2008.

The spacecraft carries seven instruments: a cosmic ray telescope (PI Harlan Spence, Boston University) to characterize the global lunar radiation environment, an IR radiometer to identify cold-traps and potential ice deposits (PI David Paige, UC Los Angeles), a laser altimeter to measure landing site slopes and search for polar ices in shadowed regions (PI David Smith NASA Goddard SFC), cameras to identify landing site hazards and illumination conditions near the poles, together with instruments to study the permanently shadowed polar craters: a Lyman-Alpha instrument, a neutron detector provided by the Institute for Space Research in Moscow and a miniature synthetic aperture radar. The imaging system comprises two narrow-angle 'push broom' cameras and a wide-angle camera that together provide the means of assessing landing site conditions in the polar region. The cameras have also remapped the Apollo landing sites at a resolution not available before and have identified the remains of old Soviet-era spacecraft and of the impact locations of several past lunar probes.

LRO's initial orbit was a 50 km altitude polar orbit; in May 2015 the spacecraft's orbit was lowered to 20 km above areas near the lunar South Pole. Its detailed mapping aims to identify safe landing sites, to locate potential resources, and, by means of the cosmic ray telescope, to characterize the radiation environment. As part of its payload, LRO has a very sensitive instrument capable of imaging permanently shadowed regions illuminated only by starlight and the glow of interplanetary hydrogen emission, the Lyman-Alpha line. Also, the laser is used to illuminate crater floors to estimate the reflectance of the surface; in the case of 19 km Shackleton Crater this was in fact found to be brighter than others nearby and consistent with the presence of small amounts of ice. There was, however, some uncertainty because, curiously, while the crater's floor was found to be bright the crater walls were even more reflective. The LRO radar penetrates to depths of a meter or two. The radar search for evidence of water ice is based on the distinctive radar polarization signature of ice compared to the surrounding material. Observations of Shackleton crater made between December 2009 and June 2010 have led to an estimate by the team that the ice content of the uppermost meter of loose material of the crater's walls is between five and ten percent ice by weight. After seven years of operations, the results of the Lunar Orbiter Laser Altimeter experiment has provided a high accuracy global geodetic reference frame for the Moon.

The Russian neutron detector observed "local regions with weak neutron emission at the lunar south pole "with very probable hydrogen-enhanced content in the regolith. Somewhat unexpectedly, these regions do not match the permanently shadowed regions (PSR) in the bottom of lunar polar craters, although it was assumed earlier that probable local regions with enhanced hydrogen content should coincide with PSRs in the pole's vicinity."[127]

A video record created by NASA Goddard SFC is recommended to the reader: *"Take a virtual tour of the Moon in all-new 4K resolution, thanks to data provided by NASA's Lunar Reconnaissance Orbiter spacecraft. As the visualization moves around the near side, far side, north and south poles, we highlight interesting features, sites, and information gathered on the lunar terrain."*[128]

2009 NASA Lunar Crater Observation and Sensing Satellite

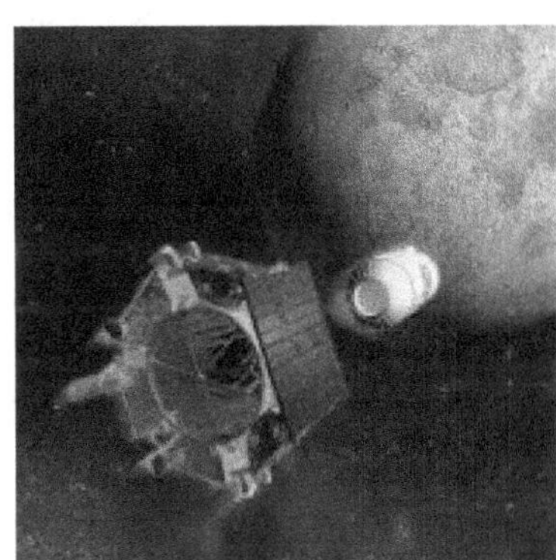

NASA's Ames Research Center's Lunar Crater Observation and Sensing Satellite (LCROSS) was conceived as the most direct means, short of soft landing a spacecraft, of assessing whether water ice remains trapped in the permanently shadowed polar craters. The approach involved directing a massive body at high speed to impact the target crater and, by means of a following instrumented 'shepherd' spacecraft, to fly through and characterize the resulting plume of debris.

The data would be returned to Earth in the few minutes before the shepherd crashed into the Moon – not unlike the Rangers of the 1960s. The impactor was the empty Centaur upper stage of the Atlas V launch vehicle; it had a mass of 2,305 kg, and an impact velocity of about 9,000 km/hour "releasing the kinetic energy equivalent of detonating approximately 2 tons of TNT".[129]

Following 6 minutes behind the shepherd spacecraft was a simple instrumented payload adapter (built by Northrop Grumman) that had attached the LRO to the Centaur. This was the least expensive way that a space mission could be conceived. Dan Andrews of NASA Ames was the Project Manager and Tony Colaprete the Project Scientist.

As noted above LCROSS launched together with LRO on an Atlas V rocket on 18 June 2009. Four and a half days later LCROSS and its attached Centaur made a swing-by of the Moon and entered into a polar Earth orbit that returned it to the Moon a month later for another maneuver to impact the target crater, Cabeus, located about 100 km from the South Pole. The temperature within its shaded region is below 100K such that water ice would persist without sublimating for billions of years. All the LCROSS impact events took place as planned on 9 October 2009; the shepherd spacecraft returned data from which the presence of water ice was confirmed; the concentration of water was determined to be 5.6+/-2.9% – much less than had been anticipated on the basis of previous orbital estimates. It had been hoped that the plume would also be detected and characterized by terrestrial telescopes but that did not happen.

2010 CNSA Chang'e 2

China's second lunar orbiter, Chang'e 2, was similar in most respects to Chang'e 1 but had significant improvements in its instrumentation. Its goals were focused on choosing a safe and scientifically interesting landing site for China's planned first descent of Chang'e 3, scheduled for a December 2013 soft landing. One of the orbiter's cameras was now capable of one-

[127] https://web.archive.org/web/20120406102630/http://l503.iki.rssi.ru/LEND-en.html
[128] https://www.camein.com/moon-nasa-4k-video-information/
[129] https://en.wikipedia.org/wiki/LCROSS

meter resolution from an orbital altitude of 100 km while a second camera would provide broader coverage with 10 meters resolution. Similarly, the laser altimeter was improved to provide 5 meters vertical accuracy.

Chang'e 2 was launched from the Xichang launch site in Sichuan on a Long March 3C booster at on 1 October 2010 and was inserted into a 12-hour elliptical lunar orbit five days later. On 9 October the orbit was lowered to its 100 km science orbit. The primary lunar science mission ended on 8 June 2011 towards the end of which the orbit was lowered to 15 km x 100 km to test the tracking ability of a new X-band ground tracking system.

In an enterprising spirit the mission controllers on 25 August 2011 directed Chang'e 2 to leave lunar orbit and move to the L2 Sun-Earth Lagrange point. Here it lingered until 15 April 2012 and then departed to fly within a few kilometers of the asteroid 4179 Toutatis on 13 December 2012.

2011 NASA GRAIL-A & GRAIL B

NASA's *Gravity Recovery and Interior Laboratory* (GRAIL) mission was designed to address fundamental questions about the Moon that remained puzzling after decades of investigation: the origin of the maria and especially the reason for the near-side/far-side asymmetry. GRAIL would attempt to determine the structure of the lunar interior from crust to core. Such an improved understanding of the Moon's interior would provide planetary scientists with a much better understanding the Moon's evolution. This, in turn, would provide new insights into the evolution of the other terrestrial planets.

Analysis of the tracking data from the NASA's Lunar Orbiter in 1968 had led to the identification of very large positive gravity anomalies associated with lunar basins. NASA's Discovery Program of low cost planetary missions (now managed by NASA's Marshall Space Flight Center in Alabama) at last provided the opportunity to make a definitive map.

The Gravity Recovery and Interior Laboratory mission was proposed by Maria Zuber of the Massachusetts Institute of Technology (MIT) and her team to make a high-resolution gravity map of the Moon using a technique much more powerful than tracking just one orbiter – an approach that had been proven for Planet Earth by the 2002 *Gravity Recovery and Climate Experiment* mission (GRACE): placing two identical spacecraft in essentially the same orbit and tracking the distance separating them by means of a super-sensitive microwave ranging system.

The GRAIL mission, eleventh in the Discovery Program, was managed by JPL; Lockheed Martin Space Systems in Denver built the spacecraft. The Project Manager was David Lehman. PI Maria Zuber of M.I.T. was assisted by Deputy PI David Smith of NASA Goddard Space Flight Center, Zuber and Smith having previously led 12 experiments to map gravity and topography of planetary bodies.

The two small spacecraft GRAIL-A and GRAIL-B were launched on 10 September 2011 aboard a single-launch vehicle: the most powerful configuration of a Delta II. GRAIL-A entered lunar orbit on 31 December 2011 and GRAIL-B followed on 1 January 2012. Thus as GRAIL-A and -B circle the Moon, areas of slightly stronger gravity affect the lead satellite first, pulling it away from the trailing satellite. Then as the two satellites continue, the trailing satellite is pulled toward the lead satellite as it passes over the gravity anomaly. The data enable the team to map the structure of the lunar crust and lithosphere, determine the subsurface structure of the impact basins and their mascons, constrain the deep interior structure of the Moon, and place limits on the size of the Moon's inner core.

The GRAIL team summarized their findings in a paper in the journal *Science* and report that the GRAIL mapping reveals features not previously resolved, including tectonic structures, volcanic landforms, basin rings, crater central peaks, and numerous simple craters. Virtually all of the gravitational signature is associated with topography, a result that reflects the preservation of crater relief in highly fractured crust. The remainder represents fine details of subsurface structure not previously resolved.

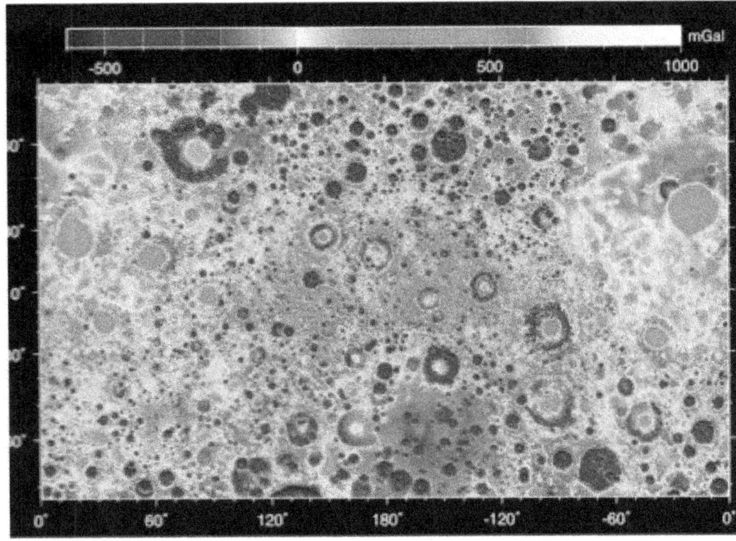

This map shows the gravity field of the moon as measured by GRAIL. The viewing perspective is a Mercator projection and, somewhat confusingly, shows the far side of the moon in the center and the nearside at either side. "The field shown resolves blocks on the surface of about 20 km and measurements are three to five orders of magnitude improved over previous data. Red corresponds to mass excesses and blue corresponds to mass deficits that create areas of lower local gravity. The map shows more small-scale detail on the far side of the moon compared to the nearside because the far side has many more small craters".

The GRAIL data are most informative regarding the hemisphere dichotomy when the data are mapped in the form of Bouguer gravity anomalies. Bouguer gravity is what remains from the gravity field when the attraction of surface topography is removed, and therefore represents mass anomalies inside the moon due to variations in crustal thickness or crust or mantle density. The difference between the lunar near side (left and right images) and the far side (center) leaps out.

The team's analyses have led them to conclude that the major difference between the near side of the Moon with its striking lava-filled basins and the dissimilar far side is the result of a difference in abundance of heat-producing uranium and thorium in the two hemispheres: temperatures on the near side were higher leading to the "vast majority of volcanic eruptions on the near side".[130]

Katarina Miljkovic of the Institut de Physique du Globe de Paris comments that "simulations indicate that impacts into a hot, thin crust representative of the early Moon's near-side hemisphere would have produced basins with as much as twice the diameter as similar impacts into cooler crust, which is indicative of early conditions on the Moon's far-side hemisphere".[131]

---

[130] https://sservi.nasa.gov/articles/nasas-grail-mission-puts-a-new-face-on-the-moon/
[131] https://www.nasa.gov/mission_pages/grail/news/grail20131107.html

The GRAIL maps show gravity gradients due to underlying mass anomalies. The data have revealed long, linear gravity anomalies, with lengths of hundreds of kilometers that crisscross the surface. These evidently indicate the presence of dikes – long, thin, vertical bodies of solidified magma in the subsurface. Jeff Andrews-Hanna, a GRAIL guest scientist with the Colorado School of Mines concludes "the dikes are among the oldest features on the moon, and understanding them will tell us about its early history".[132]

Regarding the thickness of the lunar crust, GRAIL co-investigator, Mark Wieczorek, at the Institut de Physique du Globe de Paris reports:

With our new crustal bulk density determination, we find that the average thickness of the moon's crust is between 21 and 27 miles (34 and 43 kilometers), which is about 6 to 12 miles (10 to 20 kilometers) thinner than previously thought," said. "With this crustal thickness, the bulk composition of the moon is similar to that of Earth. This supports models where the moon is derived from Earth materials that were ejected during a giant impact event early in Solar System history.[133]

The lunar mascons inevitably had the same consequence for the GRAIL orbiters as they had for the Apollo sub-satellites – namely destabilizing the low-altitude orbits enough to cause an early crash into the surface. GRAIL had to perform trajectory correction maneuvers no less than three times a week! The mission ended when two spacecraft impacted the lunar surface on 17 December 2012.

2013 NASA LADEE: Lunar Atmosphere and Dust Environment Explorer

A third spacecraft, NASA's Lunar Atmosphere and Dust Environment Explorer was to have been launched along with the GRAIL twins. Its science goals were relatively esoteric: to study the tenuous lunar exosphere and find out if the Apollo astronaut sightings of diffuse emission at tens of kilometers above the surface were sodium glow or, alternatively, dust. It had long been supposed that the Moon has a tenuous 'atmosphere' of tiny dust particles resulting from the ionization of the lunar soil particles as a result of solar UV and X-rays – such that positive charges build up and lead to repulsive forces that loft particles above the surface. LADEE was intended to acquire data that would test this theory. Perhaps as important, there was also a technology demonstration of two-way laser communication from lunar orbit to be carried out.

NASA likes to create new program names for a series of missions and LADEE was part of the Agency's 'Lunar Quest Program' described as 'Strategic roadmap-initiated lunar robotic flight' and 'Science community prioritized objectives captured in roadmaps and Scientific Context for Exploration of the Moon'. Enough already! The mission was managed by the NASA Ames Research Center with a contract to Orbital Sciences Corporation and major cooperation from NASA Goddard SFC who operated the sensor suite, technology demonstration payloads and also managed launch operations. Butler Hine of Ames was the Project Manager and Richard Elphic the Project Scientist. There were three science instruments: Ultra Violet and Visible Light Spectrometer, Neutral Mass Spectrometer, and Lunar Dust Experiment. In addition the mission included the demonstration of Lunar Laser Communication Demonstration (LLCD) experiments. The launch of LADEE was not carried out until two years after GRAIL took off from the Cape and was, in fact, quite novel in that the launch on 6 September 2013 took place from the small launch complex adjacent to NASA's Wallops Island Flight Facility on the Eastern Shore of Virginia. The Minotaur V rocket had been derived from an ICBM (the lower three solid propellant stages are those of the Peacekeeper Missile and the fourth and fifth are liquid-fueled Star 48 and Star 37 upper stages). Following a somewhat complex transit involving three phasing loops about Earth, LADEE entered lunar orbit on October 6 2013. (The phasing approach requires less propellant than the usual direct approach and provides greater flexibility in terms of launch windows and opportunities). LADEE achieved its operational 75 km circular orbit more than a month later.

---

[132] https://www.nasa.gov/mission_pages/grail/news/grail20121205.html
[133] https://www.jpl.nasa.gov/news/news.php?feature=3613

The Lunar Dust Experiment (LDEX) team noted an increase in dust that is attributed to the Geminids meteor shower. The Ultraviolet and Visible light Spectrometer (UVS) observed an increase in sodium in the exosphere in connection with the Geminid meteor shower, as well as evidence of increased light scattering due to dust. The LADEE team reports that the Moon's atmosphere is dominated by argon, helium and neon – gas species that don't form a haze. The spacecraft, orbiting at low altitude, was able to view the Moon's horizon in complete darkness and observed a yellow haze on the horizon that subsequently expanded into large diffuse glow – the zodiacal light along with a smaller measure of light coming from sun's outer atmosphere or corona. The dust instrument measured only a very tenuous dust cloud around the Moon.

The test of a pulsed laser communication system using three ground stations was successful. The LLCD team demonstrated all-optical acquisition data transfers at 10 and 20 megabits per second uplink, and transfers at 155, 311, and 622 Mbps downlink (rates higher than typical home internet streaming video). All optical contact attempts with the US ground stations at White Sands in New Mexico and Table Mountain in California were successful. The LLCD team was also able to demonstrate the ability to operate through thin clouds, as well as handing off from one ground station to another. The technology is a direct predecessor to NASA's planned Laser Communication Relay Demonstration system. The LADEE mission was short-lived because the small spacecraft lacked fuel to maintain a long-term lunar orbit. So it was intentionally allowed to decay following the mission's final low-altitude science phase impacting the Moon, on 17 April 2014.

2013 CNSA Chang'e 3

Following the successes of the Chang'e 1 and 2 lunar orbiters that provided mapping coverage for landing site selection, China launched its first lander/rover mission to the Moon, Chang'e 3, on 1 December 2013 as part of a robotic program that is planned to lead to sample return. The series of spacecraft are named after *Chang'e* who, in mythology, is the goddess of the Moon while the rover has the name Yutu meaning Jade Rabbit. The spacecraft was launched on a Long March 3B rocket from the Xichang site, reached the Moon on 6 December and landed on 14 December. In doing so it became the first lunar lander since the Soviet Luna 24 sample return mission in 1976.

The Chang'e 3 lander has a mass of 1,200 kg, carries the 140 kg Yutu rover and its design is planned to evolve to be the landing platform for the Chang'e 5 sample return mission presently planned for 2020. Like the rover, the lander is powered by solar panels and both carry radioisotope heaters to help survive the long lunar night during a mission planned to last a year. The science payload comprises seven instruments including three panoramic cameras, an extreme UV camera, a soil probe and a UV telescope. The latter will serve as the first Moon-based astronomical observatory operating in the near-UV band (245-340 nm). It is capable of detecting objects as dim as magnitude 13 and will allow extremely long, uninterrupted observations of a target.

The rover was deployed from the lander on 14 December 2013 and by the 17th the lander had deployed all of the scientific tools. Yutu was designed to explore a 3-km^2 area during its 3-month nominal mission and had a maximum range of 10 km. It is six-wheeled with a payload capacity of about 20 kg. It was designed to transmit video in real time and to perform simple analysis of soil samples. It could cope with inclines and had sensors to prevent it from colliding with other objects. Two solar panels and batteries provided energy. Instrumentation included an alpha particle X-ray spectrometer and an IR spectrometer intended to analyze the composition of lunar samples. In a fresh crater named Zi Wei, Yutu identified a different type of basaltic lunar rock – one that contains a high enrichment of titanium dioxide and olivine.

There are two panoramic cameras and two navigation cameras on the rover's mast, which stands ~1.5 meters above the lunar surface, as well as two hazard-avoidance cameras. On its underside Yutu carried a ground-penetrating radar able to measure structure down to a depth of 30 meters or more. In August 2016 "China's Yutu rover exploring the lunar surface has stopped operating after a record-setting mission of 31 months, the State Administration for Science, Technology and Industry for National Defense (SASTIND) announced".

2014 CNSA Lunar Sample Re-entry Capsule Chang'e 5-T1

China has ambitious plans for further lunar exploration. Chang'e 5 is a sample return mission planned for mid-2020 launch with the challenging goal of returning at least 2 kilograms of lunar soil and rock samples. (Recall: the last robotic sample return mission was the Soviet Luna 24 back in 1976; it returned 170 gm of lunar regolith.) In addition to the technology needed to land on the Moon, to acquire the sample and launch it back to Earth, the Chang'e 5 mission also needs to be able to re-enter the Earth's atmosphere at speeds that are in excess of those involved in returning crew from the Chinese Space Station. To have assurance of this capability the prototype lunar return capsule Chang'e 5-T1 was launched on 23 October 2014 using a Long March 3C rocket onto a free-return trajectory – one that looped behind the Moon and then returned to Earth. The return capsule underwent an unusual high-speed atmospheric skip re-entry; this technique, which calls for precision navigation, involves one or more successive skips off the atmosphere to bleed off the speed in stages to slow the spacecraft before its final entry. Meanwhile the spacecraft's "service module" remained in orbit around the Earth and was then relocated via the L2 libration point to lunar orbit for further tests.

2019 Israel AeroSpace Industries Beresheet

Beresheet ("in the beginning" from Genesis) was a small robotic lunar lander that was launched on a SpaceX Falcon 9 on 22 February 2019 and achieved lunar orbit on 4 April and then circularized its orbit. After maneuvering to carry out a landing the gyroscopes failed on 11 April 2019, the main engine shut off and the lander crashed on the Moon.

2019 India Space Agency Chandrayaan-2

Chandrayaan-2 comprised a lunar orbiter, a lander (*Vikram*), and a rover (*Pragyan*). The main scientific objective was to map surface composition, as well as to identify the location and abundance of water ice. The mission was launched on 22 July 2019 by a GSLV Mk III rocket and achieved lunar orbit on 20 August 2019. The rover was planned to land at about 70° south. On 6 September 2019 the landing sequence commenced but communications were lost and it is assumed that the spacecraft crashed.

The India Space Agency lost no time in planning to carry out make a second effort to carry out its lunar lander/rover mission. The chairman of ISRO, K. Silvan, announced the Agency's approval of Chandrayaan-3 mission on 1 January 2020 with launch as early as the end of the year. The Chandrayaan-2 orbiter remained in good health and will provide the necessary orbital support for the lander and rover.

2021 Roscosmos/ESA Luna 25

As reported by Leonard David in *Scientific American* 27 August 2020, Russia is readying to return to the robotic exploration of the Moon after many decades: flight units of 8 scientific instruments for its Luna 25 moon lander having been delivered from the Space Research Institute (IKI). Landing is to be targeted to the south polar region like other lunar lander missions seeking to determine the character of the frozen water ice in the permanently shadowed crater floors. The article in question quotes Brown University scientist Jim Head commenting on the news from Russia where he, for many decades, has maintained a close link through the regular Brown-Vernadsky microsymposia. A recent microsymposium focusing on China's space exploration was held in March 2018:

> International exploration of the Solar System has grown significantly since the beginning of the Space Age, with missions launched by many nations, and participation by scientists worldwide in planning and data analysis. Microsymposium 59, cosponsored by the Watson Institute for International and Public Affairs, and the Russian Academy of Sciences Vernadsky and Space Research institutes, is dedicated to a focus on the lunar and deep space exploration activities of the People's Republic of China. Microsymposium 59 will be highlighted by reports from a wide range of Chinese university and Chinese Academy of Sciences scientific colleagues and their descriptions of both data analysis and results of previous missions to the Moon. It will also include informal and unofficial presentations on future mission plans and activities for the Moon, Mars, asteroids and beyond. Of particular interest will be on the analysis of data for the Chang'E 1-3 missions and the concepts and landing site selection studies for the Chang'E 4 farside rover and Chang'E 5 nearside mare lunar sample return missions.

Unfortunately, because of the ongoing pandemic the planned March 2020 Microsymposium 61 had to be cancelled as was the annual Lunar and Planetary Science Conference at NASA JSC. Jack Mustard of Brown and Vlada Stamenkovic are now the organizers of the microsymposia. The focus was to have been the subsurface of the planets and moons, a subject of particular interest to the author:

> The third dimension of planetary bodies remains largely unexplored. Pioneering and revealing geophysical measurements have shown some aspects of the interiors of planetary objects. But the subsurface presents many opportunities to expand our knowledge of habitability & resources for human exploration. For example, the surprising discovering of 12-14% porosity kilometers into the Moon by the GRAIL mission is transformative for understanding subsurface habitability of Mars and Early Earth as well as other planets. While the cold, dry surface of Mars with its harsh radiation environments is widely considered to be uninhabitable, the subsurface has been hypothesized to be the longest-lived habitable environment, protected from the harsh surface conditions and a place where water could be stable. Similarly, for Europa the surface temperature and radiation conditions would not support life, yet the promise of subsurface habitability is huge. On the Earth, we are learning more and more of the vast world of Life Underground from the diverse yet largely unexplored biology to the prospects of billion-year old groundwater. And we will be needing subsurface exploration to determine the availability of useful resources on the Moon and on Mars for human exploration. The field of "Planets Underground" poses many exciting science puzzles and questions, and technological challenges to gain access to measurements relevant to these questions.[134]

2020 Chang'e 5 Lunar Sample Return

China's next lunar mission, ongoing at the time of writing, has taken the considerable step of returning samples -- something last successfully carried out by the Soviet Union in August 1976. On 1 December 2020 the Chang'e 5 lander

---

[134] http://www.planetary.brown.edu/html_pages/micro61.htm

touched down on Oceanus Procellarum, 43.1N, 51.8W in the northern hemisphere of the near side of the Moon, east of a volcanic plateau named Mons Rümker. This mission was launched out of the Wenchang spaceport on November 23 on a heavy-lift Long March 5 toward capture into lunar orbit on 28 November. The lander detached from the orbiter & the return section and carried out a successfully landing.

After collecting samples the ascent vehicle launched, rendezvoused & docked with the orbiter. The container of rocks was transferred to the Earth return element to bring home in mid-December. At the time of writing the return vehicle is expected to fire its engines to leave the Moon's orbit on December 13, "setting course for landing of the sample capsule in China's Inner Mongolia region a few days later."

A Lunar Olive Branch

Regarding China's plans for analysis of the two kilos of samples, Xu Yansong, who is Director-General of the Asia-Pacific Space Cooperation Organization and Director for International Cooperation at China's National Space Administration, has responded in an interview with Stephen Cole, host of the China Global Television Network's talk show, *The Agenda*, as follows:

"There's a great interest in the international community to study these samples because of the geographic locations," explains Xu. "This region is the newest region on the moon, some one billion years old." That's much younger than the samples brought back previously by the U.S. and Soviet Union, which were three to four billion years old.

"So these samples are very useful for telling the moon's characters – the volcanic activities and the later-stage moon activities. We will distribute the sample to the international community."

"Xu underlines the importance of nations working together as humans start to look beyond their planet again: Not so much a space race as a joint quest.

"Global cooperation is essential for exploration missions," he insists. "The moon is not far away from Earth, we can have independent missions – but to go further, we need to collect all the strength that we have on Earth."

Mentioning previous cooperation with the European Space Agency – Sweden, Germany, France and Belgium had instruments and payload on board the earlier Chang'e-4 mission – Xu says China is "looking forward to more extensive international cooperation."

That includes the U.S. now that the divisive era of President Donald Trump is over. "With the new administration, I believe that the bilateral cooperation between China and the U.S. will continue," Xu says. "Globally speaking, China is fully open for international cooperation."[135]

To his surprise this point of view from China matches that of the author. One might hope that, if the US Congress would revisit its restriction on US-China collaboration in space, then some kind of cooperation between the US and China might

---

[135] https://newseu.cgtn.com/news/2021-02-01/China-s-lunar-samples-a-symbol-of-the-new-international-cooperation-Xs4JGhGCZy/index.html

take place like that earlier demonstrated by the US and Russia following the lunar space race.   NASA, JAXA, ESA and the Canada Space Agency all collaborate in the building, supply and operation of the International Space Station. Toward this end perhaps the new NASA Administrator, whoever he or she is, might seek the advise of former Administrator Dan Goldin. Incidentally, as it happens, the Co-Chair of President Biden's Council of Advisors on Science and Technology is a lunar science expert: Maria Zuber who led the GRAIL mapping mission described in Chapter 11.

CHAPTER 12

RETURN TO MARS 2: ORBITERS & ROVERS

Mars Odyssey, Mars Express & Beagle 2, Mars Exploration Rovers Spirit & Opportunity

As described in Chapter 10, after the causes of the Mars Climate Orbiter and Polar Lander failures had been very publicly evaluated in the Young Report, Administrator Goldin acknowledged that he had been pushing too hard and a new approach to Mars exploration was needed.

Meanwhile, the 2001 launch window was approaching and, under the now-suspended Surveyor program plan, an orbiter and a lander were being built for launch. Given the still uncertain nature of the Polar Lander failure the decision to postpone the lander launch was not difficult to make. The decision to continue with the launch of the orbiter, given that the problem had just been the bizarre mix-up of navigation units, was also straightforward. Needed was a replacement for the Mars Surveyor Program. To develop a new strategy a series of workshops was held and an advisory committee – the Mars Exploration Program Analysis Group (MEPAG) – was formed (and is still active, managed by JPL).

Administrator Goldin's HQ team to develop the new policy included Ed Weiler, Associate Administrator for Space Science, Scott Hubbard Mars Program Director and Jim Garvin, a planetary scientist at nearby NASA Goddard Space Flight Center. At JPL Ed Stone was Director of the Laboratory and Firouz Naderi was the point of contact for Mars mission planning. Viking alumni Jim Martin, Gentry Lee and Michael Carr were consultants in the process. With the search for evidence of past or present life as the overarching goal, *Follow the Water* became the general Mars science strategy. Mars Sample Return was postponed indefinitely. There were to be six missions spread out over the first decade of the 21st Century. Mars Odyssey Orbiter and the two Mars Exploration Rovers were already 'on the books' and they were to be followed by a very high-powered Mars Reconnaissance Orbiter, and then a mobile Mars Science Lander ('Curiosity'). The two solar powered Mars Exploration Rovers would be launched in 2003. As W. Henry Lambright explains in his book *Why Mars*, this doubling up represented an insurance policy and was the initiative of Administrator Goldin because he regarded the success of these Mars missions to be essential in order to establish the path toward an eventual Mars Sample Return mission – and that, in turn, was an essential step in his mind toward the eventual manned exploration of Mars.[136]

The new Mars exploration plan was successfully negotiated with the Office of Management and Budget in this the last year, 2000, of the Clinton presidency (Clinton was succeeded on 20 January 2001 by George W Bush). As discussed in Chapter 10, the early years of the George W Bush Administration with Sean O'Keefe as NASA Administrator were years of potential excitement for NASA. Still, this heavy emphasis on Mars exploration in the wake of the ALH 84001 enthusiasm understandably led to some concern in the science community about the balance in the planetary exploration program.

---

[136] *Why Mars*, W. Henry Lambert, 2014, Johns Hopkins University Press, p.163-167

Mars Hydrosphere

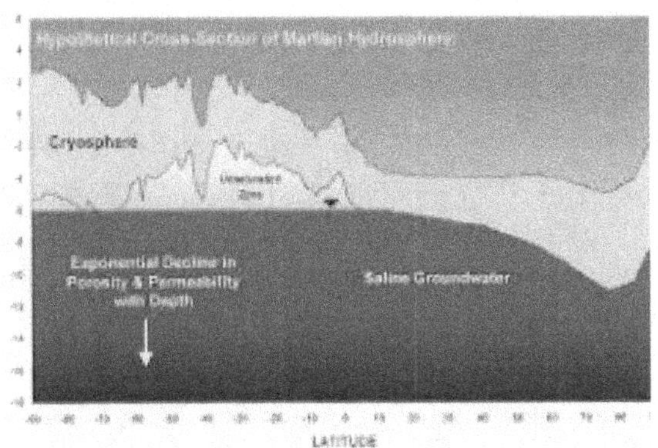

In May 2001 Steve Clifford of the Lunar and Planetary Institute and Timothy Parker of JPL published a lengthy theoretical analysis in the journal *Icarus* on the evolution of the Martian hydrosphere – a follow-up to Clifford's earlier paper on the subject.[137]

They analyzed the hydraulic conditions required to explain the origin of the Chryse outflow channels discovered by Mariner 9 assuming that, as seems clear, the water was derived from a sub-permafrost aquifer. They conclude that "throughout the planet's first billion years of evolution, as much as one third of its surface was covered by standing bodies of water and ice". Further, "the bulk of this water would have existed as an ice covered ocean in the northern plains".[138] The several orbiters and landers that were directed to further explore Mars in the next two decades would be looking for evidence to support this proposition.

2001 NASA Mars Odyssey

The Mars Surveyor 2001 Orbiter, renamed 2001 Mars Odyssey, was launched without the second launch of an accompanying Surveyor lander. The Lockheed Martin Corporation had already built this now-orphaned lander and, except for the solar arrays, the basic design was identical to that of the Mars Polar Lander. The effort and expense would not be a waste because the spacecraft was maintained in storage and became the basis of the Phoenix mission launched six years later.

Toward the end of 2001 there was a changing of the guard at NASA when Sean O'Keefe became NASA Administrator after Dan Goldin stepped down following a record ten years in that position. The new President, George W Bush appointed Mr. O'Keefe who had been Secretary of the Navy and, most significantly for NASA, had been Director of the Office of Management and Budget (OMB) through which all Federal budget submissions must pass on their way to Congress. This was an encouraging sign for the Agency that soon had to deal with major cost overruns in the construction of the International Space Station and subsequently the aftermath of the STS Columbia disaster on 1 February 2003. The 2001 Mars Odyssey orbiter would turn out to provide much needed relief to NASA, following the 1998 disasters, as a result of its great success. It was launched on 7 April 2001 and again used aero-braking to save fuel mass in order to fit on a Delta II. Remarkably, the orbiter is still operating at the time of writing in 2020 – continuing to map and monitor Mars and also serving as a relay to landed spacecraft.

Lockheed Martin built the spacecraft. The Project is managed by JPL. David Lehman is the Project Manager and Jeffrey Plaut the Project Scientist. The spacecraft is relatively lightly instrumented with just three experiments: a Thermal Emission Imaging System (THEMIS), a Gamma Ray Spectrometer (GRS) that includes a High Energy Neutron Detector (HEND) provided by Russia's IKI (the result of an implementation agreement between Rosaviakosmos and NASA), and a Mars Radiation Environment Experiment (MARIE). Principal Investigator Bill Boynton's Gamma Ray Spectrometer is based on the spare from the Mars Observer lost in 1993 and has the same elemental mapping objectives as before. The GRS/HEND measures the characteristic gamma rays and neutrons that are emitted from the Martian surface atoms as a result of the continual bombardment by high energy cosmic rays. Twenty different heavy elements plus hydrogen can be detected in this way including silicon, oxygen, iron, magnesium, potassium, aluminum, calcium, sulfur, and carbon. (Russia has provided a similar capability for the Mars Science Laboratory (Curiosity) now exploring Gale Crater.) The neutron detectors are sensitive to concentrations of hydrogen (and thus bound water and ice) in the upper meter of the surface.

---

[137] Stephen Clifford and Timothy Parker: The Evolution of the Martian Hydrosphere: Implications for the Fate of a Primordial Ocean and the Current State of the Northern Plains, *Icarus* 154, May 2001
[138] https://en.wikipedia.org/wiki/Water_on_Mars

The Thermal Emission Imaging System (THEMIS) of PI Phil Christensen is a visible and infrared camera that neatly complements the mapping of surface minerals carried out by Christensen's Thermal Emission Spectrometer on the Mars Global Surveyor (Chapter 11) – effectively increasing the spatial resolution of the spectroscopy. The IR camera has a resolution of 100 meters/pixel and operates in nine spectral bands. Both water and ice are strongly absorbing in the IR wavelength bands THEMIS uses. The visible wavelength camera has a spatial resolution of 19-meters/pixel.

Looking ahead to an era when astronauts may journey to and explore Mars, Mars Odyssey is equipped with the Mars Radiation Experiment (MARIE) to measure the space radiation environment – solar wind particles and galactic cosmic rays. Radiation damage to the central nervous system of an astronaut is the biggest potential showstopper for successfully carrying out crewed missions to Mars so its characterization is essential. NASA Johnson Space Center scientists led the experiment.

Mars Odyssey reached Mars in October 2001, was captured into a highly elliptical polar orbit and spent the next 90 days aero-braking into a Sun-synchronous orbit of 2-hour period. The spacecraft remains operational in 2020! – The longest serving spacecraft at Mars. Ample time has been available to carry out the elemental mapping of the surface including the neutron spectrometer's mapping of hydrogen. The results have been enlightening:

Ice is both widespread and abundant on the modern surface. Below 60 degrees of latitude, ice is concentrated in several regional patches, particularly around the Elysium volcanoes, Terra Sabaea, and northwest of Terra Sirenum, and exists in concentrations up to 18% ice in the subsurface. Above 60 degrees latitude, ice is highly abundant. Pole-wards of 70 degrees of latitude, ice concentrations exceed 25% almost everywhere, and approach 100% at the poles.

The science team calculates that today the water ice in the two polar regions corresponds to a water equivalent global layer of 30 meters. They estimate that originally there was 500 meters of surface water – now mostly lost to space and to the deep subsurface. The gamma ray spectrometer has mapped the abundance of other surface elements. PI Boynton comments, "We see one element, potassium, which is about twice as abundant as it is on Earth. This fact was known in advance based on meteorite studies, but was confirmed by the GRS data."[139]

This mapping of element abundances complements the mineralogical mapping that is accomplished with IR instruments like the Mars Global Surveyor's thermal emission spectrometer and Odyssey's THEMIS. A wide variety of different minerals have been mapped including different basalts, high-silica igneous rocks and quartz-bearing rocks. Olivine basalts (indicative of a dry climate) were observed on crater floors and in canyon wall layers. The quartz-bearing rocks were observed in the central uplifts of craters, perhaps once buried deep below the surface and raised by the impact event. The data from THEMIS have proved particularly valuable in selecting landing sites for probes that followed including the most recent: Mars Science Laboratory, Curiosity.

The THEMIS thermal images are used to track the advance and retreat of the Martian seasonal polar caps taking advantage of the temperature contrast between soil and ice. Rose Hayward of the USGS has used 40,000 thermal images to construct the Mars Global Digital Dune Database that initially included 550 dune fields between latitudes 65°N and 65°S and now has expanded to include the entire planet. Another use is to look for atypical pit craters that are nearly cylindrically and deeper than wide – some may provide access to underground caves.

The radiation experiment MARIE provided data for two years but, ironically, was knocked out of action by a large solar event on 28 October 2003 – likely the consequence of damage to its computer circuitry. That event in itself provides useful information about the radiation environment at Mars.

Besides its own science contribution, Odyssey has been a critical support element for NASA spacecraft on the Martian surface: it has served as the primary means of communications for Mars surface explorers – most of the images and data from Spirit and Opportunity have been linked through Odyssey. Long-lived beyond any reasonable expectation, the spacecraft now acts as a relay for UHF radio signals from Curiosity.

---

[139] http://marsmobile.jpl.nasa.gov/odyssey/news/whatsnew/index.cfm?FuseAction

2003 ESA Mars Express & Beagle 2

Following its Comet Halley mission (Giotto) and its Titan probe (Huygens) the European Space Agency entered the Mars exploration business in 2003 with a combined orbiter and lander mission: *Mars Express* and *Beagle 2* (the latter a British contribution). The orbiter carried several instruments that had been built for the lost Russian Mars 96 mission and pursued many of the same science objectives as the NASA orbiters. Mars Express also planned subsurface sounding by radar, precise measurement of atmospheric composition and study of the upper atmospheric interaction with the solar wind.

The goal of Beagle 2 was to search for signs of life, past or present. Mars Express served as the mothership for Beagle 2, named, of course, for Charles Darwin's vessel. Colin Pillinger (1943-2014) of the UK's Open University and his colleagues at the University of Leicester had conceived the mission. Astrium was responsible for program management, Martin-Baker for the entry, descent and landing system. Though small and with a modest budget Beagle 2 was well instrumented: stereo cameras, microscope, Mossbauer spectrometer, X-ray spectrometer, mass spectrometer and gas chromatograph. There was a small drill, a robotic arm and a remarkable 'mole' that could burrow into the regolith to acquire a sample from as much as 1.5 meters depth for retrieval by winching back the power cable.

2003 Beagle 2

Mars Express with Beagle 2 lifted off from the Baikonur Cosmodrome on 2 June 2003 on a Soyuz-Fregat. Beagle 2 was successfully deployed on approach to Mars on 19 December 2003 with landing planned for Christmas Day. Alas, there was no contact from the surface. The nature of the problem was only understood a dozen years later when Beagle 2 was captured in an image taken by the NASA Mars Reconnaissance Orbiter showing that, after landing successfully, two of Beagle 2's solar panels had failed to deploy – blocking the communication antenna.

This mechanical problem was, obviously, an enormous disappointment for Professor Pillinger and his team as well as those looking forward to the results of the first subsurface sampling. Landing a spacecraft on Mars is, as several space agencies (including NASA) now know to their cost, an exceptional challenge and learning experience. Given the modest cost of Beagle 2, given the clear new understanding of the technical problem, and given its unique subsurface sampling capability, a second attempt should surely be made.

2003 Mars Express

The prime contractor for Mars Express was Astrium in Toulouse, France. Rudy Schmidt is the Project Manager, Agustin Chicarro was the first Project Scientist. The spacecraft – still operational in 2020 – carries seven experiments: High Resolution Stereo Camera (PI R. Jaumann); Energetic Neutral Atoms Analyser (PI S. Barabash); Planetary Fourier Spectrometer (PI M. Giuranna); Visible and Infrared Mineralogical Mapping Spectrometer: OMEGA (PI Jean-Pierre Bibring); Mars Advanced Radar for Subsurface and Ionospheric Sounding: MARSIS (PI G. Picardi); Mars Radio Science Experiment (PI M. Patzold); Ultraviolet and Infrared Atmospheric Spectrometer (PI F. Montmessin).

The spacecraft arrived at Mars and was inserted into a highly elliptical near-polar orbit on 20 December 2003. Some months later the orbit was adjusted into a still-elliptical, 6.7-hour period orbit with a 300 km periapsis. Deployment of two 20-meter booms to create a 40-meter dipole antenna for the MARSIS experiment (the first of its kind since the Lunar Sounder of Apollo 17) proved a complex challenge after they initially failed to lock in place. The long booms are needed in order to operate the radar at very low frequency thereby to penetrate the surface and return echoes from depths of several kilometers – the first echo being from the surface of Mars. The weaker signals that are returned instants later serve to detect subsurface interfaces; the time delay between the two signals provides the measure of depth. The radar sounding was carried out only while the spacecraft was less than 800 km above the surface; this lasted about 26 minutes on each orbit. An unexpected finding was that ancient impact craters 130 to 470 km in diameter lie buried beneath the smooth, low plains of the northern hemisphere. Their origin may well be related to an ocean that once likely covered much of the northern hemisphere of Mars. MARSIS also probed the layered deposits that surround the north pole of Mars and detected two strong echoes that correspond to 1. surface reflection and 2. subsurface interface between two different materials. The MARSIS team infers that they are observing a nearly pure, cold water-ice layer that is thicker than one kilometer and that overlies a deeper layer of basaltic regolith. It had been supposed that there might be a melt zone at the base of the layered polar deposits but this observation appears to rule that out.

In further tentative evidence of an ocean in the northern hemisphere, the MARSIS team has identified a difference in the dielectric constant of the surface in the north and the south at high latitudes. Further mapping of the layered terrain of the polar regions and of subsurface ice elsewhere on Mars would await the later flight of the Mars Reconnaissance Orbiter equipped with a radar operating at shorter wavelengths: the Shallow Subsurface Radar (SHARAD) – an instrument also provided by the Italian Space Agency whose Principal Investigator is Giovanni Picardi from Universita di Roma.

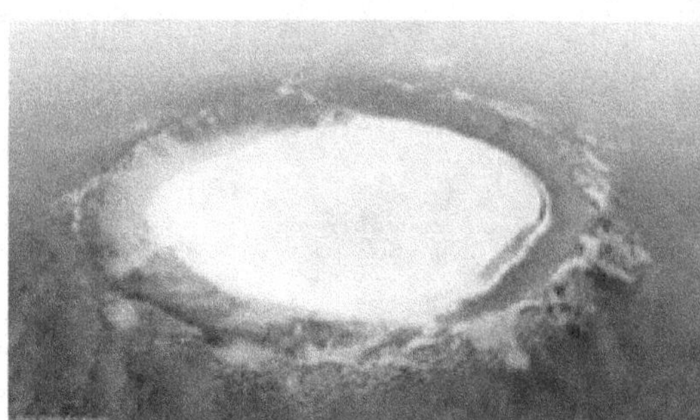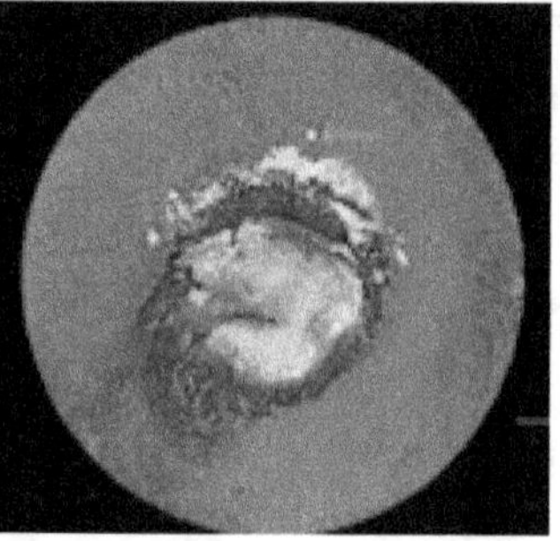

In July 2018 an important discovery by MARSIS was made namely of a subglacial lake 1.5 km below the southern polar ice cap and about 20 km wide. And in December of that year Mars Express returned images of 80 km wide Korolev Crater, located at 73°N, 165°E that is filled with ice. Water ice. The crater is located on the northern polar plain that surrounds the summer residual ice cap and its rim rises about 2 km above the plains. The Mars Express team concludes that the water ice is stable "because the crater acts as a natural cold trap."[140]

However, the subglacial lake discovery has a puzzling element as Steve Clifford points out in an email communication (January 2020) with the author:

if it is indeed a subglacial lake (and there is still an undercurrent of debate over this) – its shallow occurrence is a significant surprise. Given the local surface temperature, the most plausible estimates of the thermal conductivity of the ice and underlying crust, and the geothermal heat flow, you'd expect that you wouldn't get above freezing temperature until you were at depths of ~20 km below the ice surface. To see a subglacial lake at a depth of 1.5 km means that there is something very unusual about this spot. The MARSIS team argued for strong freezing point depressing salts, but I think they're mistaken. While some salt may get carried into the cap along with atmospheric dust, the amount would be small. And if the salt was present in sufficient quantity to cause melting, it would leach out with the meltwater as it infiltrated deeper into the crust. I think the most plausible explanation for a shallow lake is an anomalous hot spot – an area of enhanced geothermal heat flow associated with an igneous intrusion at depth or subglacial volcanism. While there is no direct evidence of such activity at this particular location, we do see evidence of it elsewhere in the nearby Dorsa Argentia Formation where there is also geomorphic evidence of subglacial melting in the form of eskers.

Steve Clifford has this additional comment on the MARSIS results:

My biggest disappoint with MARSIS is that, in lithic environments (i.e., outside of the Polar Layer Deposits and the Medussa Fossae Formation – which is believed to be a several km accumulation of high porosity dust), the penetration

---

[140] https://en.wikipedia.org/wiki/Korolev_(Martian_crater)

depth of MARSIS has been very low – maybe 200m at best. So it has not been able to penetrate the subsurface deep enough to investigate the presence of present-day sub-permafrost groundwater – which, if it exists, is most likely present at depths >4-5 km at the equator and >10-20 km at higher latitudes. Of course, with the higher geothermal heat flow expected in the past, groundwater may have existed at much shallower depths. I think the detection of methane in the Martian atmosphere could be indirect evidence of the past or present occurrence of groundwater in that, whether the methane has a biotic or abiotic origin, both require the formation of methane in the presence of liquid water. As that methane was produced, it could be vented directly to the atmosphere via fractures in the crust or (what I believe is more likely) incorporated into the cryosphere as gas hydrate – where any near-surface hydrate may become destabilized by climate and seasonal variations in temperature that result in methane release to the atmosphere.[141]

The orbiter's OMEGA instrument returned new information about the composition of the permanent ice cap at the south pole, demonstrating that, as well as carbon dioxide ice, there is water ice too in the form of permafrost made up of a mixture of soil and water ice – appearing dark in visible light images. The Mars Express team concluded that the south polar region is made up of not only the bright residual ice cap that is an 85% mixture of carbon dioxide ice and 15% water ice but also spiral scarps that are made of water ice. In addition, permafrost fields stretch for tens of kilometers away from the scarps. Mars Express carries a very sensitive Planetary Fourier Spectrometer that is used to measure the composition of the atmosphere. The team has determined that, at the level of 10 parts per billion, the atmosphere contains methane (CH_4). On Earth this gas is the result of volcanic or hydrothermal activity – but some also is the result of biologic sources (marsh gas and bovine flatulence). Michael Mumma of NASA's Goddard Space Flight Center and his colleagues also detected methane in the atmosphere of Mars using an IR spectrometer at NASA's InfraRed Telescope Facility (a 3m telescope optimized for the infrared. It was first built to support the Voyager missions and is now the US national facility for infrared astronomy) on Mauna Kea in Hawaii. "The methane was localized into clouds, or "plumes," over certain regions of the Martian surface, with the maximum density of methane reaching about 60 parts per billion."[142] As reported in *Astrobiology Magazine*: "Another ground-based observation with the Canada-France-Hawaii Telescope detected methane, but it didn't have enough spatial resolution to see plumes".[143]

Mumma's group performed their own follow-up observations in 2006. But in the three-year interim, all signs of the methane had disappeared, as reported in a 2009 *Science* paper. The implication is that the methane is a seasonal occurrence, perhaps only coinciding with summers on the Red Planet. However, the methane detection was controversial: Kevin Zahnle of NASA Ames Research Center has argued that the telescopic and orbital observations have been misinterpreted because modelling of the photo-chemistry in the Martian atmosphere leads to an estimate that a methane molecule should survive about 300 years before photochemical processes destroy it. Consequently, monthly variations of methane abundance are difficult to explain.

Incidentally, for those with long memories, the original but short-lived discovery of methane in the Martian atmosphere goes all the way back to the flyby of Mariner 7 in 1969 and the measurements made by the spacecraft's infrared spectrometer. The claim was retracted soon after by Principal Investigator George Pimentel (1922-1989) who explained that: "the anomalous lines were more likely due to CO_2 ice when it has lattice imperfections".[144]

The next opportunity to make progress on this intriguing question would be from late 2017 onwards when ESA's ExoMars Trace Gas Orbiter (which arrived in October 2016 but needed to aerobrake for months to achieve its operational orbit) began science operations with the explicit goal of mapping sources of methane and other trace gases. In doing so, the observations will help select the landing site for the ExoMars rover to be launched in July 2020.

---

[141] Email communication from Clifford to the author, 3 January 2020
[142] https://www.astrobio.net/mars/methane-debate-splits-mars-community/
[143] ibid.
[144] http://www.bibliotecapleyades.net/marte/esp_marte_34.htm

2003 NASA Mars Exploration Rovers: Spirit and Opportunity

 On 10 June 2003, eight days after Mars Express lifted off, the first of two NASA missions was launched from the Cape to Mars. These are perhaps the most successful example of Mr. Goldin's "faster, cheaper, better" philosophy. First on its way was the Mars Exploration Rover named *Spirit*. Its twin, called *Opportunity* launched on 7 July 2003. They were both elements of NASA's Mars science strategy summarized as "Follow the Water" and were intended to "search for and characterize a variety of rocks and soils that hold clues to past water activity. In particular, samples sought will include those that have minerals deposited by water-related processes such as precipitation, evaporation, sedimentary cementation or hydrothermal activity."[145] The mission of Spirit ended in 2010 while Opportunity continued till 2018.

JPL, having designed and built the rovers, managed and operated the mission. The rovers are very much bigger and more capable than little Sojourner; adopting the same airbag landing system, both benefited from that earlier success. At JPL the Project Manager was John Callas, the System Engineer Robert Nelson and the Project Scientist Matthew Golombek. The Science Team Principal Investigator was Steve Squyres of Cornell University who was supported by a large team of Co-investigators.

The mission's goal was to make progress toward determining whether life had ever gained a hold on Mars and, potentially, to help choose landing sites for an eventual sample return mission. The probes were targeted to widely separate equatorial locations. Spirit was targeted to the crater Gusev that, it was thought, had once been flooded by the wide channel, Ma'adim Valles (Ma'adim is the Hebrew name for Mars) that runs from the south into the crater. Opportunity was targeted to Meridiani Planum where grey crystalline hematite – often formed on Earth in hot springs – had been identified by the Thermal Emission Spectrometer (TES) on NASA's Mars Global Surveyor. TES showed that portions of Meridiani contain up to 20 percent gray crystalline hematite at the surface. Hematite is an iron-oxide mineral, and on Earth the gray crystalline variety forms mostly in association with liquid water.

Each cruise to Mars took place in an aero-shell attached to a cruise stage that was jettisoned shortly before atmospheric entry. Spirit and Opportunity both followed the same series of events on their way to a cushioned airbag landing: entry behind a heat shield, a parachute that was opened 6 minutes later, then, in the final stage of the descent airbags were inflated by gas generators, and a retrorocket fired to reduce speed. It had been only late in the MER development that the team realized that winds during descent could impart a surface-relative horizontal velocity to the landing system potentially sufficient to cause the airbags to rip and tear. Although the landing system could reduce horizontal velocity by pointing the final deceleration thrusters it did not have a sensor for measuring horizontal velocity. It was not feasible at this point to add a radar velocimeter but a down-looking camera was relatively easy to accommodate. The required software was quickly developed and the system field-tested in the Mojave Desert using a piloted helicopter. At an altitude of 2,400 meters, the camera took three pictures of the ground about four seconds apart and automatically analyzed them to estimate the spacecraft's horizontal speed. In the case of Spirit the wind was indeed of sufficient strength that the landing could have been jeopardized. Accordingly a secondary propulsion system fired during Spirit's descent to take out most of the horizontal speed. After the airbags were inflated the lander was released to freefall to the surface and bounce its way to a standstill about 300 meters from first impact. Spirit landed on 3 January 2004. Opportunity followed three weeks later, landing on 25 January 2004 and bounced its way into a small crater (informally known as Eagle Crater – a golfing term), just 22 meters in diameter and 10 meters deep.

Spirit and Opportunity were six-wheeled rovers like Sojourner. They were powered by solar panels feeding lithium ion batteries; each rover was 1.5 meters high, 2.3 meters wide, and 1.6 meters long, weighing 180 kilograms. Each of the six wheels had its own motor. The rovers could travel as fast as 5 cm/sec but, when moving, averaged less than 1 cm/sec. In addition to their several cameras, science instrumentation included a miniature thermal emission spectrometer (Mini-TES)

---

[145] http://mars.nasa.gov/mer/overview/

and, on an arm, a Mossbauer spectrometer, an alpha-particle X-ray spectrometer, magnets to collect magnetic dust particles and a microscopic imager. The arm also carried a rock abrasion tool to clean the surface of rocks for better analysis.

The Mossbauer spectrometer from Germany was used for close-up investigations of the mineralogy of iron-bearing rocks and soils; the alpha particle X-ray spectrometer, also from Germany, made elemental analyses. The Mini-TES was located, not on the arm, but in the body of the rover; scanning mirrors located in the Pancam Mast Assembly acted as a periscope to send light down to the instrument. Mini-TES determined the mineralogy of rocks and soils by detecting their patterns of thermal radiation (all objects emit heat but different minerals emit heat in characteristic ways).

Gusev Crater did not prove to be the ancient lakebed that had been supposed. Spirit found itself exploring a plain strewn with igneous rocks and marked with impact craters. Ma'adim Vallis, one of the largest Martian outflow channels (it is over 20 km wide and 2 km deep in some places) flows into the crater from the south and presumably did at one time deposit sediments in Gusev.

Evidently any such sedimentary rocks were subsequently covered over by volcanic materials – probably from the nearby volcano Apollinaris Mons further to the north. In some locations there were only suggestions that water may have entered cracks in the surface leading to weathering. But further along were hills where Spirit identified exposed bedrock that had suffered "extensive alteration by water". Along the way Spirit traveled to the rim of Bonneville Crater (another informal name agreed upon by the science team). Then it was on to Columbia Hills at a daily rate of about 100 meters (recall, the Martian day or sol is ~40 minutes longer than our day) while checking out the soils and rocks on the way. Spirit traveled along a path pre-planned by the science team but the rover also used its camera and on-board software to avoid obstacles. By 11 June 2004 Spirit had reached Columbia Hills and by early August had climbed to the ridge where, for the first time, it found exposed bedrock.

In 2005 Spirit found what the team considered more compositional evidence of a wet environment – but one that may have been more favorable for biology than the acidic conditions identified earlier. The rock in question was given the name Comanche and its composition was found to include a large amount of magnesium iron carbonate, a mineral that originates in wet, near-neutral conditions. Spirit identified a variety of rock types in the Columbia Hills with significantly different compositions. All show various degrees of alteration due to aqueous fluids – they are enriched in phosphorus, sulfur, chlorine, and bromine. These rocks contain varying amounts of olivine and sulfates in proportions consistent with an aqueous environment in that the olivine abundance varies inversely with the amount of sulfates.

For both rovers, sustaining enough power to provide mobility and to run the science experiments became more difficult with time as dust gradually accumulated on the solar panels. The winter months were, of course, particularly challenging so during that season they mostly remained stationary. The Martian atmosphere is rarely dust-free as was well known from the Mariner 9 and Viking experiences. In June 2007 and again in late 2008 both rovers encountered the fallout of a long-lived global dust storm that could have been fatal to their extended missions. The power output of the solar panels fell to a level where the landers had to begin to drain their batteries to maintain minimal temperatures for their electronics and other systems. So, both Spirit and Opportunity were put into a survival mode; fortunately conditions improved sufficiently that by early August it was clear that both would be able to resume their missions. Incidentally, even when Mars is largely free from the fallout of such storms, dust in the atmosphere is maintained by dust devils, one of which was beautifully captured by Spirit in a short video. On another occasion in March 2005 Spirit woke up one morning to find that its solar panels had been cleaned by just such a dust devil. This was not, obviously, something that the team could rely upon.

Having six wheels rather than, say, four proved to be a positive feature of Spirit's design. One of the front wheels proved troublesome and by March 2006 the wheel had stopped working entirely. Spirit was at the time exploring next to an eroded deposit of volcanic ash dubbed Home Plate in the Columbia Hills. At times the team drove the rover backwards, dragging the wheel. This disability created one serendipitous discovery: in March 2007, the dead wheel scraped off top soil and in doing so uncovered a bright patch of silica-rich material that on Earth would be associated with hot-springs and geysers like those at Yellowstone National Park. This was an important discovery at the end of Spirit's journey. Gusev Crater had turned out to be not a lakebed as expected but it had proved to be important in other ways.

In May 2009 Spirit became stuck in soft soil and there it would remain in spite of many months of efforts by the operations team to understand and fix the problem. By January 2010 the decision was made to accept the situation; Spirit became a stationary weather station, one that found itself in a poor solar panel orientation. Power ran down and the last communication was on 22 March 2010. It had landed six years earlier, had traveled ten times further than its design goal and had emerged from two potentially fatal dust storms.

Spirit had conducted a remarkably successful mission but Mars had not cooperated quite as much as the science team had hoped in terms of 'following the water'. By contrast Opportunity hit the jackpot immediately on landing and in its continuing operations (to everyone's amazement and satisfaction it remained operational until 2018 – clearly the rover has led a charmed life).

The airbag landing on 25 January 2004, as noted, bounced Opportunity into tiny Eagle Crater within which, remarkably, is an outcrop of exposed layered bedrock. Embedded in this layer, and scattered in the soil, are blueberry-sized gray spheres that contain hematite (the mineral form of iron oxide Fe_2O_3) – not a total surprise as orbital identification of the mineral had been the motivation for choosing to land on Meridiani Planum. The outcrop also was determined to be rich in sulfate-salt minerals, a finding indicating that the rock "had been drenched with salty water".[146]

Rounded shapes can be produced by meteor impacts and volcanic eruptions as well as by concretions that are formed when minerals come out of solution in water. The Martian spherules are typically 0.3 cm in size and are similar to ones in samples returned from the Moon by the Apollo 12 astronauts suggesting that there they were formed as a result of an impact. On Meridiani Planum Opportunity has discovered spherules both on the surface and deeper in the soil – spread evenly inside rocks. The science team interprets this distribution to be more consistent with a water-related origin rather than volcanism or impact since, in those cases, the spherules would, more likely, be stratified in layers. A rippled bedding pattern was observed "in some of the finely layered rocks, indicating that the rocks were not only exposed to water after they formed, but actually formed from sediment particles laid down in flowing water."[147]

Onward and upward: on 22 March 2004 Opportunity was able to climb out of tiny Eagle Crater without much difficulty and then ambled slowly eastwards to a more sizeable crater (130 meters diameter and 20 meters deep) given the name Endurance. It reached the crater rim on 30 April having set a one-day record driving distance of no less than 141 meters! Opportunity then spent the next half-year (mid-June to mid-December) exploring this impact crater. From the crater rim the team surveyed the interior and decided on their exploration strategy. They chose to investigate layering that they had observed at a site called *Karatepe* and were able to infer the relative ages of the layers there. On 20 March 2005 (Sol 410) Opportunity set another record for the longest single-day drive: 220 meters. This, however, was too good to be true: a month later, and for a number of weeks, Opportunity came to a grinding halt – stuck in a sand dune. Some years later, in May 2009, Spirit would become stuck in soft soil that effectively ended its mission. The Opportunity mission scientists found that all four corner-wheels were dug in more than a wheel radius in the process of attempting to climb over a dune – now designated *Purgatory Dune* – about 30 cm tall. The operations team was required to simulate the situation in a lab at JPL after which, a few centimeters at a time, they gradually propelled Opportunity to escape velocity on 4 June 2005. The rover then proceeded southward toward Erebus crater driving mostly in the firmer troughs of the dunes. Opportunity's onboard computer can be, and is now, programmed to detect that its wheels are slipping and to stop Opportunity before it digs itself further in: a learner driving lesson.

Going further into Endurance the team was tempted to study a dune field but after experiencing slippage on fine sand and being afraid of getting stuck, decided not to do so. A number of rock outcrops were investigated on the way down

---

[146] http://mars.nasa.gov/mer/newsroom/factsheets/pdfs/Mars03Rover041020.pd
[147] https://mars.nasa.gov/internal_resources/825/

further into the crater and then it was time to head back up. By investigating the cross-section of bedrock that was exposed in the crater the team concluded that there had been episodic flooding of the surface followed by dry periods. They interpreted the minerals they encountered in evaporite outcrops to imply that the water had been acidic.

The final stop inside Endurance Crater was at Burn's Cliff (named for late mineralogist Roger Burns). The cliff is steep and layered with sediments of different compositions – the deeper layers containing decreasing amounts of magnesium and sulfur suggesting to the team "a common reason for their decline – possibly dissolution in water."[148]

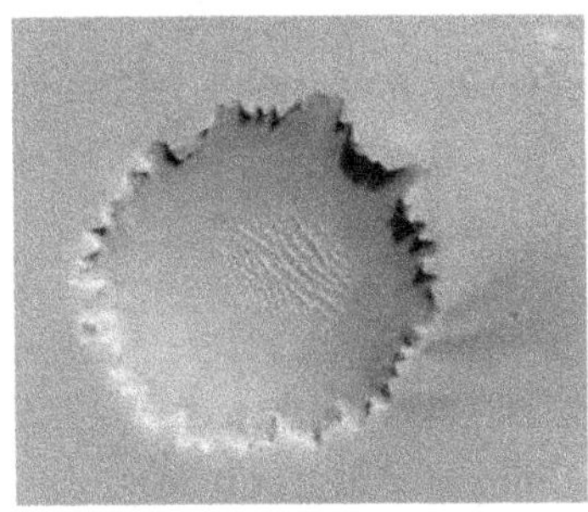

It wasn't easy for Opportunity to climb out of Endurance Crater because Burn's Cliff was too steep to climb and elsewhere there were sandy areas best avoided. So, Opportunity backed its way out the way it had entered. Leaving Endurance Crater, Opportunity drove to a man-made object that had been observed by Mars Odyssey overhead – the rover's heat shield. There, by chance Opportunity encountered and identified a meteorite – the first to be discovered on another planet.

Next stop: Victoria Crater, not named after the British monarch but, rather, for one of Ferdinand Magellan's ships. The crater is much larger – 730 meters across – than Endurance and is about 7 km from Eagle Crater.

Opportunity reached it on 26 September 2006 – a 21-month journey in all. Victoria Crater is strikingly different from the usual form of Martian and lunar impact craters: it is circular but the rim is jagged and reveals "many outcrops within recessed alcoves and promontories". The interior is unusually bland and has the appearance of a lake of sand with a field of dunes – material blown from the surrounding plain – burying any central uplift peak. Cautiously, Opportunity drove partway around the rim, mapping outcrops and looking for the safest way to enter the crater. The rover also had to sit out a dust storm for six weeks before making its way in and spending the next year investigating accessible outcrops in the steep walls. Opportunity left Victoria Crater on 29 August 2008 and headed to yet another crater – Endeavour. Endeavour crater represented another major step up in size – 22 km in diameter and 300 meters deep.

By now another NASA orbiter – Mars Reconnaissance Orbiter (MRO) – had arrived and had identified phyllosilicate-bearing outcrops along the Endeavour Crater rim. These minerals – serpentines, clays and micas – form in sheets and are hydrated with either water or hydroxyl groups attached. The crater floor shows the signature of basalt and hematite and has two dune fields. Opportunity would explore the west rim of Endeavour Crater for several years traveling just tens of meters each day identifying and analyzing rock and soil samples. Opportunity had arrived at its western edge on 9 August 2011 and, traveling north around the crater's edge, in December discovered "a vein of gypsum sticking out of the soil". It was 1 to 2 cm across and about 45 cm long.[149]

Analysis indicated its composition to include calcium, sulphur and water – almost pure calcium sulphate. This mineral vein was labelled *Homestake*; it evidently formed under more neutral conditions (a plus for biology) than the acidic conditions implied by the presence of sulphate deposits encountered elsewhere.

Opportunity continued its exploration of the northwest outcrops along the rim of Endeavour for about 2 years and then was directed south toward a hill the team named *Solander Point*, reaching the top by December 2013 near the 10-year anniversary of its landing. By March 2015 Opportunity had characterized the shallow *Spirit of St. Louis Crater* on the outer rim – it is 34 meters long by 24 meters wide and its floor is slightly darker than the surroundings. A rocky feature that rises above the rim is about 2 to 3 meters tall.

The rover was now near the entrance to a valley that cuts through the Western rim of Endeavour. *Marathon Valley* is named for the odometer reading – on 23 March 2015, NASA had announced that Opportunity had driven the full 26.2-mile distance of a marathon, with a finish time of roughly 11 years, 2 months! The rover was still searching for the clay minerals that had been the promise of the Mars Reconnaissance Orbiter's spectral observations; surprisingly, there turned out to be little confirmation. Rather, the team found that the coating of ancient iron magnesium clays observed by MRO is widespread – evidently having been broadcast far and wide by the bolide that created Endeavour Crater. These breccia

---

[148] http://apod.nasa.gov/apod/ap041115.html
[149] http://redplanet.asu.edu/?p=1147

(sedimentary rocks composed of large angular fragments over two millimeters in diameter that are bound together) have been named the *Shoemaker Formation* and "Opportunity has been driving on, through, and around it for almost her entire mission".[150]

This realization led the Opportunity scientists to conclude that the surface has only been modestly altered by water. In Marathon Valley Opportunity discovered a multitude of what are termed "red zones" all over the valley floor and along the valley's southern wall – evidently fractures along which water soluble materials had once flowed leading to chemical alteration of the rock. However, Opportunity was unable to identify any genuine bedrock. Team leader Steve Squyres commented: "All we were finding was dirt, reddish dirt. What we didn't realize until we took a close-enough look is that this stuff has been so pervasively altered, it's not bedrock. There's no solid bedrock the rover could grind with the RAT [Rock Abrasion Tool]. It's all crumbly crud now."[151]

The operational status of Opportunity at this point was still good – the rover was using 75% of the sunlight and creating 640 watt-hours of energy – about two thirds of her capability when she had landed years ago with clean solar panels. Regional dust storms elsewhere on Mars made the atmosphere more dusty than usual. The operations team had learned to cope with various minor problems – including this dusty weather and occasional 'amnesia' events & computer resets – as well as the fact that Opportunity was operating in persistent RAM mode – not using Flash memory (that retains information even when power is shut off during the rover's overnight power-conserving sleep time) – for data storage. This requires the rover to stay awake all the way up to the Ultra-High Frequency (UHF) relay passes on each sol. With Mars Odyssey now in a later orbit, relay passes come much later in the day. With winter coming and less sunlight for energy production, these late relay passes cause the rover to consume more energy from the batteries.[152]

When December 2015 came Opportunity was still inside Marathon Valley on the west rim of Endeavour Crater, positioned on steep, north-facing slopes for improved solar array energy production. The winter solstice occurred on Sol 4246 (3 January 2016) and 17 days later Opportunity celebrated an amazing 12 years on Mars. In mid-May 2016, while still in Marathon Valley the team still had not located any bedrock and decided on a brute force approach to finding out what the surface 'crud' might be hiding. By dragging one of its wheels through a 'red zone' Opportunity uncovered material that proved to have the highest sulphur content (in the form of magnesium sulphate) that had yet been encountered on Mars. Not having found smectite clays (a group of sheet silicate mineral species) as expected and, therefore, with relatively modest implications for the past Martian climate here, team member Ray Arvidson commented: "There is water and ice and mechanical transport and some degree of chemical alteration, but it's not balmy Cuba we're talking about here, more like Greenland."[153]

The plan from here was to drive Opportunity around the southern wall of the Marathon Valley and southeast downhill along the crater rim. The mission status record reported "On Sol 4438 (July 18, 2016), the plan was to continue driving. However, rover flight software stopped the drive just as the initial turn for the drive began. Evidence suggests that the terrain was causing excessive load on the right-front wheel. Being cautious, the project conducted a set of diagnostics on Sol 4439 (19 July 2016). Preliminary analysis still points to a terrain effect, but more diagnostics are planned for Sol 4440 (20 July 2016). The rover is otherwise in good health".[154]

A month later regional dust storms were limiting the power available to Opportunity and thus the distance that the rover could drive each day. It had left Marathon Valley through the newly named *Lewis and Clark Gap* and was driving through *Bitterroot Valley* on the way to the next target that the team had identified for panoramic imaging. It was not until 20 September 2016 that Opportunity arrived at the base of *Spirit Mound*. There Opportunity inspected a relatively bright outcropping of rock (one that the team calls *Gasconade*) using the microscopic imager on its arm and concluding that the

---

[150] http://www.planetary.org/explore/space-topics/space-missions/mer-updates/2016/06-mer-update-finalscience-campaign-marathon-valley.html
[151] ibid.
[152] http://www.jpl.nasa.gov/news/news.php?feature=4406
[153] http://www.planetary.org/explore/space-topics/space-missions/mer-updates/2016/06-mer-update-finalscience-campaign-marathon-valley.html
[154] http://mars.nasa.gov/mer/mission/status_opportunityAll.html

outcropping is wind-etched with angular bits of darker rock within a lighter matrix. These it is thought "may have been formed from fallout of the impact event that excavated the crater".[155]

During this time, in October 2016, ESA's ExoMars spacecraft – the combination of its Trace Gas Orbiter and Schiaparelli landing pathfinder – were closing-in on Mars and aiming at landing Schiaparelli on Meridiani Planum. This circumstance was obviously not a coincidence but simply the convenient choice of a landing site proven to be safe. It was hoped that Opportunity might, from a still considerable distance, be able to observe the final descent phase of Schiaparelli. Alas, it was not able to do so because, as is discussed later in more detail in the description of the ExoMars program, Schiaparelli's retro-propulsion system cut off too early and the vehicle crashed – a repeat of NASA's Polar Lander disaster.

Toward the end of 2016 Opportunity was preparing to drive to the East through a gully that the Mars Reconnaissance Orbiter had mapped at high resolution and judged to be carved by a fluid – presumed to be water. It is about 250 meters in length. Similar gullies have been mapped from orbit since the arrival of the Mariner 9 and Viking orbiters in the 1970s. PI Steve Squyres described the first objective of this new mission extension: "We hope to learn whether the fluid was a debris flow, with lots of rubble lubricated by water, or a flow with mostly water and less other material."[156]
The next goal was to compare rocks inside Endeavour Crater to the dominant type of rock Opportunity had examined on the plains it explored before reaching Endeavour. "We may find that the sulfate-rich rocks we've seen outside the crater are not the same inside" Squyres said. "We believe these sulfate-rich rocks formed from a water-related process, and water flows downhill. The watery environment deep inside the crater may have been different from outside on the plain – maybe different timing, maybe different chemistry." The science team summarizes: "The layers of sedimentary rock exposed in the walls of these craters contain minerals, such as sulfites and jarosite, and chemicals, such as chlorine and bromine that require considerable interaction with water to produce. The evidence suggests that Opportunity's landing site was once the shoreline of a salty sea."[157]

By June 10, 2018 when the last contact was made with opportunity the rover had travelled over 45km. The mission was declared complete in February 2019.

---

[155] https://www.nasa.gov/image-feature/jpl/pia21141/opportunity-inspects-gasconade-on-spirit-mound-of-mars
[156] http://www.jpl.nasa.gov/news/news.php?feature=6642
[157] http://science.nasa.gov/missions/mars-rovers/

CHAPTER 13

RETURN TO MARS 3:

BIGGER & BETTER

Mars Reconnaissance Orbiter, Phoenix, Phobos-Grunt, Curiosity

2005 Mars Reconnaissance Orbiter (MRO)

With the ongoing success of the 2003-launched Mars Exploration Rovers, it was now the turn of an orbiter. NASA selected the Mars Reconnaissance Orbiter (MRO) for launch in 2005. Although the Mars Global Surveyor and Odyssey orbiters had provided large quantities of excellent data, the science community's appetite had been only whetted for more. MRO's instrumentation and capability to return a great range of ultra-high-resolution data are unmatched.

This was not a grey mouse mission of the faster, cheaper, better variety. By the spring of 2019 MRO after 13 years had completed 60,000 orbits and with its Context Camera had acquired almost complete coverage of Mars with a resolution of 6 meters. The HiRISE camera with its 1 meter resolution) had achieved 3% coverage. Both accomplishments provide spectacular science results. Not surprisingly, the mission cost amounts to about the combined cost of both Spirit and Opportunity – still a bargain given the quality and quantity of data in question. Ultra-high-resolution surface imaging was a key goal of the mission because the selection of future landing sites was to be a priority objective; this requires both a very large aperture camera and a very high data rate communication system to return the images. The super-resolution image processing capability developed by Bob Kanefsky and colleagues at NASA Ames also can be applied when multiple images are acquired of special targets. The High Resolution Imaging Science Experiment (HiRISE) camera has a 0.5 meter mirror – the largest ever flown or even contemplated in the past. From 300 km altitude it has a resolution of 30 cm/pixel (twice as good as Google Earth maps). HiRISE is thereby capable of resolving objects smaller than a meter – including landed spacecraft; indeed, there is now nowhere to hide on Mars! The images are acquired in three colors: blue green, red and the near infrared; this, combined with IR spectroscopic data allows minerals to be mapped in detail, not just identified. Ball Aerospace (who had provided the Viking Orbiter cameras) built the HiRISE instrument for the University of Arizona (Principal Investigator Alfred McEwen). HiRISE is targeted to observe surface features that the team identifies and prioritizes – targeting that, as always, evolves with time as new information is acquired – not only by HiRISE and the other MRO instruments but also from the other Mars orbiters and landers. Precision targeting allows surface feature changes to be monitored. Researchers at the Mullard Space Science Laboratory, University College London have improved the spatial resolution of some targets by a factor of five using the software of Kanefsky et al. In a paper presented at the 47th Lunar and Planetary Science Conference (2016) Y. Tao and colleagues describe a ten-step process involving their "automated 3D reconstruction/DTM generation and super resolution restoration, both based on the use of the 5th generation of an adaptive least squares correlation matcher, called Gotcha".[158] All very straightforward!

This one instrument alone creates a flood of data and MRO's communication system (combined with NASA's DSN) is designed to cope with it: the antenna is 3 meters in diameter and operates at X-band frequencies at 8 GHz. The system can send back data at up to 6 Mbps, an order of magnitude higher than for earlier Mars orbiters. HiRISE is not the only instrument churning out enormous volumes of data. The Context Camera (CTX) serves to provide the broader coverage needed to interpret the targeted HiRISE images and the results from other instruments – the imaging spectrometer and the ground penetrating radar. The context camera (another Malin Space Science Systems special) is itself a high-resolution imager providing 6 meters/pixel spatial resolution for general-purpose mapping: mosaics of large areas, stereo coverage, and potential landing site maps.

The Compact Reconnaissance Imaging Spectrometer for Mars (CRISM) operates at visible and near IR wavelengths. Being able to detect light in these wavelength ranges enables identification of a broad range of minerals. At infrared wavelengths CRISM has much greater sensitivity to iron-containing minerals as well as sulfates, carbonates, hydroxyl, and water that is incorporated into mineral crystals that formed in hot springs, thermal vents, lakes, or ponds. The Principal Investigator is Scott Murchie of Johns Hopkins University Applied Physics Lab.

MRO's Shallow Subsurface Radar (SHARAD) experiment is designed to probe the permanent polar caps and the layered sediments upon which they sit. More generally SHARAD explores planet-wide, identifying subsurface layers of ice and rock in up to 1 km of crust. This instrument operates at higher frequencies than the MARSIS instrument on Mars Express and therefore penetrates less deeply. It operates in a 15 to 25 MHz band and is sensitive to changes in the electrical reflectance properties of rock, sand and water ice in particular. Its horizontal resolution is between 0.3 and 3 km and has a vertical resolution of 15 meters. Both MARSIS and SHARAD were provided by the Italian Space Agency and provide complementary data. Roberto Seu of University of Rome La Sapienza is the Principal Investigator.

MRO carries yet a third CCD camera, less high powered than the others: the Mars Color Imager (MARCI) whose purpose is to produce a global map each day that serves as a daily weather report of dust storms, growth and retreat of the polar caps, condensate clouds, and surface streaks resulting from winds moving dust grains around. The camera's fisheye lens with seven color filters – five visible light filters and two in the UV – produces images that are 1 to 10 km in resolution. Malin Space Science Systems again produced this camera and provides its science team. The MRO payload also includes an IR spectrometer to study the dynamic atmosphere of Mars – an updated version of an experiment that had been on both the ill-fated Mars Observer and on the ill-fated Mars Climate Orbiter. The indomitable team, as before, was made up of scientists and instrument builders at JPL – led by Dan McLeese – and, at Oxford University, led by Fred Taylor. On the positive side of history, there had been significant developments since the creation of the original instrument in the late 1980s in that cooling of the sensors was no longer necessary; a much lighter, compact and less expensive spectrometer was the result – named Mars Climate Sounder. The instrument views the atmosphere vertically and also horizontally to the limb of Mars at wavelengths mostly in the far IR (12 to 50 microns) serving to measure temperature, humidity and dust content as a function of height. Analysis of data leads to a daily global weather map that nicely complements the one created by the Mars Color Imager.

Lockheed Martin, acting as prime contractor to JPL, built MRO. It is a big, heavy spacecraft – 1,031 kg dry mass, 2,180 kg at launch – and even though it planned to use aerobraking to save fuel, MRO still required an Atlas V launcher. Launch took place on 12 August 2005. MRO arrived at Mars in March 2006 and was inserted into a highly elliptical polar orbit where it spent the next five months (to September 2006) gradually losing energy in the uppermost atmosphere before reaching its nearly circular science orbit (250 to 316 km above the surface with a period of 112 minutes). Mars orbit was beginning to get crowded: Mars Global Surveyor, Mars Express and Mars Odyssey were still busy. Spirit and Opportunity were in their extended missions. NASA's Deep Space Network – also tracking the Cassini orbiter at Saturn – was stretched to the full.

---

[158] http://www.hou.usra.edu/meetings/lpsc2016/pdf/2074.pdf

The deployment of the SHARAD antennas proved uneventful this time but throughout the many years of its operations the team has had occasional problems to deal with – but none of an existential nature. Huge volumes of science data from all the instruments have been returned with endless discoveries about the processes – and the role of water and sedimentation in particular – that have shaped Mars since its formation. The SHARAD science team has focused its analysis on the Utopia Planitia region in the mid-northern hemisphere where data have been acquired from hundreds of orbits. They have identified a deep deposit of water ice mixed with dirt – "more extensive than the state of New Mexico".[159] The deposit is between 80 and 170 meters thick and is buried 1 to 10 meters beneath the surface.

CRISM has returned data indicating that billions of years ago Mars was extensively covered in ice. The evidence is provided by minerals and landforms that show that volcanoes erupted beneath such ice sheets. Sisyphi Montes in southern Mars is a region studied with flat-topped mountains that have caught investigators' attention because these domes resemble terrestrial volcanoes that erupt underneath an ice – the 2010 eruption of ice-covered Eyjafjallajökull in Iceland is an example. On Earth characteristic minerals resulting from such sub-glacial volcanism on Earth include zeolites, sulfates and clays and Co-investigator Sheridan Ackiss of Purdue University reports these are what CRISM with a resolution of 18 meters/pixel has identified in the Sisyphi Mons.

The HiRISE experiment has returned 50,000 images, many of which are spectacularly beautiful. The team has created a website with a somewhat overwhelming amount of science analysis and information. They have prepared many dozens of PowerPoint slides summarizing their discoveries – 'nuggets'[160] too numerous to cover here except with a few examples selected in part for the striking appeal of the images.

• Present-Day Gully Formation on Mars

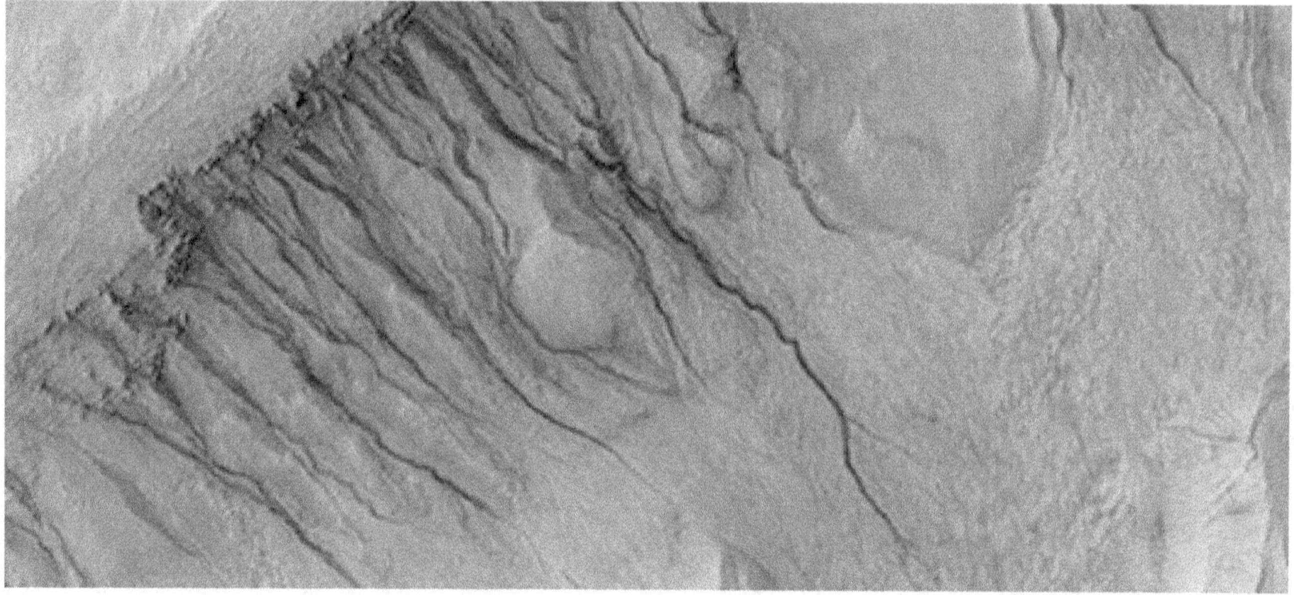

---

[159] https://astronomynow.com/2016/11/23/martian-ice-deposit-holds-as-much-water-as-lake-superior/
[160] http://www.uahirise.org/epo/nuggets/

This latest step in resolving power provided by HiRISE has taken our appreciation of the dynamic nature of Martian geologic processes, past and present to a new level. Among the many fascinating observations one astrobiology question stands out: does liquid water still make an appearance at the surface of this frigid planet where low atmospheric pressure (below the triple point except at the very lowest locations) makes water unstable? The CRISM and HiRISE teams have been systematically acquiring high-resolution compositional data from more than 100 gully sites across the planet. These data have been correlated with images from the two camera systems and, to some surprise, the team finds no mineralogical evidence for liquid water being the agent that carves them. CRISM and HiRISE data have been combined in this example of highly incised gullies on the rim of the Hale crater (about 150 km across and located at 36°S, 323°E). Unaltered mafic material from the crater rim is carved and transported down-slope but no hydrated minerals are observed. There is thus no evidence for the role of liquid water. The freeze and thaw of carbon dioxide ice is an alternative being considered.

• Slope Streaks

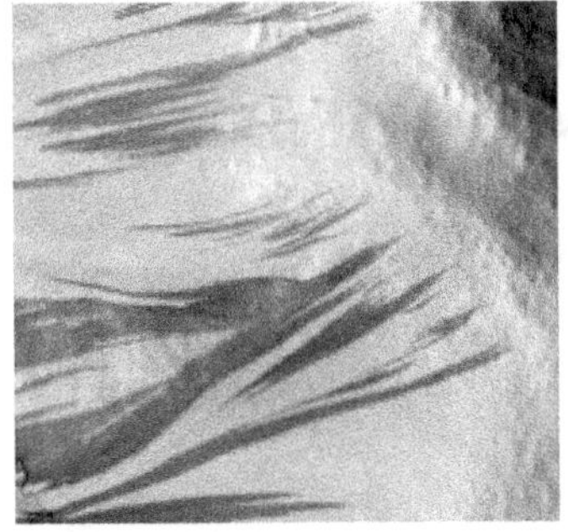

There are many types of slope streaks on Mars. They were observed first by the Viking orbiters and are typically observed as dark features on slopes in the equatorial regions on relatively steep terrain such as along escarpments and crater walls. They may be tens of meters in length and gradually fade away with time as new ones form. The most common explanation is that they are formed by dust avalanches that expose underlying dark materials.

• Spectral evidence for hydrated salts in recurring slope lineae

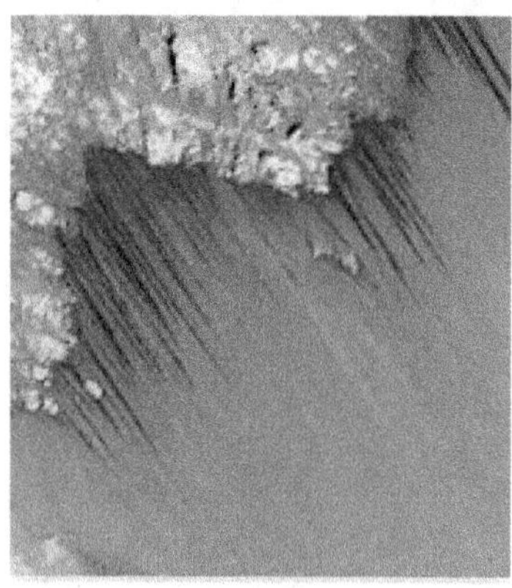

HiRISE has discovered a new type of slope streak called recurring slope lineae. These are dark flows up to a few meters wide on steep, rocky slopes. They are smaller in size than the common slope streaks and are notable for their seasonal behavior – they appear to be active only in summer. PI McEwen and colleagues report, "They behave like salty water flows in terms of temperature dependence, seasonality, and growth patterns, but the origin of the water is not known. They were previously reported in southern middle latitudes and now are known to be abundant in equatorial regions, especially deep in Valles Marineris. They follow the sun: active on N- facing slopes when subsolar latitude is to the north; active on S-facing slopes when sun is to the south. Shallow water may be surprisingly abundant near the surface in equatorial regions of Mars."[161]

Recurring slope lineae that are observed repeatedly by HiRISE are active flows on sun-facing slopes. One of the most active is in the central peaks of Hale Crater. The Hale features are unusual because they began activity much earlier than most RSL sites in the middle southern latitudes – they were well developed in the early spring. If seeping water causes these features they would have to be rich in salts to significantly lower its freezing point. Research on the origin of these features continues and most recently (2020) the consensus is that they're actually caused by dark sand sliding down inclines. Curious! A time lapse image of the phenomenon can be viewed at https://www.jpl.nasa.gov/spaceimages/images/largesize/PIA17606_hires.jpg.

• Large "Mud Cracks"

Polygonal patterns are very common on Mars and often attributed to be the result of thermal contraction. HiRISE coverage has shown that various surface cracking patterns cannot be explained by a thermal contraction alone. HiRISE observed large polygonal surface features resembling giant mud cracks in many locations – in areas characterized by minerals that may have formed in ancient lakes and rivers. When these dried out these striking features remained. Sites like these would be strong candidates for in-situ exploration and the search for paleo-organic materials.

• Mud Volcanoes

HiRISE has returned images of what appear to be mud volcanoes in a region in the northern plains called Acidalia Planitia. They are between 20 and 500 meters in size and more than 18,000 have been mapped! Astrobiologists are especially intrigued because muddy sediments spewed from underground might contain organic materials that could be biosignatures of past or even present life. Dorothy Oehler, a research scientist at the Astromaterials Research and Exploration Science Directorate at NASA's Johnson Space Center comments, "Mud volcanoes themselves are an indicator of a fluid-rich subsurface, and they bring up material from relatively deep parts of the subsurface that we might not have a chance to see otherwise."[162]

---

[161] https://www.uahirise.org › epo › nuggets › RSL-VM-nugget
[162] http://www.astrobio.net/mars/mud-volcanoes-on-mars/

• Marsquakes and water-lain sediments

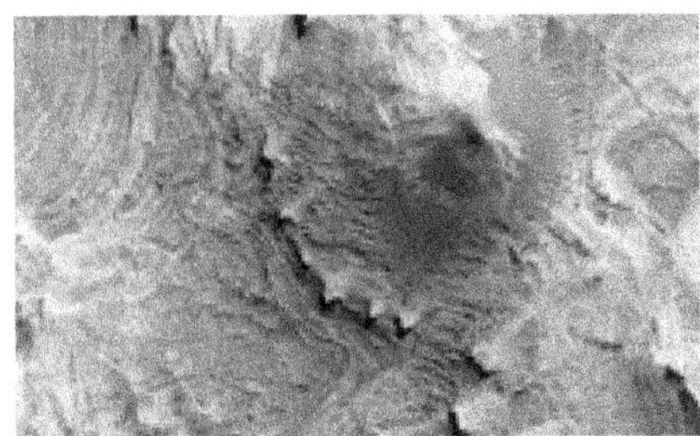

HiRISE data have revealed evidence of past marsquakes and water-lain sediments in the west Candor Chasma region of Valles Marineris as in this image that is about 1 km across. The orientations of the sediment layers argue that these rocks formed by sand and dust that were blown in by the wind and became trapped in shallow playa lakes. Conical hill features geologists call injectite megapipes are also observed – they formed by underground movement of the water-lain sediments in response to ground shaking (marsquakes) from several large fault zones in the area. These injectites served as reservoirs for ground water and thereby would have hosted potentially habitable environments in the Martian subsurface approximately 3 to 3.5 billion years ago.

• Ancient Northern Ocean

The puzzle presented by the northern plains and the possibility that this is where an early ocean of water was to be found appears to be increasingly credible bearing out the previously mentioned analysis of Clifford and Parker. In May 2016 a HiRISE science team led by Alexis Rodriguez of the Planetary Science Institute and including Victor Baker and Virginia Gulick has published a paper in *Nature* that supports the northern ocean model and explains why no complete shoreline is to be found.[163]

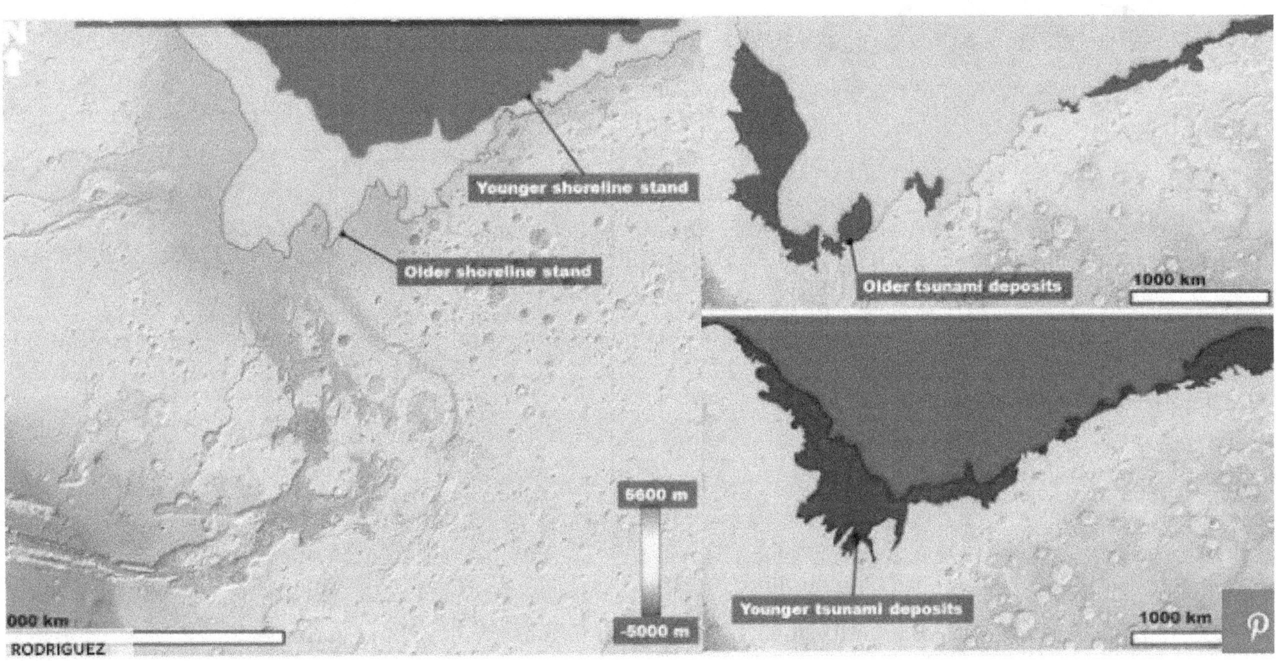

Based on geomorphic and thermal image mapping in the circum-Chryse and north-western Arabia Terra regions of the northern plains, in combination with numerical analyses, we show evidence for two enormous tsunami events possibly triggered by bolide impacts, resulting in craters ~30 km in diameter and occurring perhaps a few million years apart.[164]

This interpretation of data acquired by the Mars Reconnaissance Orbiter argues that a giant asteroid impact into the northern ocean took place ~3 billion years ago and led to an enormous tsunami that served both to sweep away parts of the then-existing shoreline and, also, deposit huge quantities of debris – in the form of lobes tens to hundreds of kilometers in extent. The stratigraphy at the edge of the northern plains leads the team to argue that the Martian climate

---

[163] http://www.nature.com/news/giant-tsunamis-washed-over-ancient-mars-1.19916
[164] ibid.

subsequently cooled and that the ocean had substantially iced over when a second asteroid struck creating another huge tsunami and more debris lobes, this time ice-rich in form.

The terrestrial asteroid strike that wiped out the dinosaurs and other species at the time of the Cretaceous-Tertiary (K/T) extinction event – 66 million years ago – produced an enormous tsunami in what is now the Gulf of Mexico. Richard Cowen of UC Berkeley reported in 1999:

Around the Caribbean and at sites up the Eastern Atlantic coast of the United States, existing Cretaceous sediments were torn up and settled out again in a messy pile that also contains glass spherules of different chemistries, shocked quartz fragments, and an iridium spike. All this implies that a great tsunami or tidal wave affected the ocean margin of the time, washing fresh land plants well out to sea and tearing up seafloor sediments that had lain undisturbed for millions of years. The resulting bizarre mixture of rocks has been called the Cretaceous-Tertiary cocktail.[165]

So, the MRO team's analysis is certainly plausible and appears to have solved one of the biggest Martian mysteries.

Water Ice Cliffs

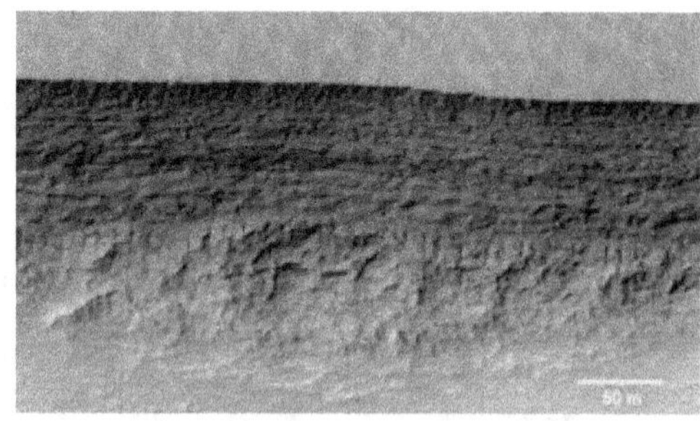

Another remarkable discovery from the steady accumulation of HiRISE images was made by Co-I Colin Dundas of the USGS and his colleagues in 2018 – namely of thick layers of water ice that are exposed in the sides of steep cliffs at a number of different sites. The observations (enhanced color here) were of "water ice that can be >100 meters thick, extending downward from depths as shallow as 1 to 2 meters below the surface. The scarps are actively retreating because of the sublimation of the exposed water ice. The ice deposits likely originated as snowfall during Mars' high obliquity periods and have now compacted into massive fractured and layered ice".[166]

These cliffs will provide a stratigraphic record of past Martian climates – an opportunity for which new technology will be required.

MRO Climate Sounder Experiment CRISM

On the question of water on early Mars, the CRISM experiment has added to the weight of evidence that has been accumulating regarding minerals that are hydrous weathering products: phyllosilicates, carbonates, hydrated silica and hydrated sulfates. And, as noted at Opportunity's Eagle Crater, the iron oxide hematite can precipitate out of standing water.

After its seemingly endless wait to reach Mars orbit, the Climate Sounder experiment (Principal Investigator Dan McCleese of JPL) has achieved almost all of its goals as "unfortunately the water vapor measurement was not a success".[167]

The instrument scans the horizon to observe the atmosphere in 5 km vertical slices creating profiles that are combined into daily global weather maps. Its measurements are made at all times of day in contrast to previous experiments limited to daytime observations.

Existing global circulation models have been confirmed and the physics of observed water ice condensate cloud formation has been clarified: they form in association with a strong temperature inversion, the result of cooling at altitudes above 20 kilometers. During northern summer, cloud formation in the tropics is driven by solar radiation and the associated thermal tide. This leads to cloud formation above the summits of the Tharsis volcanoes first observed by Mariner 9.

---

[165] Cowen, R. 1994. *History of Life*. 2nd edition, 460 pp. Blackwell Scientific Publications, Cambridge, Massachusetts
[166] https://science.sciencemag.org/content/359/6372/199
[167] Fred Taylor, personal communication, December 2019

Widespread Shallow Water Ice at High Latitudes and Mid-latitudes

In an important discovery Sylvain Piqueaux of JPL and colleagues have used more than a dozen years of Climate Sounder night-time temperature data to infer the presence of frozen water ice at shallow depths over large areas of Mars. This ability comes about because "frozen water is a very strong heat conductor compared to typical Martian regolith. As a result, near-surface ice measurably influences seasonal surface temperature trends, and the depth of the H2O table controls the amplitude of the effect." The large amount of thermal data has allowed the team to model the physics in question. The "results are consistent with widespread water ice at latitudes as low as 35°N/45°S buried sometimes a few centimeters below sand-like material, with high lateral ice depth variability, and correlated with periglacial features."[168]

Since October 2012 the Mars Reconnaissance Orbiter has been in its second extended mission "Targeting new locales of interest, following up on earlier discoveries, and characterizing inter-annual variability in atmospheric and surface processes; and providing relay support for Mars Science Lab and Mars Exploration Rovers."[169]

2007 NASA Phoenix

The next Mars launch window after that of the Mars Reconnaissance Orbiter was in August 2007; in August 2003 NASA had selected the Phoenix mission proposed by the University of Arizona in competition with other proposals. It was in fact the first selection made under yet another named Mars Program – the *Mars Scout Program* where each project was to cost less than US$485 million. The program was short-lived as only two missions were selected (Phoenix and MAVEN) before the program was retired in 2010. In terms of cost, aptly named Phoenix had a major advantage because the lander that had been built for the 2001 launch was in storage at Lockheed Martin's facility. Several instruments from the planned 2001 mission were available along with some instrument spares from the lost Polar Lander. Peter Smith of the University of Arizona's Lunar and Planetary Laboratory was Principal Investigator; there were 24 Co-Investigators. There was unusually wide participation in the mission from US universities and NASA Centers. Also helping with cost, international contributions were provided by the Canadian Space Agency; the University of Neuchatel, Switzerland; the universities of Copenhagen and Aarhus Denmark; the Max Planck Institute, Germany; and the Finnish Meteorological Institute.

JPL provided project management with Barry Goldstein as Project Manager and Leslie Tamppari as the Project Scientist. The science instruments and operations were the University of Arizona's responsibility. The instruments included stereo-cameras, a volatiles-analysis instrument, a trench-digging robot arm, wet chemistry laboratories, optical & atomic force microscopes and a suite of meteorological instruments.

Phoenix was launched from the Cape on August 4, 2007 on a Delta and arrived at Mars on 25 May 2008. It entered the Martian atmosphere in an aero-shell, descended under a parachute and, using retrorockets, touched down safely. Like the Vikings, Phoenix used descent thrusters in the final seconds before landing on its three legs. The lander used radar to continually assess the spacecraft's vertical and horizontal motion during the final minutes of descent – continually adjusting the descent accordingly. Compared with the Spirit and Opportunity descents through the atmosphere, Phoenix separated from its parachute at an altitude 100 times greater and its velocity at touchdown was one tenth as great. In spite of their three earlier successes it seems unlikely that NASA will use the heavy bouncing airbag system in the future.

Where the lost Polar Lander had aimed at a site near the South Pole, because of the season, Phoenix was targeted to a northern lowland site at 68.35°N, one that had the informal name *Green Valley*. The season was late spring and at that high latitude the sun was above the horizon all day – convenient for the solar-powered spacecraft but, inevitably, the mission would be of limited lifetime once the autumn arrived and the sunlight hours were too few to maintain power and temperature. Thus the first sunset for Phoenix was experienced at the start of September 2008 and the mission ended on 2 November 2008.

---

[168] Geophysical Research Letters, 10.1029/2019GL083947
[169] http://mars.jpl.nasa.gov/mro/mission/timeline/

The panoramic images acquired by Phoenix – the fifth (following the two Vikings, Pathfinder, Spirit and Opportunity) from the surface of the Martian northern lowlands – revealed a bland, flat landscape lacking in any features, e.g. dunes and, unlike other northern sites, with only a few small rocks and pebbles. This may well be the surface that once lay at the bottom of a deep ocean.

The flat surface in Green Valley observed *in situ* and from orbit is shaped into polygons 2 to 3 meters across bounded by troughs a few tens of cm deep. These are much like the periglacial patterned ground that characterizes cold-climate regions on Earth. The polygonal shapes are the result of ice in the soil expanding and contracting with major temperature changes. The soil itself is composed of both flat particles – thought to be a type of clay – and rounded particles.

The robotic arm was deployed to scoop up soil that was then delivered to an oven where the emitted gases were analyzed; also to a microscope and to the Wet Chemistry experiment. This laboratory analyzed three samples, two from the surface and one from a depth of 5 cm revealing a slightly alkaline soil and low levels of salts including ~0.6% by weight calcium and magnesium perchlorate (ClO_4). This key discovery of perchlorate has raised important questions about the soil analyses carried out by the Viking landers decades earlier. The complete absence of organic molecules had been a bit of a puzzle because, independent of any Martian biology, meteorites would have been a source of organic molecules. Because perchlorates break down organics when heated it remains possible, indeed likely, that the Viking soil samples did contain carbon-based molecules.

"This doesn't say anything about the question of whether or not life has existed on Mars, but it could make a big difference in how we look for evidence to answer that question," is the comment of Chris McKay of NASA's Ames Research Center who co-authored a study published online by the *Journal of Geophysical Research – Planets*, re-analyzing results of Viking's tests for organic chemicals in Martian soil.[170]

In another important discovery, Phoenix determined that water-ice is located just a few inches under the surface. When exposed to the atmosphere the ice slowly sublimes at a rate consistent with water rather than CO_2. Bright material exposed by digging on 15 June and still present on 16 June had vaporized by 19 June.

Phoenix carried an elaborate meteorological package from Canada that included a LIDAR (Light Detection and Ranging) laser system – and measured the daily weather throughout the mission. This experiment monitored wind speed, pressure and temperature as well as the vertical distribution of dust, ice, fog and clouds – all of which is grist for the improvement of general circulation modeling. Wind speeds were between 11 and 58 km/hour, typically 36 km/hour. The highly successful Phoenix mission ended in August 2008; there was a last brief communication on November 2 as the spacecraft's solar power diminished.

---

[170] https://www.youtube.com/watch?v=nAthdQpjFwg

2011 Russia's Phobos-Ground (*Fobos-Grunt*)

The Russian planetary science community, together with Chinese partners, attempted to carry out another major Mars mission with a launch on 8 November 2011. They proposed to return samples from the inner Martian moon, Phobos, and also to further monitor the atmospheric dynamics of Mars by means of a Chinese orbiter, Yinghuo 1. *The Planetary Society* was also a participant with an innovative experiment that involved sending microorganisms on a three-year interplanetary round-trip. The spacecraft was a largely new design but with inheritance from the successful Luna sample return missions. Alas, decades of work came to nothing when, after an initially successful launch on a Zenit-2SB, the upper stage failed to perform so that the spacecraft was stranded in orbit. Another tragedy.

China subsequently began an independent Mars project, Tianwen-1, for launch in 2020 – an orbiter, lander and rover being readied for launch at the time of writing – Chapter 15.

2011 NASA Mars Science Laboratory: *Curiosity*

NASA's 2011 Mars mission – the Mars Science Laboratory: *Curiosity* – launched on 26 November 2011 a few weeks after Fobos-Grunt and was, by contrast, a big success. It is a flagship science mission (still operational at the time of writing in early 2020) that had been conceived in the heady days when meteorite ALH84001 led many to believe mistakenly that they had discovered evidence of ancient life on Mars and that a Mars Sample Return mission could soon be initiated. The science objectives were broad: a further investigation of Martian habitability, continued study of geology and climate, anticipation of sample return and manned exploration. Science instrument proposals had been solicited in 2004 and eight had been selected. Curiosity (whose name in a contest was suggested by a 12-year-old schoolgirl) carries the most advanced suite of instruments yet sent to the Martian surface and has the capability to analyze soil samples and core samples drilled from rocks. These provide a record of Martian climate and geology necessary to assess the potential past habitability of Mars.

The mission is managed by JPL. During development Pete Theisinger was the Project Manager and John Grotzinger the Project Scientist. During operations Jim Erickson is the Project Manager and Ashwin Vasavada the Project Scientist. Curiosity is loaded to the gills with instrumentation from several countries. A US-French consortium led by NASA's Goddard Space Flight Center developed Curiosity's powerful Sample Analysis at Mars (SAM) instrument suite; this takes up more than half of the payload and serves to analyze organics and gases from both atmospheric and solid samples. The Principal investigator is Paul Mahaffy. SAM includes a Quadrupole Mass Spectrometer to detect gases sampled from the atmosphere or released from solid samples by heating. A gas chromatograph heats soil and rock samples until they vaporize then separates individual gases into their molecular components with the resulting gas flow analyzed in the mass spectrometer. A tunable laser diode measures oxygen and carbon isotope ratios in carbon dioxide (CO_2) and methane (CH_4) in the atmosphere of Mars as a means of distinguishing between different origins: geochemical or biological.

The PI of the X-ray diffraction and fluorescence instrument called CheMin is David Blake of NASA's Ames Research Center that identifies and measures the abundances of various minerals – including those that had been discovered by earlier missions, namely, olivine, pyroxenes, hematite, goethite and magnetite, all of which indicate the environmental conditions when they formed. Of obvious importance is the further study of the role that water has played. Kenneth Edgett of Malin Space Science Systems is PI of the Mars Hand Lens Imager that is mounted on an arm and takes extreme close-up pictures of rocks and soil, revealing details smaller than the width of a human hair (this is the camera that takes self-portraits of Curiosity). Also on the arm is the Alpha Particle X-ray Spectrometer provided by the Canadian Space Agency; this instrument determines the relative abundances of different elements in rocks and soils. Ralf Gellert of the University of Guelph is PI.

Curiosity's Mast Camera is mounted at about human-eye height and records images in high-resolution stereo and color. It also can take and store high-definition video sequences. Michael Malin is the PI. ChemCam uses a laser to vaporize thin layers of material from targets up to 7 meters away and includes a spectrometer to characterize the vapor and a telescope on the rover's mast. The instrument also serves as a passive spectrometer to measure composition of the surface and atmosphere. Roger Wiens of Los Alamos National Laboratory is the PI. The rover's Radiation Assessment Detector makes measurements that are needed to plan eventual astronautic exploration of Mars. The PI is Donald Hassler of Southwest Research Institute.

Curiosity has meteorological instrumentation provided by Spain's Ministry of Education and Science – the Environmental Monitoring Station that measures atmospheric pressure, temperature, humidity, winds and also ultraviolet levels. The PI is Javier Gómez-Elvira of the Center for Astrobiology in Madrid. The Russian-provided Dynamic Albedo of Neutrons (DAN) instrument detects hydrogen beneath the rover to a depth of one meter (at the rover's very dry study area on Mars, the detected hydrogen is mainly in water molecules bound into minerals). Igor Mitrofanov of the Space Research Institute (IKI) in Moscow is the Principal Investigator.

Rock sampling can also be carried out by a drill that delivers powdered sample to the SAM suite of instruments for analysis of organics. Powdered samples can be delivered to the CheMin instrument for X-ray diffraction analysis of the mineral composition. As noted, such analysis has identified minerals, including feldspar, pyroxenes and olivine similar to the kind of weathered basalts found in the soils on Hawaiian volcanoes. Landing site selection started with 60 candidates where the goal was to choose a site most promising in terms of its potential to support microbial life – a site that demonstrated both mineralogical and geological evidence that water had been present in the past. Both orbital spectroscopy and the results of Spirit and Opportunity had provided encouragement through their detection of hydrated minerals such as clays, sulfate salts and, also, the iron oxide hematite that is deposited in aqueous environments.

A series of workshops was held to successively reduce the list of candidate sites; the 154 km diameter Gale Crater emerging as the choice. It is located just 5 degrees south of the equator at an elevation of -4.4 km and was formed by a sizable impact 3.5 to 3.8 billion years ago. In the center of the crater is a mountain, Aeolis Mons (in honor of deceased Caltech geologist Robert Sharp (1911-2004) it is informally named Mount Sharp) that is 5.5 km high. Layering of this central feature suggests that the crater was filled with a sequence of sediments that were laid down in a lake over, perhaps, 2 billion years. Then evidently, after the water was slowly lost to space, the sediments were eroded by wind action over another billion years. Information about lake levels in the past might be provided by the fans and deltas within Gale Crater. Each layer "helps tell the story about how Mars changed from being more Earth-like – with lakes, streams and a thicker atmosphere – to the nearly airless, freezing desert it is today."

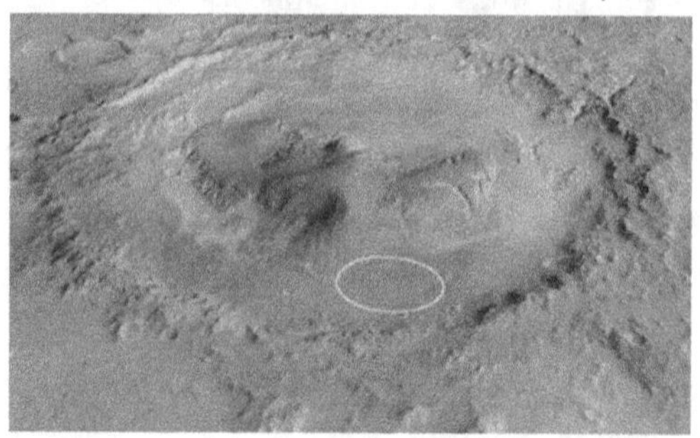

The Mars Science Laboratory is a large mobile lander powered by a radioisotope thermal generator (RTG) fueled by 4.8 kg of plutonium-238 oxide supplied, as always, by the US Department of Energy. It produced 125 watts of electrical power at the start of the mission, a level that will slowly reduce to 100 watts after 14 years. It was launched from the Cape on 26 November 2011 on an Atlas V rocket with a Centaur upper stage. The spacecraft arrived at Mars on 6 August 2012 and successfully carried out the landing by means of JPL's innovative 'sky-crane' approach, an advance in safety over Viking and Phoenix-style retrorocket 3-leg landing and, much more so, over the airbags used by Spirit and Opportunity. Curiosity benefited from a new high-accuracy entry, descent-and-landing (EDL) system that served to reduce the size of the landing ellipse to 20 x 7 km from the previous 150 x 20 km targets.

The EDL phase lasted a tense 7 minutes during which the entry capsule first slowed down to Mach 1.7 and the supersonic parachute was deployed at 10 km altitude. At 1.8 km altitude the rover and its descent stage were ejected from the aero-shell. Radar altimetry served to measure speed and altitude to the rover's flight computer while Curiosity was lowered on three nylon tethers 7.6 meters beneath the descent stage. The altimeter determined when the moment had arrived for Curiosity to gently touch down at which time the tethers were cut and the descent stage flew off to crash land some 650 meters away. The landing was not only gentle but highly accurate: less than 2.4 km from the center of the planned landing ellipse. The Curiosity team named their landing site Bradbury Landing in honor of the American science fiction writer Ray Bradbury (1920-2012) who had died on 5 June just a few months earlier. Famous among his many well-known books was *The Martian Chronicles* written in 1950. The coordinates of the landing site are: 4.5895°S 137.4417°E. The 6-wheeled, car-size laboratory has the two front and two rear wheels on individual steering motors that allow the rover to make arching turns.

A lot of attention has been paid to its design by the team at JPL that had been designing and testing Mars rovers since the 1970s: the rocker-bogie suspension allows Curiosity to climb over obstacles that are 25 cm in size (this is the wheel diameter). The cleats on the wheels provide the needed grip to make progress on soft sand. The hazard-avoidance program stops Curiosity for a safety check every 10 seconds – so Curiosity is not designed for speed – tops is 5 cm/sec. Like Spirit and Opportunity, average speed is designed for about 1 cm/sec. The team at JPL has provided a detailed description of Curiosity's travels and findings at a Wikipedia site from which the summary below has mainly been drawn.[171]

Following instrument and mobility tests, Curiosity began to assess its surroundings on 19 August 2012 using the ChemCam laser (30 shots in 10 seconds!) to analyze a conveniently nearby rock christened Coronation; as anticipated, it turned out to be a basalt. During nearly four years on Mars, ChemCam had inspected multiple points on more than 1,400 targets by detecting the color spectrum of plasmas generated when laser pulses zap a target – more than 350,000 total laser shots at about 10,000 points in all. Then, ten days later, it began what was expected to be a year-long journey to the base of Mount Sharp while analyzing the rocks and soil on the floor of Gale Crater. The first major stop, just 400 meters away to the east, was named *Gleneig* (presumably with the encouragement of a scientist with West Highland ancestry) in an area that the team subsequently called *Yellowknife Bay*. Likely everyone on the science and operations team was given an opportunity to provide an informal name (later to be adopted by the IAU) for the many dozens of locations where Curiosity stopped on its journey south and upward. It didn't take long (27 September 2012) before evidence was found of an ancient streambed "suggesting a 'vigorous flow' of water". And a month later Curiosity carried out its first X-ray diffraction analysis of the soil – finding feldspar, pyroxene and olivine, minerals similar to the weathered basaltic soils of the Hawaiian volcanoes. It was not until February 2013 that the Curiosity team plucked up their courage to use the drill for the first time. The analysis of the 5 cm deep samples would take a long time to digest and would prove to be very illuminating.

The immediate results were exciting: water, carbon dioxide, oxygen, sulfur dioxide and hydrogen sulfide were all detected leading the team to conclude that "the geochemical conditions in Gale Crater were once suitable for microbial life". The results were "consistent with the presence of smectite clay minerals" – these form in the presence of water. Analyses of a number of rock samples also provided evidence of mineral hydration. In addition, the Russian pulsed neutron generator instrument (DAN) "provided evidence of subsurface water, amounting to as much as 4% water content, down to a depth of 60 cm in the rover's traverse from the Bradbury Landing site to the Yellowknife Bay area". Further analysis concluded that Curiosity had identified a fine-grained sedimentary rock known as a mudstone, one that along with shale, makes up about two thirds of all sedimentary rocks on Earth.

Mudstones would be identified at most of the sites that Curiosity investigated as it worked its way to Mt. Sharp – evidence that Gale Crater had indeed been lake-filled in the past. By September a year later (2013), the team had sufficient confidence to report that they had detected abundant, easily accessible water (1.5-3 weight percent) in the soil. Also reported: that perchlorate salts had been detected just as had been the case at the Phoenix north polar landing site – suggesting that these salts are distributed globally. If so then, because of the way in which they destroy organics when soil is heated, such widespread dissemination may make it difficult to detect life-related organic molecules.

Through most of April 2013 Mars lay behind the Sun as seen from Earth (solar conjunction) leading the operations team to bring the rover to a halt and hand over control to Curiosity itself (the concern being the possibility of uplink data corruption). Therefore, during this period of stationary autonomous operation, science activities were limited and routine. Not

---

[171] https://upload.wikimedia.org/wikipedia/commons/0/01/PIA19067-MarsCuriosityRover-TraverseMap-Context Sol0817-20141123.jpg

long after, on 5 June 2013 the Curiosity team was ready to begin the lengthy journey to the base of Mt. Sharp with stops for analysis at various points on the way. In planning the journey and in deciding where to stop for experiments the Curiosity team greatly benefitted from the very high-resolution MRO imaging and spectral data. Along the way Curiosity systematically analyzed the atmospheric composition noting, in July 2013, the marked absence of any methane, a trace component that had been detected by ground-based telescopes years earlier.

There were inevitably some technical alarms along the way. Because of a problem with Curiosity's active computer it was necessary on 28 February 2013 to switch to the backup; this became fully operational on 19 March 2013. Also, there was an electrical problem involving the rover's RTG power supply – apparently an internal short that caused "an unusual and intermittent decrease in a voltage indicator".

Discovery of Organic Molecules

By the first anniversary of the rover's landing, 5 August 2013, Curiosity had discovered convincing evidence of "ancient environments suitable for life". It had driven over 1.6 km, had transmitted 190 gigabits of data to the DSN including 36,700 full-frame images and the laser had been fired 75,000 times at 2,000 targets. In December 2013 the team further reported in the journal *Science* that their analysis of the 5 cm drill samples had identified organic molecules – comparable to samples of sedimentary rock rich in organics on Earth, included thiophenes, methylthiophenes methanethiol and dimethylsulfide – that they were confident were not terrestrial contamination. These could be from meteorites. "The carbon could be from organisms, but this has not been proven". It was concluded that the core samples "were probably once mud that for millions to tens of millions of years could have hosted living organisms. This wet environment had neutral pH, low salinity and variable redox states of both iron and sulfur species".[172]

The age of the mudstone sample was judged to be "Late Noachian/Early Hesperian or younger" indicating that "clay mineral formation on Mars extended beyond Noachian time; therefore, in this location neutral pH lasted longer than previously thought". The Noachian era is 4.1 to 3.7 billion years ago when many of the large impact craters were formed and when the atmosphere was denser and Mars was warmer. The Hesperian era followed and saw widespread volcanism and the flooding that carved the huge outflow channels.

Curiosity traveled on to a pre-selected waypoint called the *Kimberley* on the 589th Martian day (2 April 2014). It was selected because of the diversity of rock types distinguishable in orbital images, "exposed close together at this location in a decipherable geological relationship to each other".[173]

Onward. Curiosity had already fulfilled its prime mission goal: Mars certainly was habitable in the deep past. Fourteen productive months after leaving Yellowknife Bay, Curiosity reached the slopes of Mount Sharp on 11 September 2014 and began the climb upwards. The planned route would take it first to visit a geologic unit that Mars Reconnaissance Orbiter

---

[172] https://en.wikipedia.org/wiki/Timeline_of_Mars_Science_Laboratory
[173] https://www.jpl.nasa.gov/spaceimages/details.php?id=PIA18073

mapping had identified as including the iron-bearing mineral hematite followed by a clay-rich unit and subsequently a sulfate unit – each providing sequential information about the history of the mountain.

Sampling of rocks that have evidently been laid down in water has confirmed that Mars has (and had in the distant past) chemistry – carbon, hydrogen, oxygen, phosphorus and sulfur – that would have been hospitable to microbial life. A sample acquired by Curiosity from inside a rock has confirmed the presence of clay minerals and, based on the limited amount of salts, the likelihood that the water had been reasonably fresh. Curiosity also acquired images of rock surface texture that, in comparison with those acquired earlier by Opportunity, argue for a habitable environment in Gale Crater. The sedimentary rock imaged by Curiosity reveals a very fine-grained surface that was probably deposited under water that cemented the sediments. White lines running through the rock are evidently sulfate minerals formed in fractures. The Curiosity team interprets the image and associated chemistry and mineralogy measurements to mean that the aqueous environment was habitable in terms of neutral pH, low salinity and chemical ingredients. The carbon content of powdered rock samples included organic carbon – another basic ingredient of biology.

By August 2014, after two years on Mars, less time had been allocated to stopping and exercising the analytical instruments – instead the goal was to drive on and reach the base of Mount Sharp, 3 km away, after carefully navigating a path through a series of sandy valleys with just a couple of stops on the way. Now the panoramic images were interpreted to be tilted beds of sandstone most likely formed by streams of water flowing down and carrying sediments from higher elevations – indicative of precipitation to form a shallow lake in the crater. By September 2015 Curiosity had arrived at the base of the mountain in the middle of Gale Crater and began to observe a different kind of rock formation – finely layered mudstones.

The layering suggests that the environment had not been constant but, rather, episodic as a result of changes in the climate and in the supply of water to the lake. Curiosity's drill was used a number of times to acquire samples for mineral analysis; these indicated that the formation had been created under fresh water. Environmental conditions eons ago evidently would have been favorable for microbial life. Much harsher and drier conditions followed to create the planet we see today.

By now Curiosity had spent two Martian years (each of 687 Earth days) on the surface of Mars while monitoring atmospheric conditions – not just the weather but also the background concentration of methane that, as previously noted, had registered a large spike (7 ppm) during the first southern hemisphere autumn. This spike was not repeated one Mars year later when the concentration measured varied between 0.3 and 0.8 ppm. Curiosity has, however, observed more subtle changes that may follow a distinct seasonal pattern along with pressure, temperature, UV illumination and water vapor concentration. Gale Crater is, of course, extremely cold and dry with daytime temperatures above freezing and at night reaching 90 below.

Now it traveled onward following a path planned using the high-resolution coverage from the Mars Reconnaissance Orbiter. Curiosity provided some wonderful images of the Bagnold dunes (informally named after British military engineer Ralph Bagnold (1896-1990) who had pioneered the physics of how such dunes are created on our own planet) that form a band along the north-western flank of Mount Sharp. Not surprisingly, Curiosity avoided entering the dune field but skirted its edge. The Mars Reconnaissance Orbiter had studied these dunes in some detail by means of its Imaging Spectrometer whose team had concluded that the mineral composition of the dunes is not evenly distributed. MRO's HiRISE camera had monitored the dunes' movement showing that edges of individual dunes move as much as one meter in an Earth year.

On Mars sand dunes are widespread as Spirit and Opportunity had found to their cost and as might be expected on a planet that has regular global-scale dust storms. They accumulate where a supply of sand grains is found and where wind speeds suffice to transport the grains by saltation – sand grains lifted by the wind and then, falling back, knock others into the air. The sand gets deposited where the winds weaken. Martian dune formation differs from the terrestrial case because of the much thinner atmosphere and much reduced gravity. The NASA Ames Research Center has a Planetary Aeolian Laboratory that has been used since the 1970s for experiments and simulations of Aeolian processes under different planetary atmospheric environments, including Earth, Mars, and Venus.

Now, at close range in a location called the Pahrump Hills, Feb. 4, 2015, scientists could directly compare terrestrial and Martian dune formation. In addition to the large sand dunes and small (~30 cm) sand ripples, Mars also has mid-sized ripples, about 3 meters apart – more like sand ripples that form under moving water on Earth. The Bagnold Dune Field's local supply of sand is evidently the result of the winds that for eons sandblasted the sedimentary rock in Gale Crater – a conclusion the Curiosity team bases on the measurements of its SAM mass spectrometer. This measures isotopes of helium, neon and argon in the rock that are produced by cosmic rays. "The fewer of these isotopes they find, the more recently the rock has been exposed near the surface. The 4-billion-year-old lakebed rock drilled by Curiosity was uncovered between 30 million and 110 million years ago by winds which sandblasted away 2 meters of overlying rock". Other characteristic observations from the Pahrump Hills site near the base of Mount Sharp included a small ridge (about 1 meter across) that clearly does a better job of resisting wind erosion than the flatter plates nearby – and would have been very punishing to Curiosity's wheels if the rover had driven over it.

Close-by to Pahrump Hills on Curiosity's route is Marias Pass where the DAN instrument detected an unusual hydrogen-rich patch of material that the team concluded was related to "water or hydroxyl ions in rocks within three feet beneath the rover". [174]

This was not announced until December 2015 after a collaborative analytical effort by the DAN team and the ChemCam team who had identified a region of high silica in the same area. By the time that the investigators realized that both instruments had acquired unusual measurements, Curiosity had moved on and it had been necessary to return to the site in question to make further observations. These determined that the Marias Pass site had large amounts of the quite rare silica mineral tridymite. This is a high-temperature form of silica (SiO_2) – colorless crystals sometimes found in cavities in igneous rocks rich in feldspar and quartz. Since there is no indication of local volcanic activity the origin of the tridymite remains a puzzle.

---

[174] https://www.nasa.gov/jpl/msl/pia19809/curiosity-finds-hydrogen-rich-area-of-mars-subsurface

Curiosity's next major stop in September 2016 would be to observe some buttes – steep isolated hills somewhat like the ones in Monument Valley, Arizona – that had been identified from orbit by the MRO cameras. Murray Buttes (named to honor Bruce Murray who had died in 2013) is located at a gap in the Bagnold dune field – so their characterization could be conveniently carried out en route to the base of Mount Sharp.

The buttes are up to 15 meters in height and the way in which they formed is no doubt similar to those on Earth, namely the erosion of a sedimentary rock formation that is capped with denser, erosion-resistant rock. The composition of the terrain changed significantly as Curiosity slowly climbed and showed a marked difference after reached the Murray Buttes. In December 2016 the team reported:

"This series of pie charts shows similarities and differences in the mineral compositions of mudstones at 10 sites where NASA's Curiosity Mars rover collected rock-powder samples and analyzed them with the rover's Chemistry and Mineralogy (CheMin) instrument.

"The charts are arrayed in chronological order, with an indication of relative elevation as the rover first sampled two sites on the floor of Gale Crater in 2013 and later began climbing the crater's central mound, Mount Sharp. The pie chart farthest to the right and uphill shows composition at the "Sebina" target, sampled in October 2016. Five non-mudstone rock targets that the rover drilled and analyzed within this time frame are not included.

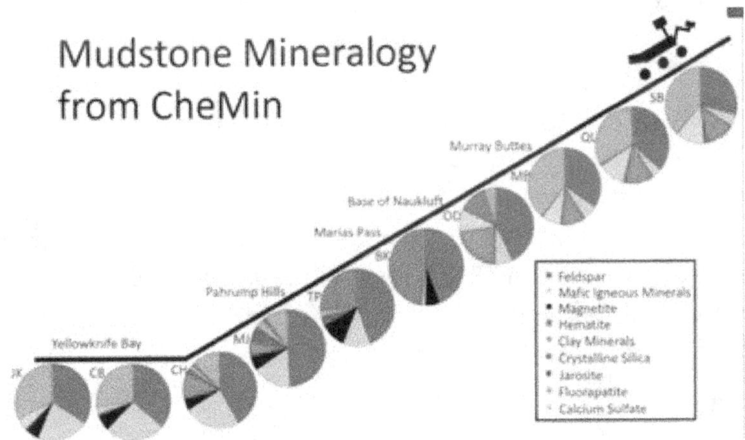

"The mineralogical variations in these mudstones may be due to differences in any or all of these factors: the source materials deposited by water that entered lakes, the processes of sedimentation and rock forming, and how the rocks were later altered. One trend that stands out is that the mineral jarosite – shown in purple – was more prominent in the "Pahrump Hills" area of lower Mount Sharp than at sites examined either earlier or later. Jarosite is an indicator of acidic water. Mudstone layers uphill from Pahrump Hills have barely detectable amounts of jarosite, indicating a shift away from acidic conditions in these overlying – thus younger – layers. Clay minerals, shown as green, declined in abundance at sites midway through this series, then came back as the rover climbed higher."[175]

Curiosity makes direct measurements of atmospheric chemistry as it travels along using its tune-able laser spectrometer (part of the SAM instrument) A remarkable tenfold spike in the abundance of methane in the atmosphere was observed over a two-month period. "Sample measurements taken 'a dozen times over 20 months' showed increases in late 2013 and early 2014, averaging 7 parts of methane per billion in the atmosphere. Before and after that, readings averaged around one-tenth that level". This phenomenon is still a mystery.[176]

Other sensitive measurements of atmospheric composition by the Quadrupole Mass Spectrometer instrument of the Sample Analysis at Mars instrument suite has determined that, as expected from the Viking measurements, hydrogen, carbon and argon are enriched in their heavier isotopes – a circumstance that confirms an earlier determination that Mars has lost most of its original atmosphere and water. In March 2015 new results of the SAM experiment were reported concerning the first detection of nitrogen – in the form of nitric oxide – released following the heating of sedimentary samples. Such nitrogen is useable by organisms and, therefore, adds to the conclusion that in the past the surface of Mars would have been habitable (the deep subsurface may well be habitable today).

The discovery of tridymite (a high-temperature form of silica) was repeated as Curiosity climbed further up the mountain. And yet another form of silica – Opal-A was identified. This is a hydrated form of silica i.e. $SiO_2.nH_2O$ deposited at a relatively low temperature in rock fissures and amorphous in structure rather than crystalline.

The Los Alamos National Laboratory members of the laser ChemCam team made an important announcement at a press conference on 14 December 2016 – that they had identified the element boron (boron has two stable isotopes 10-B nucleus has 5 protons and 5 neutrons 11-B has 5 protons and 6 neutrons) in mineral veins. Patrick Gasda of LANL commented, "If the boron that we found in calcium sulfate mineral veins on Mars is similar to what we see on Earth, it would indicate that the groundwater of ancient Mars that formed these veins would have been 0-60 degrees Celsius [32-140 degrees Fahrenheit] and neutral-to-alkaline pH. The temperature, pH, and dissolved mineral content of the groundwater could make it habitable".[177]

Boron is a relatively rare element in spite of its low atomic number – 5 – because it is not created by stellar nucleosynthesis but is made in supernovae. It is, however, essential to life as boron compounds serve to strengthen the cell walls of plants and for humans is an essential nutrient.

In the US boron is most notably associated with Death Valley in California where the mineral borax – sodium borate – was found in quantity and thereby became available as a commercial cleaner. Death Valley is, of course, one of the hottest places in the world in summer and in the past this led to the repeated evaporation of the wide shallow lake that seasonally filled the lowest spot in the valley. Earlier, during the Quaternary period that ended about 10,000 years ago, Death Valley was filled with a lake 160 km long and 180 meters deep.

The extensive sodium borate deposits are the result of the evaporation of this large lake. However, the environmental implications of the boron found by Curiosity are still open to debate. "Scientists are considering at least two possibilities for the source of boron that groundwater left in the veins: It could be that the drying out of part of Gale lake resulted in a boron-containing deposit in an overlying layer, not yet reached by Curiosity. Some of the material from this layer could have later been carried by groundwater down into fractures in the rocks. Or perhaps changes in the chemistry of clay-bearing deposits and groundwater affected how boron was picked up and dropped off within the local sediments".[178]

Following this first discovery of boron, Curiosity has discovered a systematic variation in the way that the composition changes with elevation with the quantity of both clay and boron increasing. "These and other variations can tell us about

---

[175] https://mars.nasa.gov/resources/mudstone-mineralogy-from-curiositys-chemin-2013-to-2016/
[176] http://www.astrobio.net/news-exclusive/mystery-methane-on-mars-the-saga-continues
[177] http://www.youtube.com/embed/D29hjfiSHzc
[178] http://www.lanl.gov/discover/news-release-archive/2016/December/12.13-first-detection-of-boron.php

conditions under which sediments were initially deposited and about how later groundwater moving through the accumulated layers altered and transported ingredients. Groundwater and chemicals dissolved in it that appeared later on Mars left its effects most clearly in mineral veins that filled cracks in older layered rock. But it also affected the composition of that rock matrix surrounding the veins, and the fluid was in turn affected by the rock".[179]

The Mars Hand Lens camera on the rover's arm is used regularly to check the condition of the wheels. Holes and tears in the wheels had worsened significantly during 2013 as Curiosity crossed terrain studded with sharp rocks. Development simulations at JPL had established that when three of the zigzag-shaped treads on any wheel have broken, that wheel has reached about 60 percent of its life. This wear and tear is the price that has to be paid in order to reach key geological units on lower Mount Sharp one of which is known from orbital mapping to contain hematite, an iron oxide that, on Earth, is laid down in liquid water. A band rich in clay minerals lies just above and then another layer of sulfate minerals. Gaining access to these layers was an important part of choosing the Gale Crater landing site in order to assess for how much time the environment on Mars would potentially have supported microbial life.

Curiosity's journey thus far:

The terrain through which Curiosity is traveling is extremely rugged as the MRO HiRISE images show and one is struck that such a journey would make a daunting challenge for an astronaut crew if they should ever set foot on Mars (the lunar surface as experienced by the Apollo astronauts and Lunakhods is relatively benign). So, the role of robotic mobile landers like Curiosity will surely not end should that day eventually arrive (the author will long be 6 feet under by then so he will never know). Astronauts would be limited to explorations that take place in the general vicinity of their landing site – a considerable restriction considering how many different regions on Mars invite in situ exploration. Today, as opposed to the situation decades ago, the most obvious benefit that might be provided by astronauts at a Mars base would be the *real-time* control of mobile robots by professional geologists in the crew – potentially a useful increase in productivity. However, perhaps in a relatively short time from now Mars surface rovers will be endowed with a useful degree of artificial geologic and astrobiology intelligence so that even real-time control by astronauts will not be a significant consideration. There is now an enormous amount of imaging coverage and associated spectroscopic data from a number of Mars rovers that can be used to 'train' such an AI Mars astrobiologist.

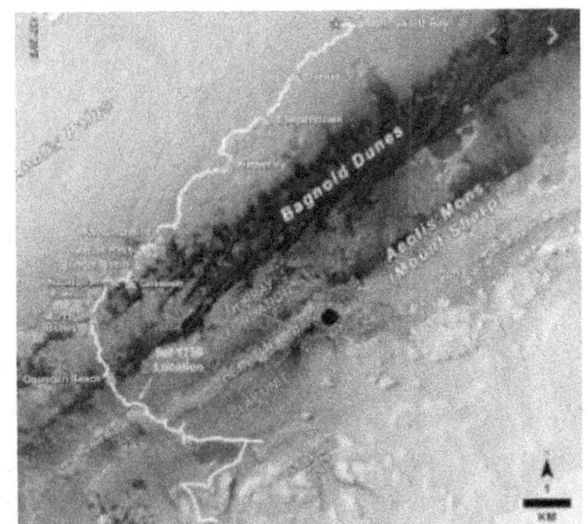

In the light of the Covid-19 lockdown, readers who may wonder about the continuing operation of Curiosity by the dispersed JPL team in Pasadena should know that operations continue. The flight teams at JPL have encountered many problems over the years and have dealt with them:

"For people who are able to work remotely during this time of social distancing, video conferences and emails have helped bridge the gap. The same holds true for the team behind NASA's Curiosity Mars rover. They're dealing with the same challenges of so many remote workers - quieting the dog, sharing space with partners and family, remembering to step away from the desk from time to time - but with a twist: They're operating on Mars. On March 20, 2020, nobody on the team was present at NASA's Jet Propulsion Laboratory in Southern California, where the mission is based. It was the first time the rover's operations were planned while the team was completely remote. Two days later, the commands they had sent to Mars executed as expected, resulting in Curiosity drilling a rock sample at a location called *Edinburgh*."

---

[179] https://en.wikipedia.org/wiki/Timeline_of_Mars_Science_Laboratory

"The team began to anticipate the need to go fully remote a couple weeks before, leading them to rethink how they would operate. Headsets, monitors and other equipment were distributed (picked up curbside, with all employees following proper social-distancing measures). Not everything they're used to working with at JPL could be sent home, however: Planners rely on 3D images from Mars and usually study them through special goggles that rapidly shift between left- and right-eye views to better reveal the contours of the landscape. That helps them figure out where to drive Curiosity and how far they can extend its robotic arm".[180]

JPL also responded to the Covid-19 emergency in a quite different way as reported in the *New Yorker* of May 19, 2020. In a lengthy article in the magazine's *Coronavirus Chronicles* entitled BREATHING ROOM, James Sommers describes the way in which engineers at the Jet Propulsion Laboratory led by David Van Buren, Roger Gibbs and Michelle Easter took on a new goal: "build the simplest possible, easy-to-manufacture ventilator, made from readily supplied parts, that was capable of treating all but the most complex cases of COVID-19, and get it approved by the FDA". "The project had attracted more than a hundred participants" and "for almost forty days straight they worked from sunup to sundown – brutal days that were a relief, in their way, from the ennui of lockdown." The design requires about 300 parts "carefully chosen to avoid siphoning supplies from medical-grade ventilator manufacturers". It has been tested and received an Emergency Use Authorization from the F.D.A. The author, given his JPL background is, naturally, very pleased to read of this life-saving spin-off from our exploration of the Solar System.

News from the Project Office, 12 July 2020: "After more than a year in a clay-rich region, Curiosity is making a mile-long journey around some deep sand so that it can explore higher up Mount Sharp. NASA's Curiosity Mars rover has started a road trip that will continue through the summer across roughly a mile (1.6 kilometers) of terrain. By trip's end, the rover

will be able to ascend to the next section of the 3-mile-tall Martian (5-kilometer-tall) mountain it's been exploring since 2014, searching for conditions that may have supported ancient microbial life".

"Stitched together from 28 images, NASA's Curiosity Mars rover captured this view from 'Greenheugh Pediment' on April 9, 2020, the 2,729[th] Martian day, or sol, of the mission. "In the foreground is the pediment's sandstone cap. At center is the "clay-bearing unit"; the floor of Gale Crater is in the distance. the 2,729[th] Martian day, or sol, of the mission after the rover ascended a steep slope, part of a geologic feature called "Greenheugh Pediment." In the foreground is the crusty sandstone cap that stretches the length of the pediment, forming an overhanging ledge in some parts. At center is the "clay-bearing unit," a region with a unique story to tell about the history of water on Mount Sharp, the 3-mile-tall (5-kilometer-tall) mountain Curiosity has been ascending since 2014. In the distance at the top of the image is the floor of Gale Crater, which is 96 miles (154 kilometers) wide."[181]

Curiosity's mission, one that has been more successful than anyone hoped, continues.

---

[180] https://www.jpl.nasa.gov/news/news.php?feature=7638&utm_source=iContact&utm_medium=email&utm_campaign=nasajpl&utm_content=weekly20200417-5

[181] https://www.jpl.nasa.gov/spaceimages/details.php?id=PIA23974

CHAPTER 14

RETURN TO MARS 4:

INDIA JOINS IN

Mangalyaan, MAVEN, ExoMars Trace Gas Orbiter & Schiaparelli Lander INSIGHT

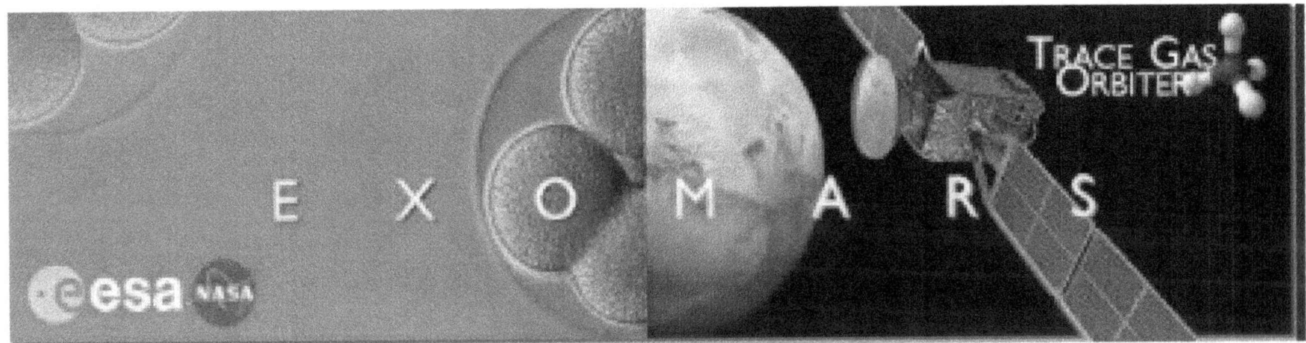

Driven by the vision of Vikram Sarabhai, considered the father of the Indian space program, the objective of the Indian Space Research organization, ISRO, is "to use space technology and its application to various national tasks". India has been active in space since the 1970s, initially using Soviet launch vehicles. In 1980, the first satellite was launched to Earth orbit by a small domestic launch vehicle. Two larger rockets were developed for launching numerous communications and Earth observation satellites – into polar orbits and into geostationary orbits. In October 2008, as related above, ISRO successfully inserted Chandrayaan-1 into lunar orbit and then turned its attention to Mars. On 24 September 2014 India succeeded on its first attempt to orbit Mars.

2013 ISRO Mars Orbiter Mission (Mangalyaan)

The Indian Space Research Organization (ISRO) launched its first interplanetary mission – the Mars Orbiter Mission (also called Mangalyaan) on 5 November 2013. The ISRO control center is in Bangalore and communications are at S-band frequencies to the Indian Deep Space Network antennae in Byalalu. The ISRO had been successful in carrying out a lunar orbiter mission in 2008, Chandrayaan-1, making a Mars orbiter the next challenge. The mission serves as a demonstrator for the many technologies required to successfully carry out deep space missions.

Moumita Dutta was the Project mission manager, Nandini Harinath the Deputy Operations Director of Navigation., Ritu Karidhal the Deputy Operations Director of Navigation, BS Kiran the Associate Project Director of Flight Dynamics, V Kesava Raju the Mission Director and V Koteswara Rao the ISRO scientific secretary.

The orbiter carries five science instruments to address atmospheric composition including methane abundance, particle environment and visible and IR surface studies. These are: Mars Colour Camera, Thermal Infrared Imaging Spectrometer, Methane Sensor for Mars, Mars Exospheric Neutral Composition Analyser and Lyman Alpha Photometer.

The launch took place at the Satish Dhawan Space Center on a Polar Satellite launch vehicle. The probe spent about a month in Earth orbit while making a series of seven propulsive maneuvers to increase orbital apogee before a final burn to send the spacecraft on to Mars. It reached Mars after a 298-day journey and went into orbit successfully on 24 September

2014, an impressive feat for India's first attempt. The MOM orbit is highly elliptical with a period of 72 hours 52 minutes, periapsis of 420 km and apoapsis of 77,000 km.

On 24 September 2015, ISRO released its "Mars Atlas", a 120-page scientific atlas containing images and data from Mangalyaan's first year in orbit. After the completion of one year of primary mission (September 2015), the orbiter reduced its periapsis altitude from 400 km to 260 km to benefit the Mars Exospheric Neutral Composition Analyzer that successfully measured the profiles of CO_2 and O_2. At the time of writing, December 2019, the spacecraft was operating successfully and so the Indian Space Agency may develop a lander mission for launch between 2020 and 2022.[182]

MOM has completed six years in orbit about Mars, a considerable success given that its designed mission life was only six months. MOM's Mars Colour Camera has captured more than a thousand images. The probe is still in good health and continues to work nominally with enough propellant for about another year of operations: a huge success for India's space agency. In contrast to the ultra-high-resolution images of the Mars Reconnaissance Orbiter, Magalyaan provides a synoptic view of Mars that is ideal for observing the changing weather on Mars – particularly dust storms that cover large swaths of the planet for days, weeks and months.

2013 NASA MAVEN

The year 2013 also saw the launch of NASA's Mars Atmosphere and Volatile EvolutioN Mission (MAVEN) designed to learn how the Martian atmosphere and surface water were lost to space over time: billions of years. The mission was, following Phoenix, successful in winning its place in the low-cost Mars Scout series (this program was subsequently discontinued). The mission is a five-part collaboration among: NASA's Goddard Space Flight Center, University of Colorado Boulder, University of California Berkeley, JPL, and Lockheed Martin Space Systems. The Principal Investigator is Bruce Jakosky of the University of Colorado's Laboratory for Atmospheric and Space Physics. Janet Luhmann of UC Berkeley is Deputy PI. Lockheed Martin built the MAVEN spacecraft, a design based on the Mars Reconnaissance Orbiter and Mars Odyssey. David Mitchell of NASA Goddard is the Project Manager. MAVEN was launched on an Atlas V Centaur on 18 November 2013 and was inserted into Mars orbit on 22 September 2014.

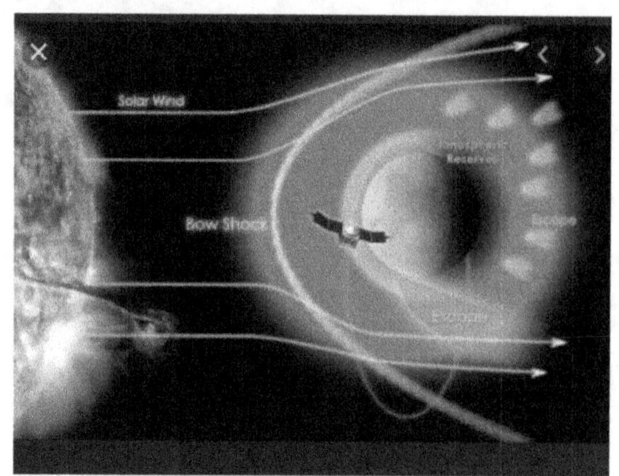

The orbit is of high inclination (75°) but is somewhat unusual in having a very low periapsis altitude of 150 km while apoapsis is 6,200 km and the period is 4.5 hours. As a result the spacecraft passes through the uppermost atmosphere 6 times each day and spends most of the time exposed to the solar wind whose intensity varies. The mission science goals focus on a fundamental problem in understanding the evolution of Mars, one that was first addressed directly by the mass spectrometer measurements made by the Viking entry vehicles. Specifically MAVEN proposed to make definitive measurements of the ratios of stable isotopes in the atmosphere and the loss of volatiles to space. Also, MAVEN would measure the current state of the upper atmosphere, the ionosphere and their various interactions with the solar wind. The current rates of escape of neutral gases and ions and the processes controlling them would be determined. And the orbiter would also serve as a backup relay for the several mobile landers on the Martian surface. To achieve these science goals MAVEN was equipped with a

---

[182] https://www.sciencemag.org/news/2017/02/india-eyes-return-mars-and-first-run-venus

comprehensive instrument payload: a particles and fields package built by the University of California, Berkeley measures solar wind and ionosphere electrons, solar wind and magneto-sheath ion density and velocity, thermal ions, the impact of solar energetic particles on the upper atmosphere. An Imaging UV Spectrometer remote sensing package built by the University of Colorado's LASP measures the upper atmosphere and ionosphere. A Neutral Gas and Ion Mass Spectrometer package built by NASA's Goddard Space Flight Center measures the composition and isotopes of neutral gases and ions.

A press release by the Project explains the way in which MAVEN in its eccentric orbit samples both the uppermost atmosphere and the solar wind environment:

On each orbit around Mars, MAVEN dips into the ionosphere – the layer of ions and electrons extending from about 75 to 300 miles above the surface. This layer serves as a kind of shield around the planet, deflecting the solar wind, an intense stream of hot, high-energy particles from the Sun. Scientists have long thought that measurements of the solar wind could be made only before these particles hit the invisible boundary of the ionosphere. MAVEN's Solar Wind Ion Analyzer, however, has discovered a stream of solar-wind particles that are not deflected but penetrate deep into Mars' upper atmosphere and ionosphere. Interactions in the upper atmosphere appear to transform this stream of ions into a neutral form that can penetrate to surprisingly low altitudes. Deep in the ionosphere, the stream emerges, almost Houdini-like, inion form again. The reappearance of these ions, which retain characteristics of the pristine solar wind, provides a new way to track the properties of the solar wind and may make it easier to link drivers of atmospheric loss directly to activity in the upper atmosphere and ionosphere. New insight into how gases leave the atmosphere is being provided by the spacecraft's Supra-thermal and Thermal Ion Composition (STATIC) instrument. Within hours after being turned on at Mars, STATIC detected the "polar plume" of ions escaping from Mars. This measurement is important in determining the rate of atmospheric loss. As the satellite dips down into the atmosphere, STATIC identifies the cold ionosphere at closest approach and subsequently measures the heating of this charged gas to escape velocities as MAVEN rises in altitude. The energized ions ultimately break free of the planet's gravity as they move along a plume that extends behind Mars.[183]

MAVEN is providing a definitive understanding of how the Martian atmosphere is gradually being stripped away by the solar wind – a process that has been ongoing since Mars lost its magnetic field as the interior cooled around 4 billion years ago. Solar storms are particularly destructive. Surface water that was at one time plentiful on Mars would have evaporated into the atmosphere to be split by the solar radiation into hydrogen and oxygen atoms. The hydrogen is easily lost and is thought to have done so at a fairly constant rate – a rate, however, that is greatly influenced by distance from the Sun.

MAVEN's observations through a full Martian year show that the escape rate is highest when Mars is closest to the Sun and only one-tenth as great when it is at its most distant. PI Bruce Jakosky reports the team's conclusion that "most of the gas ever present in the Mars atmosphere has been lost to space. Specifically, about 65 percent of the argon that was ever in the atmosphere has been lost in this way".[184]

As of early 2020 MAVEN was fully operational and well into its extended mission. In time MAVEN may function for several years as a tele-communications orbiter for future landed spacecraft.

2016 ESA ExoMars Trace Gas Orbiter and Schiaparelli Lander

ESA's ExoMars program, as its name suggests, seeks evidence of past and present life on Mars by a combination of remote sensing from orbit and, later, by investigations using a capable mobile surface laboratory. Monitoring from orbit will serve to detect trace gases – methane in particular. As earlier noted, in 2003 astronomers at NASA's Goddard Space Flight Center first detected this gas in the Martian atmosphere – which on Earth is associated with both biological and geological processes. In 2004 there were further reports – from ESA's Mars Express orbiter and from telescopic monitoring. The

---

[183] http://www.nasa.gov/press/goddard/2014/december/nasa-s-maven-mission-identifies-links-in-chain-leading-to-atmospheric-loss
[184] https://mars.nasa.gov/news/2017/nasas-maven-reveals-most-of-mars-atmosphere-was-lost-to-space

concentration of methane was just a trace – parts per billion – and variable from year to year. As noted much earlier, the variability of methane in the Martian atmosphere is puzzling because its destruction by solar UV would take place on a timescale of about 300 years. Another process evidently dominates removal of atmospheric methane – a process that must be hundreds of times more efficient than photochemical destruction. There also must be a very active replenishing source. One possibility is that the methane condenses and evaporates seasonally from clathrates where the hydrogen-bonded framework is contributed by water.

The Mars Science Laboratory, Curiosity, has been monitoring the level of methane in the atmosphere since it landed in 2011 and, as reported in the previous chapter, has measured occasional spikes in its concentration. The simultaneous monitoring from orbit will provide a powerful means to characterize the phenomenon and, it is to be hoped, that it will be possible to locate the source(s). Non-biological processes include water-rock reactions, radiolysis of water, and pyrite formation: these produce hydrogen that could generate methane with CO and CO_2. Methane could also be produced by a high-temperature subsurface reaction involving water, carbon dioxide, and olivine (a common Martian mineral).

ESA ExoMars Trace Gas Orbiter & Schiaparelli Lander

The ExoMars Program that is being carried out by ESA with Russian support has a complex and difficult history, only touched on here, that goes back to discussions that began in late 2008 between NASA and ESA toward collaboration in exploring Mars. ESA's Ministerial Council had recommended that the Agency seek international cooperation to complete its two proposed ExoMars missions: 1) a Trace Gas Orbiter plus an Entry, Descent and landing demonstrator Module (EDM), known as *Schiaparelli*, to be launched in March 2016 and 2) an exobiology rover to be launched in 2020. Meanwhile, NASA was reassessing its Mars Exploration Program plan in the wake of a decision to delay the launch of the Mars Science Laboratory (*Curiosity*) from 2009 to 2011. Unfortunately, NASA now found itself in big financial trouble as the real cost of building the James Webb Space Telescope was becoming clear. This flagship astronomy mission was of higher priority to NASA than ExoMars so, given a lack of forgiveness by President Obama's Office of Management and Budget, in early 2012 NASA was embarrassed to have to pull out of the planned joint Mars missions with ESA.

Not to be discouraged, ESA found another partner for the ExoMars program, namely Russia. On 15 March 2012, the ESA's ruling council announced it would press ahead with its ExoMars program in partnership with the Roscosmos whose contribution was to be the launchers – Protons – plus an additional entry, descent and landing system for the 2020 rover mission. Project management is somewhat complex: "the ExoMars prime contractor is Thales Alenia Space-Italy, in conjunction with Astrium UK, and with the support of ESA, the Italian space agency ASI and Russian industry led by NPO Lavochkin, including Krunichev and the Space Research Institute (IKI). Also contributing were European organizations participating in Mars science programs, including the University of Padua and the Astronomic Observatory of Capodimonte".[185]

With its new partner ESA remained committed to its ExoMars Program aimed at investigating the Martian environment and demonstrating new technologies "paving the way for a future Mars sample return mission in the 2020s".[186]

The development of the Trace Gas Orbiter and Schiaparelli lander (built in Italy as pathfinder for the originally intended 2020 landing) proceeded as planned toward their Proton launch out of the Baikonur Cosmodrome on 14 March 2016. The two spacecraft traveled as a single combination. The cruise phase in 2016 was relatively short and arrival at Mars took place in the middle of October. Schiaparelli was detached three days before arrival, coasted the remainder of the way and entered the Martian atmosphere on 19 October 2016 behind its heat shield at 21,000 km/hr. Alas, the attempted landing on Meridiani Planum was not a success. The heat shield had functioned as planned, decelerating the lander to 1,650 km/hour and an altitude of 11 km, measuring conditions as it did so. Then, after slowing its initial entry through the atmosphere, the module deployed a parachute and was to complete its landing on retrorockets by using a closed-loop

---

[185] https://www.thalesgroup.com/en/worldwide/space/press-release/press-info-exomars-status
[186] http://exploration.esa.int/mars/46048-programme-overview/

guidance, navigation and control system based on a Doppler radar altimeter sensor and on-board inertial measurement units.

Throughout the descent, various sensors recorded a number of atmospheric parameters and lander performance. The plan was that at 7 km in altitude the front heat shield would be jettisoned and the radar altimeter turned on, then at 1.3 km altitude above Mars the rear heat cover and parachute would be jettisoned. The final stages of the landing were to be performed using pulse-firing liquid-fuel engines or retrorockets. About two meters above ground, the engines were designed to turn off and let the platform land on a crushable structure, designed to deform and absorb the final touchdown impact… Contact was lost with the module 50 seconds before the planned touch down… By October 21, 2016 the ESA, after studying the data, said it was likely that things went wrong when the parachute released early, the engines then turned on but then turned off after too short of time… Initial analysis of returned telemetry data (600 megabytes) suggests that the heat shield and parachute deployment worked as expected, but the parachute was released too soon. In addition, the subsequent rocket thrusters' firing lasted about 3 seconds instead of the expected 30 seconds, ceasing operation while the lander was still 2-4 km above the surface, leading to an impact on the Martian surface at near terminal velocity.[187]

The nature of the failure was much like that experienced by NASA's Mars Polar Lander in December 1999. So, Mars had claimed as a victim one more lander whose impact point, like that of Beagle 2 would be imaged by the Mars Reconnaissance Orbiter. The immediate *science* loss was minimal because Schiaparelli, lacking solar panels, was expected to survive on the surface of Mars for only a short time.

Because the planned 2020 ExoMars mission is based on landing using a near-duplication of Schiaparelli (the ExoMars Rover is to be delivered by the instrumented ExoMars surface platform), the crash jeopardized that follow-on mission. However, based on a preliminary report on the malfunction that was presented at the December 2016 ESA ministerial meeting, the decision was made to proceed as planned to the 2020 mission. Evidently, the nature of the problem was judged to be adequately understood and sufficient progress was felt to have been made toward a successful entry, descent and landing system. A further consideration was that the 2016 TGO orbiter is to provide the communications relay for the 2020 rover and it does not have an unlimited life. Overall, a gutsy call in the author's opinion! Moreover, at the time of writing in early 2020, ESA had experienced a test failure with the deployment of the 35 meter main parachute and the issue had not been resolved. The launch of the lander – named for Rosalind Franklin – has now been delayed to 2022. The communication link for the lander will be provided by the Trace Gas Orbiter.

2016 ESA Trace Gas Orbiter TGO

Happily TGO was inserted into orbit successfully on 19 October 2016. From its initial elliptical orbit TGO took more than a year to aero-brake into a 400 km circular orbit, concluding in April 2018. From there science activities began. TGO's instrumentation consists of an Atmospheric Chemistry Suite (ACS), a Colour and Stereo Surface Imaging System (CaSSIS), a Fine Resolution Epithermal Neutron Detector (FREND) and a Nadir and Occultation for Mars Discovery (NOMAD) experiment. The Atmospheric Chemistry Suite, provided by Russia, has three infrared spectrometer channels. The Colour and Stereo Surface Imaging System was provided by Switzerland and are used to build surface elevation models and characterize potential landing sites with a resolution of 4.5 meters. The Fine-Resolution Epithermal Neutron Detector, FREND, provided by Russia is sensitive to the presence of hydrogen in the top meter of the surface (similar to the DAN instrument on board Curiosity). The Nadir and Occultation for Mars Discovery experiment provided by Belgium has two IR and one UV spectrometer channels.

The science team is in the process of "characterising spatial & temporal variation, and localisation of sources for a broad list of atmospheric trace gases". They anticipate that "if methane CH_4 is found in the presence of propane C_3H_8 or ethane C_2H_6 that would be a strong indication that biological processes are involved. However, if methane is found in the

---

[187] https://en.wikipedia.org/wiki/Schiaparelli_EDM_lander

presence of gases such as sulfur dioxide SO_2 that would be an indication that the methane is a byproduct of geological processes".[188]

Methane mystery plot thickens

The two complementary instruments also started their measurements of trace gases in the Martian atmosphere. Trace gases occupy less than one percent of the atmosphere by volume, and require highly precise measurement techniques to determine their exact chemical fingerprints in the composition. The presence of trace gases is typically measured in 'parts per billion by volume' (ppbv), so for the example for Earth's methane inventory measuring 1800 ppbv, for every billion molecules, 1800 are methane.

Methane is of particular interest for Mars scientists, because it can be a signature of life, as well as geological processes – on Earth, for example, 95% of methane in the atmosphere comes from biological processes. Because it can be destroyed by solar radiation on timescales of several hundred years, any detection of the molecule in present times implies it must have been released relatively recently – even if the methane itself was produced millions or billions of years ago and remained trapped in underground reservoirs until now. In addition, trace gases are mixed efficiently on a daily basis close to the planet's surface, with global wind circulation models dictating that methane would be mixed evenly around the planet within a few months.

Reports of methane in the Martian atmosphere have been intensely debated because detections have been very sporadic in time and location, and often fell at the limit of the instruments' detection limits. ESA's Mars Express contributed one of the first measurements from orbit in 2004 at that time indicating the presence of methane amounting to 10 ppbv.

Earth-based telescopes have also reported both non-detections and transient measurements up to about 45 ppbv, while NASA's Curiosity rover, exploring Gale Crater since 2012, has suggested a background level of methane that varies with the seasons between about 0.2 and 0.7 ppbv – with some higher level spikes. More recently, Mars Express observed a methane spike one day after one of Curiosity's highest-level readings. The new results from TGO provide the most detailed global analysis yet, finding an upper limit of 0.05 ppbv, that is, 10–100 times less methane than all previous reported detections. The most precise detection limit of 0.012 ppbv was achieved at 3 km altitude.

As an upper limit, 0.05 ppbv still corresponds to up to 500 tons of methane emitted over a 300 year predicted lifetime of the molecule when considering atmospheric destruction processes alone, but dispersed over the entire atmosphere, this is extremely low.

"We have beautiful, high-accuracy data tracing signals of water within the range of where we would expect to see methane, but yet we can only report a modest upper limit that suggests a global absence of methane," says ACS principal investigator Oleg Korablev from the Space Research Institute, Russian Academy of Sciences, Moscow.

"The TGO's high-precision measurements seem to be at odds with previous detections; to reconcile the various datasets and match the fast transition from previously reported plumes to the apparently very low background levels, we need to find a method that efficiently destroys methane close to the surface of the planet."

"Just as the question of the presence of methane and where it might be coming from has caused so much debate, so the issue of where it is going, and how quickly it can disappear, is equally interesting," says Håkan.

"We don't have all the pieces of the puzzle or see the full picture yet, but that is why we are there with TGO, making a detailed analysis of the atmosphere with the best instruments we have, to better understand how active this planet is – whether geologically or biologically."[189]

---

[188] https://en.wikipedia.org/wiki/Trace_Gas_Orbiter
[189] http://www.esa.int/Science_Exploration/Human_and_Robotic_Exploration/Exploration/ExoMars/First_results_from_the_ExoMars_Trace_Gas_Orbiter

2018 NASA INSIGHT

The characterization of the interior of Mars – the size, thickness, density and overall structure of the planet's core, mantle and crust, as well as the rate at which heat escapes from the planet's interior – is long overdue. The Mars exploration strategy of "Follow the Water" has managed until now to leave out this one key planetary science discipline. Most of what little we know about the Mars interior comes from the detailed mapping of the magnetic field carried out by the Mars Global Surveyor (briefly discussed earlier). This determined that the magnetic field strength is very low in the northern lowlands, the impact basins and in the Tharsis volcano province. In the part of the ancient southern crust that lacks both impact basins and volcanism the magnetic field strength is high. Evidently Mars did once have an interior dynamo but it shut down billions of years ago. Only the rocks that formed before the shutdown (and were not subsequently heated above their Curie point) still retain their magnetization.

In 1976, the two Vikings each had a seismometer attached to one of the lander legs. This instrument, unfortunately, failed to unlatch on the Viking 1 lander. The Viking seismometer collected approximately 640 hours of data but detected no marsquakes. Because the one operational seismometer was poorly located on the lander's leg, it lacked the sensitivity required to detect low-magnitude or distant seismic activity. The Viking PI, Don Anderson of Caltech, estimated that a well-designed seismometer experiment could be up to a thousand times more sensitive than the leg-mounted Viking unit. A distributed array of seismometers, deployed and monitored for many years, would be ideal to permit the triangulation of marsquake locations. INSIGHT alone, however, will be a good start. Depending on the findings perhaps a seismic array can be built up in the coming years. Quite recently, as has been described in the previous chapter, MRO HiRISE imaging data have provided evidence of past marsquakes in the west Candor Chasma region of Valles Marineris. The science team observes conical hills of the kind that are formed by the underground movement of water-lain sediments in response to ground shaking from several large fault zones in the area.

The InSight spacecraft was manufactured by Lockheed Martin based on the 2008 Phoenix design. The mission is managed by JPL and most of its scientific instruments were built by European agencies. The mission launched on 5 May 2018 on an Atlas V and successfully landed on Elysium Planitia on 26 November 2018 about 600 km north of Gale Crater. After landing, the mission took some months to deploy and commission the geophysical science instruments.

The first faint seismic signal was recoded on 6 April as a subtle rumble. The team has concluded that the Martian crust is like "a mix of the Earth's crust and the Moon's" where "drier crusts like the Moon's remain fractured after impacts, scattering sound waves for tens of minutes rather than allowing them to travel in a straight line. Mars, with its cratered surface, is slightly more Moon-like, with seismic waves ringing for a minute or so, whereas quakes on Earth can come and go in seconds".[190]

By late 2019 the seismometer had made great strides. In a 13 December 2019 paper in the journal *Nature* the lander has detected more than 300 quakes and traced some back to their source. The team reports in the article that most of the marsquakes are tiny, much smaller than anything that would be felt on Earth. But a couple have been big enough – up to nearly magnitude 4 – for scientists to be able to trace them back to their source.

"Two of the biggest marsquakes came from a geologically active area known as Cerberus Fossae, which lies about 1,600 kilometres east of InSight. The quakes there might have been caused by the build-up of stress along geological faults in the Martian crust, and then released in a marsquake. Other early findings from the mission include mysterious magnetic pulses that appear around midnight each night around the lander. But one of InSight's main goals — to hammer a heat probe 5 metres into the Martian ground — remains frustratingly out of reach. The probe,

---

[190] https://mars.nasa.gov/news/8430/nasas-insight-detects-first-likely-quake-on-mars/?site=insight

dubbed 'the mole', has encountered more friction in the soil than scientists had expected. In October, it even unexpectedly backed out of its hole."[191]

In an article by Alexandra Witze in *Scientific American Nature* 18 December 2019 the science team reports: The biggest discoveries so far have come from the ever-expanding catalogue of marsquakes. InSight's highly sensitive seismometer hunts for quakes at night, after the winds that shake the ground during the day die down. The marsquakes come in two types. The most common shakes the ground at high frequencies. Less common is a type that is detectable at lower frequencies. The high-frequency signals might be coming from quakes that rupture the shallow Martian crust, whereas the low-frequency ones might be traveling from deeper within the planet, in its mantle, said Domenico Giardini, a seismologist at the Swiss Federal Institute of Technology in Zurich.[192]

Two of the biggest marsquakes hit in May and July 2019. Both were the low-frequency type. Team members were able to trace the seismic energy back to Cerberus Fossae. This area is home to recent geological activity, including faults that seem to have moved in the past ten million years.

Before InSight launched, researchers had predicted it might be able to detect quakes coming from Cerberus Fossae. The faults there could build up stress at their ends, said Alice Jacob, a planetary scientist at the Paris Institute of Earth Physics. An analysis she led suggests that this could be the source of the marsquakes picked up by InSight. The rate of quakes has been increasing, Banerdt said – from a few sporadic tremors reported after InSight landed to the current pace of two a day. Mission scientists aren't sure why.

Equally mysterious are the magnetic pulses that show up every night. InSight measured them with its magnetometer, and they are thought to be related to something happening in the space environment around Mars. One idea is that they are created when charged particles from the solar wind slam into Mars.[193]

Steve Clifford whose research focuses on the search for evidence of a deep Martian hydrosphere adds this comment in a January 4 2020 email communication with the author: "InSight has provided some mixed results. The Q values (Seismic Quality Factor Q is a measure of seismic attenuation with distance) deduced from recent seismic events seem more consistent with the Moon than Earth, suggesting the deeper crust is currently dry. This could indicate that whatever groundwater that once existed on Mars has been subsequently cold-trapped into a deepening cryosphere as the planet's heat flux has declined with time, or it may simply indicate that groundwater is absent in the immediate vicinity of the lander".

"This is the problem with having just one seismometer. A more logical investigation would have been with the four seismic station originally proposed by the CNES for the Netlander mission that was supposed to fly in 2007, but which got cancelled due to funding issues. However, InSight also has a magnetometer whose initial results suggest the presence of a highly conductive layer at depth – consistent with the presence of saline groundwater. So, who knows?"

"Robert Grimm has been looking at various possible electromagnetic investigations of the Martian subsurface that could sound down to depths of several to as much as 20 km. He has published a number of papers and abstracts on this. These are also techniques that are being actively investigated by JPL for possible future missions".[194]

On 28 February 2019, the Heat and Physical Properties Package probe started to drill into the surface. The probe and its digging mole were intended to reach a maximum depth of 5 m in about two months after just a week the mole paused its digging after penetrating only 35 cm. Evidently, the soil on Mars does not provide necessary friction for drilling: "causing the mole to bounce around and form a wide pit around itself rather than dig deeper".[195] The Project reports, July 2020, "Akin to a 16-inch-long (40-centimeter-long) pile driver, the self-hammering mole has experienced difficulty getting into the Martian soil since February 2019. It's mostly buried now, thanks to recent efforts to push down on the mole with the scoop on the end of the robotic arm. But whether it will be able to dig deep enough - at least 10 feet (3 meters) - to get an accurate temperature

---

[191] https://www.nature.com/articles/d41586-019-03796-7
[192] ibid.
[193] https://www.scientificamerican.com/article/marsquakes-reveal-red-planets-hidden-geology/
[194] Steve Clifford email communication to author 13 January 2020
[195] https://www.thestar.com/news/world/us/2019/10/01/nasa-lander-captures-marsquakes-other-martian-sounds.html

reading of the planet remains to be seen. Images taken by InSight during a Saturday, June 20, hammering session show bits of soil jostling within the scoop - possible evidence that the mole had begun bouncing in place, knocking the bottom of the scoop".

In another recent report (INSIGHT NEWS February 24, 2020) the team observes: Mars trembles more often – but also more mildly – than expected. SEIS has found more than 450 seismic signals to date, the vast majority of which are probably quakes (as opposed to data noise created by environmental factors, like wind). The largest quake was about magnitude 4.0 in size – not quite large enough to travel down below the crust into the planet's lower mantle and core. Those are "the juiciest parts of the apple" when it comes to studying the planet's inner structure, said Bruce Banerdt, InSight Principal Investigator at JPL.

Scientists are ready for more: It took months after InSight's landing in November 2018 before they recorded the first seismic event. By the end of 2019, SEIS was detecting about two seismic signals a day, suggesting that InSight just happened to touch down at a particularly quiet time. Scientists still have their fingers crossed for "the Big One."

"Mars doesn't have tectonic plates like Earth, but it does have volcanically active regions that can cause rumbles. A pair of quakes was strongly linked to one such region, Cerberus Fossae, where scientists see boulders that may have been shaken down cliffsides. Ancient floods there carved channels nearly 800 miles (1,300 kilometers) long. Lava flows then seeped into those channels within the past 10 million years — the blink of an eye in geologic time."

"Some of these young lava flows show signs of having been fractured by quakes less than 2 million years ago. "It's just about the youngest tectonic feature on the planet," said planetary geologist Matt Golombek of JPL. "The fact that we're seeing evidence of shaking in this region isn't a surprise, but it's very cool."[196]

The INSIGHT lander experiments have proven to be challenging and mysterious in a number of different ways. First, in September 2019 the team reported unexplained magnetic pulses and magnetic oscillations. Then they found that the soil on Mars does not provide enough friction for drilling – causing the mole to "bounce around and form a wide pit around itself rather than penetrating deeper". This is important information for planning future missions. A summary of results after a year on the surface showed that Mars does have active quakes and magnetic pulses. Curiously, the magnetic field at the landing site is about 10 times stronger than had been expected and it fluctuates rapidly.

The seismometer, radio experiment and the weather instruments continue to operate and the surface mission has been extended by two years, until end of December 2022.

---

[196] https://www.sciencedaily.com/releases/2020/02/200224151502.htm

CHAPTER 15

NEW AND FUTURE MARS MISSIONS

NASA Mars 2020, CSA Tianwen-1 2020, Emirates Mars Mission 2020, ESA EXOMARS 2022, Mars Sample Return, Deep Subsurface Exploration

Three missions to Mars were launched in the July 2020 window by NASA, China and a new player – the United Arab Emirates. These are described below and their goals, in the case of NASA, follow upon the earlier recommendations of the National Academy of Sciences and the internal Mars Exploration Advisory Committee, MEPAG.

An Astrobiology Strategy for the Search for Life in the Universe (2018)

In future searches for evidence of past or present microbial life on Mars, the National Academy report notes that "biological ecosystems thrive in specific niches, commonly thought of as spatially large and long-lived. This need not be the case universally as the situation in Chile's hyper-arid Atacama Desert provides a case in point. In the hyper-arid core of the desert, photo-synthesizing microbial communities are to be found on undersides of translucent rocks. The rocks provide protection from desiccation and from solar ultraviolet solar radiation while their translucence allows the microbes sufficient light for photosynthesis" … "Similarly, endolithic communities are found sparsely distributed throughout terrestrial hot spring systems". Evidently, in the case of Mars "The search becomes not only for habitats that may have become inhabited as or after life emerged, but also for locations where life or signs of extinct life may persist despite widespread loss of surface habitability". So, although conditions on Mars today are such that life could not emerge there on the surface it is possible that viable cells on Mars could survive and maybe flourish for short periods in isolated, transient habitable niches at the surface. Recently, the ongoing Curiosity rover mission in Gale Crater has illuminated the potential for ancient sediments to provide nutrients and energy for life; combined with the presence of water, Gale Crater could have served as an ephemeral habitat for life. That mission continues. More generally, the search for evidence of past or present life on the Martian surface is one that should continue based on the increased understanding that ongoing and near future missions provide.

Regarding the possibility of life at depth the report notes: "Another example of slow-growing cells to take into account when considering planetary habitability are those entities residing in Earth's crust or deep sediments. Deep-sea sediments host viable cells down to depths of nearly 2 km. In oligotrophic environments, the low availability of both nutrients and water severely limits cell viability. Cells in this environment have a remarkably slow metabolic turnover of about one cell division per thousand years. The fractured crust is a somewhat different environment, where water that has infiltrated deep fractures can entrain microbes over long periods of time at depths up to 3 to 4 km. Each fracture system, some as small as 1 cm, is an oasis for life and the microbial biomass in these subsurface habitats is low compared to that at the surface. Autotrophy and methane cycling are the main means of metabolic support for these deep microorganisms, with cell turnover times that range from one to hundreds of years".

The report makes the following recommendation: *NASA's programs and missions should reflect a dedicated focus on research and exploration of subsurface habitability in light of recent advances demonstrating the breadth and diversity of life in Earth's subsurface,*

NASA Mars 2020: *Perseverance*

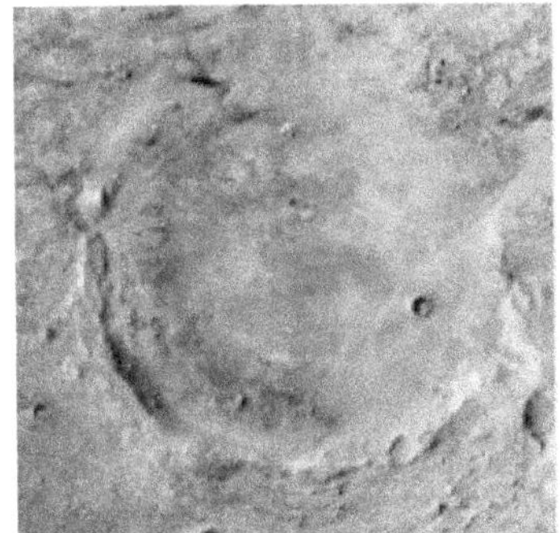

Building on the continuing success of *Curiosity's* exploration, the *Perseverance* rover with a new science payload was launched on an Atlas V on 30 July 2020 toward a landing in February 2021.

NASA has chosen Jezero Crater as the landing site for the 2020 rover after a five-year search, during which details of more than 60 candidate locations were considered. The crater is located on the western edge of Isidis Planitia, a giant impact basin just north of the equator. The 45-km, 500-meter-deep crater was once home to an ancient river delta when, billions of years ago, water and sediments flowed into the crater to become a huge lake. Water entering from a channel to the west deposited sediments to form a delta —

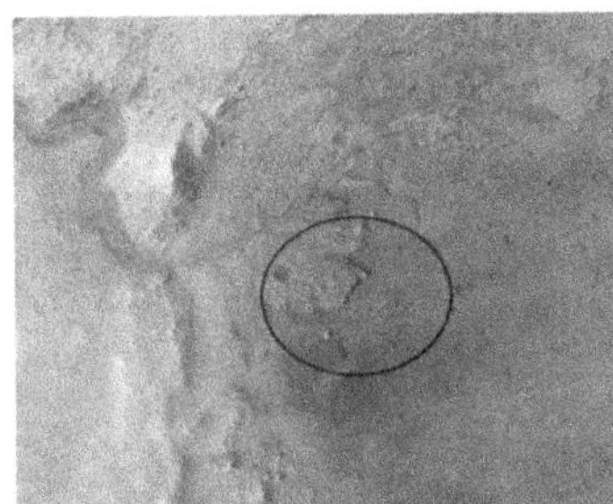

the landing target. Orbital observations have identified there a range of rock types, including clays and carbonates — these may have preserved organic molecules that would an important clue regarding possible past life. "Particularly enticing is the "bathtub ring" of sediments laid down at the ancient lake's shoreline. It's here that Perseverance could find stromatolites, ancient fossilized microbial mats". Deputy project scientist Katie Stack Morgan notes "they leave behind very thin layers, with concentrations of particular elements and organics at repeated intervals. We'll be looking for those fine laminations, looking for chemistry and textures you wouldn't expect if these things were just abiotic, or didn't involve life".[197]

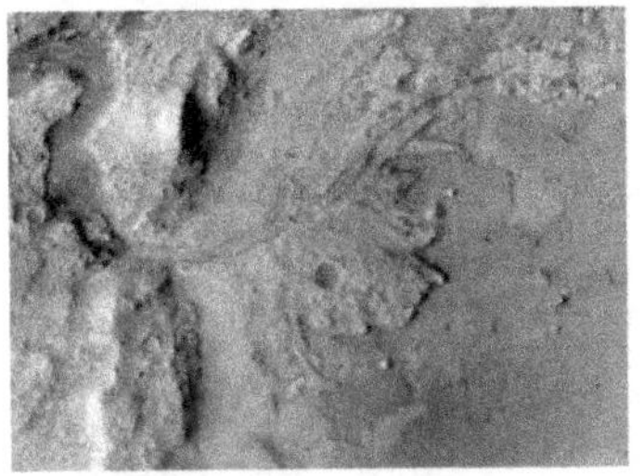

Perseverance Mars 2020 rover science and technology payload

The payload has been selected: an X-ray fluorescence spectrometer; a ground-penetrating radar able to image tens of meters beneath the rover; a set of environmental sensors (temperature, pressure, wind speed etc.; an advanced version of Curiosity's ChemCam to analyze rocks and regolith from a distance using a laser; a stereoscopic imager with zoom lens; a UV spectrometer and fine-scale imager with a UV laser to identify minerals and detect organics; and a small device that will produce oxygen from atmospheric carbon dioxide toward future astronaut life support and use as rocket fuel.

Perseverance is a major evolution of the very successful Curiosity rover and carries a broad suite of instruments. It does not include any biology experiments of the kind that the Viking landers carried out but, instead, its astrobiology investigation will consist of the collection of multiple core samples that will subsequently be returned to Earth for analysis. The science payload is as follows:

* Mastcam-Z, an advanced camera system with panoramic and stereoscopic imaging capability with the ability to zoom. The instrument also will determine mineralogy of the Martian surface and assist with rover operations. The principal investigator is James Bell, Arizona State University in Tempe.

* SuperCam, an instrument that can provide imaging, chemical composition analysis, and mineralogy. The instrument will also be able to detect the presence of organic compounds in rocks and regolith from a distance. The principal investigator is

---

[197] https://www.bbc.co.uk/news/science-environment-51544476

Roger Wiens, Los Alamos National Laboratory, Los Alamos, New Mexico. This instrument also has a significant contribution from the Centre National d'Etudes Spatiales, Institut de Recherche en Astrophysique et Planétologie (CNES/IRAP) France.

* Planetary Instrument for X-ray Lithochemistry (PIXL), an X-ray fluorescence spectrometer that will also contain an imager with high resolution to determine the fine-scale elemental composition of Martian surface materials. PIXL will provide capabilities that permit more detailed detection and analysis of chemical elements than ever before. The principal investigator is Abigail Allwood, NASA's Jet Propulsion Laboratory (JPL) in Pasadena, California.

* Scanning Habitable Environments with Raman & Luminescence for Organics and Chemicals (SHERLOC), a spectrometer that will provide fine-scale imaging and uses an ultraviolet (UV) laser to determine fine-scale mineralogy and detect organic compounds. SHERLOC will be the first UV Raman spectrometer to fly to the surface of Mars and will provide complementary measurements with other instruments in the payload. The principal investigator is Luther Beegle, JPL.

* Mars Environmental Dynamics Analyzer (MEDA), a set of sensors that will provide measurements of temperature, wind speed and direction, pressure, relative humidity and dust size and shape. The principal investigator is Jose Rodriguez-Manfredi, Centro de Astrobiologia, Instituto Nacional de Tecnica Aeroespacial, Spain.

* The Radar Imager for Mars' Subsurface Exploration (RIMFAX), a ground-penetrating radar that will provide centimeter-scale resolution of the geologic structure of the subsurface. The principal investigator is Svein-Erik Hamran, the Norwegian Defence Research Establishment, Norway.[198]

Perseverance has improved computational capability using RAD750 radiation hardened computers manufactured by BAE Systems Electronics. Nevertheless, autonomous operations on Mars still require humans-in-the-loop because of the limitations of the computer processors. This will change in the near future when high performance, multi-core radiation-hardened processors become available to provide a hundred times computational capacity using no more power than at present. To take advantage of this advance Mars geologists have been labelling Curiosity images to 'train' future rovers toward AI surface operations.

---

[198] https://mars.nasa.gov/news/nasa-announces-mars-2020-rover-payload-to-explore-the-red-planet-as-never-before/

2020 Mars Technology Demonstrations:

Rotorcraft Ingenuity

Mars 2020 will feature a remarkable new technology experiment in the form of a small rotorcraft that will be transported to Mars on the belly-pan of the rover. Larry Young and colleagues at the NASA Ames Research Center have been developing Mars rotorcraft technology for many years in chambers that can simulate the low atmospheric pressure at the Martian surface and JPL has been doing so since 2013.[199]

Ingenuity was built by a collaboration between AeroVironment Inc. and JPL. The experimental rotorcraft weighs just 2 kg and has twin counter-rotating blades operating at ~3,000 rpm!

Given the rugged nature of the Martian surface there could be many applications for a rotorcraft in support of a rover. A lead role for a larger instrumented rotorcraft could include carrying out stratigraphic exploration of the steep walls of the Valles Marineris.

Mars Oxygen ISRU Experiment (MOXIE)

Moxie will produce oxygen from Martian atmospheric carbon dioxide. The principal investigator is Michael Hecht, Massachusetts Institute of Technology, Cambridge, Massachusetts.

Planetary Protection

Mars 2020 is to be the leading half of the first Mars sample return mission – carried out together with ESA. The mobile lander will collect and seal drill samples – more than 40 – in metal tubes. This coupling of the 2020 mission with subsequent sample return – another 'flagship' mission with much new technology required – makes for some complication. Before samples can be returned to Earth a facility would have to be created in which to analyze the samples under quarantine conditions more complex than even for Bio-Safety Level 4 (the highest level) facilities that the US Centers for Disease Control and Prevention have established. The reason for this is the possibility of introducing of extra-terrestrial organisms into Earth's biosphere (the scientific consensus is that the potential for large-scale effects, either through pathogenesis or ecological disruption, is extremely small). NASA has from the beginning of lunar and Mars exploration taken seriously its task of protecting Planet Earth from any possible contamination by extra-terrestrial organisms and has appointed a series of Planetary Protection Officers with this responsibility. A central tenet is the presumption of ignorance. The responsibility is enshrined in law – a treaty signed by the US and the Soviet Union in 1967. See the website of the Lunar and Planetary Institute for more information.

The possible danger of extra-terrestrial contamination was brought to public attention in 1969 by the publication of Michael Crichton's novel *The Andromeda Strain*. This was made into a movie in 1971 and in 2008 into a television miniseries of the same name. So, NASA is well aware that this is an issue it must address with rigor. Recall that one of the Principal Investigators on the Viking Biology team, Dr Gil Levin whose Labelled Release experiment in 1976 produced a positive result (in spite of Viking's failure to find organics in the Martian soil) and he remains convinced that his experiment *did* discover primitive life on Mars.

When dealing with samples from Mars it is essential to protect the samples from terrestrial contamination – *forward* contamination – to avoid false positives. Mars landers undergo sterilization. So, *two-way* protection – both forward and

---

[199] https://rotorcraft.arc.nasa.gov/Research/Programs/mars_helicopter.html

backward – is required. At high bio-safety levels, precautions to protect the environment include airflow systems, multiple containment rooms, sealed containers, positive-pressure personnel suits, established protocols for all procedures, extensive personnel training, and high levels of security to control access to the facility. It would therefore seem ideal to have a two-way containment arrangement for Mars samples – like the evacuated walls of a giant vacuum flask – between the chamber containing the samples and the laboratory where the analysts function. Such a laboratory would require a large degree of automation as glove-box manipulation like that of the Apollo samples would not be possible. In NASA's planning, however, a simpler approach is proposed to allow glove box operations in which "the Facility uses successive series of 5-6 negative pressure levels to the outside in 25 to 50 Pa Increments. This ensures the protection of the population."[200]

A major undertaking before Mars samples are returned to Earth will be the location and creation of this unique facility – a multi-year project in itself. It seems safe to say that this problem is not yet fully understood and will require much more attention before 'selling' this part of the MSR project to the public. Work would need to be started on designing, locating, constructing and testing such a facility as much as a decade before the launch of a sample return mission. The cost of this will be significant.

The team at JPL, learning from their Curiosity experience, added important elements of capability to the Perseverance landing system to increase its chances of success:

First, using what is called *Terrain-Relative Navigation*, Perseverance, acquired images as it descended that were used to estimate its location with an accuracy of about 40 meters. This allowed it to maneuver, avoid potential hazards and land closer to the target area of greatest scientific interest – a valuable form of artificial intelligence.

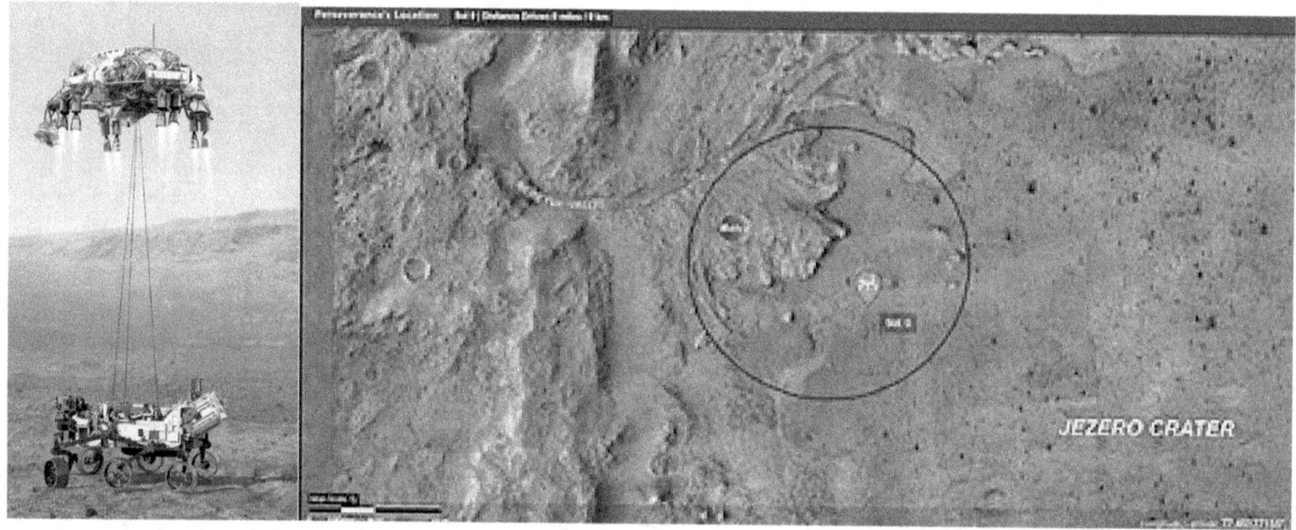

Perseverance, using its sky crane touched down safely on 18 February 2021 as planned and immediately returned images of the surrounding terrain. Not surprisingly, given the effort to avoid hazards, the site appears very bland.

---

[200] https://www.lpi.usra.edu/pss/presentations/200803/04-Atlas-PPSonMSR.pdf

At the time of writing in February 2021 the landed mission was just getting underway. The reader will no doubt have followed its progress in JPL news releases – along with the continuing progress of Curiosity in Gale Crater.

2020 CSA Tianwen-1

Having been frustrated by the failure of the Russia-China Phobos-Grunt mission in 2011 but having successfully landed a spacecraft with an instrumented rover on the Moon, the Chinese Space Agency was ready in 2020 to do the same on Mars – a much bigger challenge. The mission name is Tianwen-1 – translation *Questions to Heaven*. The orbiter, lander and rover have been developed by the China Aerospace Science and Technology Corporation (CASC), and managed by the National Space Science Centre (NSSC) in Beijing. Simulated landings have been performed by the Beijing Institute of Space Mechanics and Electricity.

The orbiter is instrumented with cameras, magnetometer, mineralogy spectrometer, subsurface radar and charged particle detectors.

The rover is powered by solar panels and is instrumented with a radar to probe the subsurface. In addition to a camera, magnetometer and meteorology instrumentation, it is instrumented to analyze the chemistry of the soil looking for bio-molecules and biosignatures.

The mission was successfully launched from the Wenchang Spacecraft Launch Site on 23 July 2020 on a Long March 5 heavy-lift rocket. Its stated objectives are to search for evidence of both current and past life, and to assess the planet's environment. Mars orbit insertion took place on 15 February 2021 and landing is anticipated to take place in May or June after the orbiter has carried out a survey of potential landing sites.

2020 United Arab Emirates Mars Mission: Hope

The United Arab Emirates is the latest newcomer to space exploration with plans to launch an orbiter to Mars during the 2020 launch window. The mission is effectively an international collaboration with the US and Japan as partners but with the Emirates funding the enterprise. According to the New York Times (front page article 21 February 2020) the Emirates "are planning for the future when petroleum no longer flows as bountifully, to invest its current wealth in new knowledge-based industries." Sarah al-Amiri, the minister of state for advanced sciences who leads the mission's science asks rhetorically "How do you develop highly skilled people that are able to take on higher risks? – That was the reason to go to space exploration." The orbiter has been built in collaboration with Emirati engineers at the University of Colorado where several investigators have long experience in building instrumentation for NASA's Mars orbiters.

The science focus is on better understanding the mechanisms driving the gradual loss of the Martian atmosphere – a study presently being carried out by NASA's MAVEN mission (described in the previous chapter) whose PI is Bruce Jakosky of the University of Colorado. Hope will seek to understand the connection between current Martian weather and the ancient climate of Mars, will track the behavior and escape of hydrogen and oxygen, and determine how the lower and upper levels of the Martian atmosphere are connected. The Project Manager is Omran Sharaf. Sarah Amiri is Deputy Project Manager and Lead Science Investigator. The spacecraft instrumentation comprises (PI Phil Christensen of the University of Arizona, an IR spectrometer, a UV spectrometer and a camera that is a collaboration between the University of Colorado and the Mohammed Bin Rashid Space Centre in Dubai.

Hope was launched out of the Tanegashima Space Center in Japan on a Mitsubishi H-IIA launch vehicle on 19 July 2020 and reached Mars at the time of the UAE's celebration of the 50th anniversary of its independence. It performed a successful capture on 9 February 2021 into an elliptical, roughly 22,000 x 44,000 km orbit with a period of 55 hours and a 25 degree inclination.

ESA/Roscosmos EXOMARS 2022

ExoMars 2022 (postponed from planned 2020 launch) is a cornerstone mission in ESA's exploration program with ministerial support confirmed in December 2016. The mission is similar in many respects to NASA's Mars 2020 mission. The plan is for the European rover and its Russian landing 'surface platform' to be launched on a Proton rocket out of Baikonor. The rover, named for English chemist and X-ray crystallographer Rosalind Franklin (1920-58) has been built by Thales Alenia Space with several space-engineering facilities in the UK.

This photograph shows the completed ExoMars rover, named "Rosalind Franklin," lifted in preparation for shipment from a facility in Stevenage, England. Credit: Max Alexander/Airbus. The solar-powered robot weighs about 300 kilograms. It is equipped with a drill to acquire samples from up to 2 meters depth and, also, with a range of analytical instruments. Its lander, named *Kazachok* ("Little Cossack") will be the platform for a number of geophysics experiments. The choice of landing site, of course, will be of critical importance: "a site with high potential for finding well-preserved organic material, particularly from the planet". Several kilometers of travel are anticipated.[201]

An ESA carrier module will contain the surface platform and the rover within a single aeroshell. A descent module provided by Russia will separate from the carrier shortly before reaching Mars and will directly enter the atmosphere. The descent and landing will follow the same approach as that tested on the 2016 Schiaparelli Pathfinder: heat shield, parachutes, thrusters, and damping systems. It is very much to be hoped that the lessons learned from that Pathfinder make the landing a success this time. Site selection is a process where NASA's Mars 2020 debate should be helpful as both missions' primary goal is to explore a site with the potential for finding organic material from the early history of Mars.

Rather like Spirit and Opportunity, after landing, Rosalind Franklin will egress from the platform to start its science mission. Like those NASA rovers, but unlike Curiosity, the ExoMars 2022 will be solar powered. The very big difference from Spirit and Opportunity is the drill. An IR spectrometer will serve to characterize the mineralogy in boreholes. Once collected, a sample is to be delivered to the onboard instrumentation for analysis – looking in particular for evidence of organic substances. The ExoMars Trace Gas Orbiter will support both up and downlink communications. The Rover Operations Control

---

[201] https://www.airbus.com/newsroom/news/en/2017/11/ExomarsRoverMission.html

Centre (ROCC) will be located in Turin, Italy. Commands to the Rover will be transmitted through the Orbiter and the ESA space communications network operated at ESA's European Space Operations Centre (ESOC).

First Application of Artificial Intelligence

Signaling a significant advance in deep space operations by robotic spacecraft:

"Victoria Da Poian and Eric Lyness (both at NASA's Goddard Space Flight Center), have trained artificial intelligence systems to analyze hundreds of rock samples and thousands of experimental spectra from the Mars Organic Molecule Analyzer (MOMA), an instrument that will land on Mars within the ExoMars Rosalind Franklin Rover in 2023. MOMA is a state-of-the-art mass-spectrometer-based instrument, capable of analyzing and identifying organic molecules in rocks samples. It will search for past or present life on the Martian surface and subsurface through analysis of rock samples. The system to be sent to Mars will still transmit most data back to Earth, but later systems for the outer solar system will be given autonomy to decide what information to return to Earth.

First results show that when the system's neural network algorithm processes a spectrum from an unknown compound, this can be categorized with up to 94% accuracy and matched to previously seen samples with 87% accuracy. This will be further refined until being incorporated into the 2023 mission".[202]

2022 ExoMars Kazachok

A veritable Christmas tree of eleven science experiments will be carried out on the Kazachok lander:

* Russian seismometer
* Radio science experiment developed in Belgium "will study the internal structure of Mars, will help to understand the sublimation/condensation cycle of atmospheric CO_2, and will make precise measurements of the rotation and orientation of the planet by monitoring two-way Doppler frequency shifts between the lander and Earth. It will also detect variations in angular momentum due to the redistribution of masses, such as the migration of ice from the polar caps to the atmosphere".
* Swedish instrument package "will investigate the amount of water vapour in the atmosphere, daily and seasonal variations in ground and air temperatures, and the UV radiation environment".
* Meteorological package developed by Russia, Spain and Finland will incorporate a pressure and humidity sensor, radiation and dust sensors and a sensor to measure magnetic fields.

---

[202] https://phys.org/news/2020-06-nasa-life-mars.html

* Magnetometer developed by Russia and the Czech Republic
* Russian cameras
* Russian IR Fourier spectrometer for atmospheric studies
* Russian radiation detector
* Russian multi-channel diode laser spectrometer
* Russian radio thermometer for soil studies
* Russian dust particle suite of instruments
* Russian Gas Chromatograph Mass Spectrometer for atmospheric analysis

The combination of the Rosalind Franklin rover and the Little Cossack lander promises to make the ESA/Roscosmos ExoMars mission a big step forward.

Prospects for 2026 NASA ESA Mars Sample Return

NASA's plans for the continuation of its ongoing Mars exploration missions and the relatively near term return of samples from Mars have received a major blow from the submission of the Administration's NASA budget for 2021 based on a report by Leonard David *in Scientific American* on April 30, 2020:

> "NASA's Mars Exploration Program is in calamitous straits. Cuts to the program in President Donald Trump's budget proposal for the 2021 fiscal year (FY) could pull the plug on the space agency's ensemble of orbiters, as well as its only active Mars rover, Curiosity, which has been prowling the Red Planet since 2012. If unchanged, the budget numbers would, in this calendar year, shutter an aged but functional communications relay and science orbiter, Mars Odyssey, which has operated at the planet since 2001. They would also curtail Curiosity just as it reaches new heights in its ongoing science investigations on Mount Sharp in Gale Crater. The funding shortfall would close out the rover's work late next year, before it can explore a major transition in the ancient climate of Mars that is thought to be recorded in rocks higher on the mountain.
>
> Furthermore, the FY 2021 budget reduces the science sleuthing of NASA's Mars Reconnaissance Orbiter (MRO) by 20 percent. It cuts the number of targeted observations MRO can execute in half, purging most of the special data products associated with them. Like Mars Odyssey, MRO is a dual-purpose orbiter, serving as a crucial data relay while also providing high-resolution imagery of potential future landing sites."

The Planetary Society has been following the budget fight in Congress and reports that: "The congressional budget did restore a number of programs slated for cancellation by the White House: including the Nancy Grace Roman Space Telescope, the Odyssey Mars orbiter, NASA's STEM Outreach program, and several Earth Science missions.

The Planetary Society was generally pleased with the outcome. The final legislation funded our top priorities, including operations for all existing Mars missions, the start of a Mars Sample Return mission, and called for a 2025 launch of the NEO Surveyor mission, which would help identify potentially hazardous asteroids."

NASA's Mars 2020 plan has been to use a drill to acquire shallow core samples from rocks and soil. This and their storage are considered the mission's biggest risk in terms of readiness to launch on schedule. Until recently plans called for a follow-up NASA mission to return the samples requiring a second, precise landing on Mars, a surface rendezvous and transfer of the samples, ascent to Mars orbit, rendezvous with an orbiter, and transfer of the samples after which the orbiter will then leave Mars and return to Earth. Given the additional requirement for building a unique sample analysis facility, the author has been inclined to think that it may be some time before Mars sample return is actually carried out. Recently there has been a very encouraging development in the form of a proposed collaboration between NASA and ESA. NASA and ESA signed a "statement of intent" in April 2018 to jointly work on a Mars sample return program. In its FY2021 NASA budget submission to Congress there is $250M identified for Mars Sample Return.

The Mars sample return mission, if approved by both NASA and ESA, would pick up rock and soil samples that had been collected and stored in tubes by NASA's Mars 2020 rover. The planning calls for two launches in 2026 that would face particularly challenging performance requirements. The first launch would be a NASA lander, followed in the fall by an ESA orbiter. Both would take non-standard trajectories to get to Mars, with the orbiter getting there in about a year with the assistance of solar electric propulsion, The lander would be targeted to the location of NASA's Mars 2020 rover (that may or may not still be operational) and would send out a small ESA 'fetch' rover to collect samples (stored in metal tubes) acquired by NASA's 2020 rover. The NASA lander's ascent vehicle would then carry the soccer ball-sized sample container into Mars orbit where it would be transferred to ESA's orbiter for return to Earth using solar electric propulsion. All of this would be carried out by the autonomous spacecraft. There the samples, about 500 gm in total, in an Earth entry vehicle would descend to a test range in Utah a decade from now in 2031.

This ambitious plan would establish an entirely new level of international collaboration beyond that of the NASA-FRG Galileo Jupiter orbiter and the NASA-ESA Cassini-Huygens mission to Saturn and Titan (Chapter 22). Management of such an intimate collaboration will be a challenge but one that has a sturdy base to build upon given the many international successes in the past including that of the International Space Station.

Automated drill development for future Mars Rovers

Looking further ahead, it will be necessary for astrobiologists to analyze samples acquired from the top few meters below the surface and, eventually, from km depths. NASA's development of an automated Mars drill capable of sampling at meter depths began in the mid-1990s as a multi-Center project in NASA's Astrobiology Technology & Instrument Development Program involving NASA Ames, NASA Johnson, Baker-Hughes Inc, and UC Berkeley. The author was Principal Investigator. The principal goal has been to develop a low-mass (~20 kg) drill that will be operated without drilling fluids and at very low power levels (~100 watts) to access and retrieve subsurface samples from permafrost regions of Earth and Mars at depths that provide protection from cosmic radiation and solar UV.

A two-meter drill was designed and built as a joint effort by NASA JSC (led by Jeff George and Brian Derkowski) and Baker-Hughes. It took the form of a down-hole unit attached to a cable so that it could, in principle, be scaled easily to reach significant depths. The base of the drill is anchored to the surface and a pipe-like drill module follows the electrically powered bit down into the surface. The module is pulled to the surface every few cm in order to remove the core and cuttings. This is a slow process. The UC Berkeley contribution (led by Professor George Cooper) has been a laboratory effort to characterize the physics of dry drilling under Martian conditions of pressure, temperature and atmospheric composition. Data from the UC Berkeley and NASA JSC laboratory experiments were used as input to a drill simulation program to provide autonomous control. The first Arctic field test was organized by Wayne Pollard of McGill University and took place at 80°N in May 2004 near the Canadian Weather Station in Eureka on Ellesmere Island.

Since the author's retirement from NASA Ames, development has continued by a team led by Howard Cannon from Ames, the University of Oklahoma, the Centro de Astrobiologia Spain and Honeybee Robotics – the Mars Astrobiology Research and Technology Experiment (MARTE). In September 2005 the team carried out an experiment in Minas de Riotinto in SW Spain using a 10 meter dry auger coring drill, a core sample handling system, onboard science and life detection experiments all mounted on a simulated lander platform. Their lengthy report – MARTE: Technology Development and Lessons Learned from a Mars Drilling Mission Simulation – is available online at https://ti.arc.nasa.gov.

The ongoing effort combines a drill with a mobile, instrumented platform operating in Chile's hyper-arid Atacama Desert – an environment that resembles Mars but in a different way than Ellesmere Island. NASA's Atacama Rover Astrobiology Drilling Studies project – PI Brian Glass of NASA Ames – is in its second season of test – demonstrated that roving, drilling and life detection can all be combined effectively. The rover carries a lightweight, low-power, two-meter drill together with a robotic sample transfer arm and three life-detection instruments – including JPL's Wet Chemistry Laboratory that flew on the 2007 Phoenix mission to Mars and also the Signs of Life Detector, contributed by Spain's Center for Astrobiology. PI Glass reports "The drill, rover and robot arm combination behaved beautifully in the field. It was a steady platform that enabled us to go deeper than we expected."[203]

Probably, to reach much deeper in order to penetrate to below the permafrost it will be necessary to adopt a different technology entirely – using an electrically heated probe to melt down through rock. This capability is not presently under development.

A Deep Mars Hydrosphere?

The modelling of the Martian cryosphere by Steve Clifford has been discussed in previous chapters implying a possible deep (10+ km) hydrosphere – with profound implications. On Earth, microbial life flourishes at great depth. Most recently, scientists led by Robert Grimm at Southwest Research Institute have continued the modeling of the secular retention of ground water and ice in the circumstance where tropical ground ice undergoes long-term sublimation and likely exospheric escape. The model deals with a variety of complexities and they conclude: "If Mars post-Noachian crustal H_2O inventory was a few hundred meters global equivalent layer or more, then the modest loss since then implies that the cryospheric seal has been maintained following to down freezing and that ground water likely exists globally today".[204]

Orbital radars cannot penetrate to such depths and the technology to drill or melt our way down has yet to be developed and would be a significant challenge. As a start an approach to characterizing the deep subsurface in a variety of locales is provided by the steep cliffs of Valles Marineris and elsewhere where the MRO HIRISE cameras have recorded the presence of thick layers of water ice. On Earth such cliffs form along shorelines or valley walls and provide an opportunity to decipher the history of rocks, minerals, fossils, structures at these locations over hundreds of millions of years. The same should be true for Mars and would seem to be an obvious, though very challenging, place to start. Technology to characterize the cliffs would seem to require a capable rover landed on the Valles floor and equipped with an advanced version of the rotorcraft carried by NASA's 2020 Perseverance lander.

Electromagnetic Induction Sounding

To detect liquid water at depths of 10 or more km on Mars a landed EM sounding instrument is needed that is capable of characterizing the subsurface at depths not accessible to orbital radar. The electromagnetic induction method described online by Geophysical Survey Systems Inc (GSSI) is based on the measurement of the change in mutual impedance between a pair of coils on or above the earth's surface. These coils are electrically connected and are separated by a fixed distance. The transmitter coil is used to generate an electromagnetic field at a specific frequency – the primary field that causes electrical currents to flow in conductive materials in the subsurface. The flow of eddy currents in the subsurface generates a secondary magnetic field, which is sensed by the receiver coil. The magnitude of the secondary field depends upon the type and distribution of conductive material in the subsurface. Both the induced secondary field, along with the primary field, are detected at the receiver coil.

Whereas an orbital radar in a polar orbit can map the whole planet, a landed EM instrument can only make a local measurement – but to greater depth. So, to characterize the Martian subsurface to many km depth an array of landers will be required targeted to a range of different sites – presumably at as low elevation as possible e.g., the floor of Valles Marineris and of the large basins like Hellas. Such a project will likely take many years to complete but will provide the potential basis for deep penetration sampling that can tell us if microbial life may still flourish at depth on Mars – if so, the detection of such life would surely be the jewel on the crown of Solar System exploration.

---

[203] https://www.nasa.gov/feature/ames/mars-rover-tests-driving-drilling-and-detecting-life-in-chile-s-high-desert
[204] https://agupubs.onlinelibrary.wiley.com/doi/pdfdirect/10.1002/2016JE005132

CHAPTER 16

ASTRONAUT EXPLORATION OF MOON AND MARS:

THE FUTURE

After decades of soliciting advice from blue ribbon panels that NASA, with international partners, should soon return astronauts to the Moon and follow with a first mission to Mars, specific development steps are finally underway toward creating a lunar base in this present 2020s decade. The first astronaut missions to Mars are also being reconsidered for the decade of the 2030s.

Astronaut Return to the Moon: The Artemis Program

The present (2020) NASA plan for human lunar exploration has been given the name of Apollo's sister *Artemis.* "With the Artemis program, NASA will land the first woman and next man on the Moon by 2024, using innovative technologies to explore more of the lunar surface than ever before. We will collaborate with our commercial and international partners and establish sustainable exploration by the end of the decade. Then, we use what we learn on and around the Moon to take the next giant leap – sending astronauts to Mars".[205]

The news in mid-2020 about the giant Space Launch System that will serve Artemis is that "the SLS core stage for the Artemis 1 un-crewed test flight is currently at the Stennis Space Center for a "Green Run" static-fire test scheduled for later this year, while the Orion spacecraft for that mission is wrapping up testing at NASA's Plum Brook Station. The SLS core stage should arrive at the Kennedy Space Center in late summer or early fall, allowing teams to begin integrating for a launch hopefully in the mid '21 timeframe, mid to late '21 timeframe for Artemis 1."[206]

NASA has estimated that the Artemis Program will cost an additional $35 billion over the next four years in order to develop the lunar landing system to achieve the Administration's 2024 goal. Not highlighted are the many, many billions that have been spent by NASA since 2005 in the development of Ares and SLS launchers intended to recreate the equivalent of the Saturn V half a century earlier. Announced in February 2020, NASA's proposed fiscal year 2021 budget calls for a 12 percent increase over the previous year – $25.2 billion. The biggest increase is proposed for the landing system, $3.37 billion in FY 2021 alone. The next step will be for Congress (where in 2020 the President's party does not have a majority in the House of Representatives) to authorize NASA's budget and pass an appropriation of funds. Then the reality of the President's plan will become clear.

The Administration's 2021 budget request notably makes cuts across most science agencies (including the National Science Foundation, the National Institutes of Health and the Department of Energy) – a request that in the past Congress has repeatedly rebuffed. At the time of writing the coronavirus emergency was in full swing around the world and NASA Centers were under lockdown with negative consequences yet to be understood.

---

[205] https://www.nasa.gov/specials/artemis/
[206] https://spacenews.com/first-sls-launch-now-expected-in-second-half-of-2021/

Led by NASA, Artemis partners include ESA, JAXA and the Canadian Space Agency, the elements of the program include some that have been under development: the 6-person crew vehicle Orion with a Service Module provided by ESA, a lunar space station called the Lunar Gateway and along with a still-to-be-developed lander and ascent vehicle. Gateway will be solar powered and will serve as a communications hub, a science laboratory and a short-term habitation module. Orion is solar powered and has an automated docking system; it comprises a Crew Module built by Lockheed Martin and a Service Module built by Airbus Defense and Space.

The SLS launcher will take the crew in Orion to the Gateway orbiting station. Commercial launchers will suffice to serve the other program elements. Mobile robotic landers targeted to the lunar south-polar region will support the detailed mapping of the distribution and concentration of water ice in the permanently shadowed crater floors and, also, the testing of approaches to its exploitation. An early planned mission of this kind is VIPER – Volatiles Investigating Polar Exploration Rover – that is managed by the Ames Research Center.

Mobile surface vehicles for crew will be similar to those that the Apollo astronauts used. They may be prepositioned on the surface and could also be operated remotely from the orbiting Gateway. JAXA is evaluating a large pressurized rover for use of landed crew. It is expected that a pressurized surface outpost for use by astronauts will be required and that it would be sent to the Gateway orbiter and from there to a safe polar surface location. This would be the beginning of an International Lunar Base. No doubt China will be developing similar plans.

As a point of comparison, the first Antarctic research station was established soon after World War 2. The US set up the Amundsen-Scott South Pole Station in 1957. There are, in fact, today numerous national research stations in Antarctica located, not surprisingly, mainly near the coast. Many are staffed only during the summer months. With the successful building and operation of the International Space Station over several decades, the prospects for a similar Lunar International Research Station seem promising. Such a station will no doubt fully consume any realistic NASA budget into the indefinite future.

Meanwhile, Mars exploration will continue to be the domain of robotic spacecraft of ever-increasing capability as will, no doubt, be the exploration of the other planets, minor planets, asteroids and comets. All will, doubtless, feel major budget squeezes as other national and international priorities are weighed in the aftermath of the Covid-19 pandemic.

It would not be surprising if the operation of an International South Polar Lunar Base will comprise the deep space activity of astronauts for many decades to come while the robotic exploration of the Solar System continues.

The Artemis Accords

It is encouraging to learn that NASA's plans for the international and private sector participation in Artemis include establishing a set of essential principles – the *Artemis Accords* – that will guide the activities. "International space agencies that join NASA in the Artemis program will do so by executing bilateral Artemis Accords agreements, which will describe a shared vision for principles, grounded in the Outer Space Treaty of 1967, to create a safe and transparent environment which facilitates exploration, science, and commercial activities for all of humanity to enjoy".[207]

Thus:

* International cooperation on Artemis is intended not only to bolster space exploration but to enhance peaceful relationships between nations. Therefore, at the core of the Artemis Accords is the requirement that all activities will be conducted for peaceful purposes, per the tenets of the Outer Space Treaty.

* Transparency is a key principle for responsible civil space exploration and NASA has always taken care to publicly describe its policies and plans. Artemis Accords partner nations will be required to uphold this principle by publicly describing their own policies and plans in a transparent manner.

* Providing emergency assistance to those in need is a cornerstone of any responsible civil space program. Therefore, the Artemis Accords reaffirm NASA's and partner nations' commitments to the Agreement on the Rescue of Astronauts, the Return of Astronauts and the Return of Objects Launched into Outer Space. Additionally, under the Accords, NASA and partner nations commit to taking all reasonable steps possible to render assistance to astronauts in distress.

* NASA has always been committed to the timely, full, and open sharing of scientific data. Artemis Accords partners will agree to follow NASA's example, releasing their scientific data publicly to ensure that the entire world can benefit from the Artemis journey of exploration and discovery.

* Protecting historic sites and artifacts will be just as important in space as it is here on Earth. Therefore, under Artemis Accords agreements, NASA and partner nations will commit to the protection of sites and artifacts with historic value.

* The ability to extract and utilize resources on the Moon, Mars, and asteroids will be critical to support safe and sustainable space exploration and development. The Artemis Accords reinforce that space resource extraction and utilization can and will be conducted under the auspices of the Outer Space Treaty, with specific emphasis on Articles II, VI, and XI.

* Specifically, via the Artemis Accords, NASA and partner nations will provide public information regarding the location and general nature of operations which will inform the scale and scope of 'Safety Zones'. Notification and coordination between partner nations to respect such safety zones will prevent harmful interference, implementing Article IX of the Outer Space Treaty and reinforcing the principle of due regard.

NASAWATCH.COM

Plans for future crewed missions and the SLS launcher evolve all the time so the best place to stay informed is the NASAWatch website – nasawatch.com – where Keith Cowing reports all the latest news about what is going on in the Agency and Congress.

On February 18, 2021 NASAWATCH reported that:

NASA's acting administrator said Wednesday evening that the goal of landing humans on the Moon by 2024 no longer appears to be feasible. "The 2024 lunar landing goal may no longer be a realistic target due to the last two years of

---

[207] https://www.nasa.gov/specials/artemis-accords/index.html

appropriations, which did not provide enough funding to make 2024 achievable," the acting administrator, Steve Jurczyk, told Ars. "In light of this, we are reviewing the program for the most efficient path forward."

Keith Cowing followed this by his own perspective:

"This is, of course true - but it is not the whole story. It was widely assumed within NASA that when Vice President Pence suddenly advanced the Artemis lunar landing date to 2024 that it would be rather hard to make that happen. But NASA had to accept that challenge - and they did and worked hard to make it happen. But it did not happen. The prime reason for the problems lay at the feet of the chronic delays and cost overruns for SLS and its associated ground support systems. Even when NASA got the money it needed it still fell behind year after year as both the Government Accountability Office (GAO) and Office of Inspector General (OIG) noted with consistent regularity.

Then, of course, there was the ever-changing Gateway which added and then discarded features faster than the NASA graphic artists could update the pretty website imagery. And the lunar lander grew larger and more complex every time NASA mentioned it. So ... Jurczyk is right - he is just not fessing up to the whole story. It is mea culpa time for NASA.

NASA is as much to blame for the Artemis quagmire as past Congresses and White House Administrations are. Now, a new Administration has thrown a hopeful lifeline to the Artemis program albeit a vague one. The pandemic, a crashing economy, exploding government debt, and dysfunctional politics is going to force every program - in every agency - to redouble its explanation as to why it needs to be done.

The Biden Administration's slogan "Build Back Better" should be something that everyone at NASA pays attention to. Artemis is going to change - and be fixed - for the "better". A reformatted Artemis may well accomplish much of its original intent - but NASA may also be directed to focus human spaceflight efforts elsewhere as well. But refocusing of human spaceflight at NASA - regardless of what that ends up being - is only going to work out well if NASA stops the whole smoke and mirrors, shift the blame, give-us-what-we-want-because-we-say-so, tactics and openly admits that it did things wrong with Artemis.

Moreover, instead of being an outlier when it comes to overall national priorities, NASA needs to start becoming more of a "whole of government" player. Otherwise it may just find itself standing there with an empty, outstretched hand. NASA is also going to have to learn to let go of some things and adopt other novel approaches in the process of building back Artemis better. As soon as the new TBD NASA Administrator arrives the agency needs to hit the ground running."

Crewed Mars Mission to a Martian Moon

Assuming that a crewed mission to Mars can ever be justified, and recognizing the perennial problem of cost alone, a creative approach proposed for such future exploration of Mars has emerged from planners at JPL led by Firouz Naderi. His team, with support from the Aerospace Corporation has sought to develop a plan that would be fiscally more plausible than the Mars Design Reference Mission described in Chapter 7. It would build upon NASA's Artemis program plans for the present 2020s decade with crewed missions to Mars beginning in the following decade – first to the inner moon Phobos and only later to the surface of Mars.[208]

---

[208] https://www.nasa.gov/sites/default/files/files/Naderi_JPL_Study_of_Humans_to_Mars_NAC_Final_TAGGED.pdf

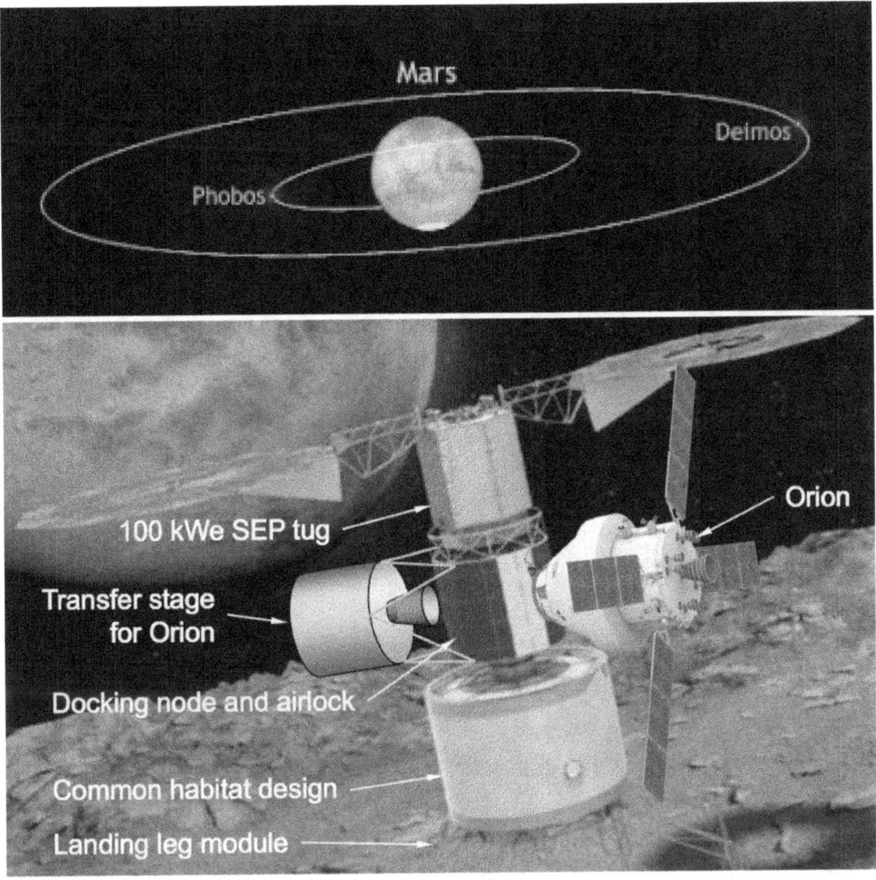

First a crew of four would be transported to low equatorial Mars orbit to rendezvous with and land on the inner moon, Phobos. Obviously, this would be very much less of a challenge than that of atmospheric entry and landing on Mars and would also make the return to Earth simple to carry out. Phobos is in an 8-hour synchronous orbit about Mars (i.e. with one side always facing Mars) thereby providing radiation protection from both below (Phobos) and some from above (Mars). Though not ideal for continuous communication and control of mobile assets on the Martian surface, communication satellites would provide the needed links.

The outer moon Deimos in its 30-hour synchronous orbit would be an almost ideal location for direct communications with robotic Mars landers but, importantly, a landed Deimos crew would not benefit as much from the radiation shielding that Mars would provide to a crew of a Phobos lander.

In comparison to today's time-delayed control of the Curiosity and Perseverance rovers (where the round-trip light-time between Earth and Mars can be as much as 50 minutes) the Phobos crew's ability to explore many sites on Mars would have the great benefit of real time communications with pre-positioned mobile robots deployed at a number of priority science sites. In time, a crew on Phobos might operate a Mars lander equipped with a drill (or with a melting probe) designed to acquire samples from beneath the cryosphere. Unlike the case of Apollo where there were real-time communications with scientists at Mission Control, the crew would need to include professional scientists – astrobiologists and geologists – to take advantage of the opportunity provided by real time VR control of pre-positioned assets.

Incidentally, "When NASA scientists want to follow the path of the Curiosity rover on Mars, they can don a mixed-reality headset and virtually explore the Martian landscape."[209] This already-available ability for anyone to take themselves to Mars raises the question of whether sending a crew to land on and travel about the Martian surface will continue to excite the imagination of tax payers sufficiently to support the considerable expense involved.

---

[209] https://www.nasa.gov/feature/jpl/take-a-walk-on-mars-in-your-own-living-room

Private Sector Plans for Activity in Deep Space

The US private sector's participation in space activities to date has been mainly in the form of building launch vehicles and spacecraft for various government and commercial organizations as well as elements of the International Space Station. In the future, commercial space activities may include tourism in low Earth orbit, processing lunar water-ice and mining of the Near Earth Asteroids. In time the establishment of a commercial spacecraft refueling facility at the South Pole of the Moon would not be an outlandish possibility. What does seem to the author to lie in the realm of science fiction, however, are the ambitions of two of Planet Earth's richest men: Jeff Bezos and Elon Musk. Not without good reason, both are focused on mankind's prospects for continuing to live on Planet Earth in the face of a variety of existential threats. An article in the November 2019 issue of *The Atlantic* magazine reports the vision of Mr. Bezos, founder of the Amazon Company – a vision that derives directly from that of Princeton physicist Gerard K. O'Neill who, in his 1976 book *The High Frontier*, made the case for moving mankind into enormous space stations in Earth orbit. They were to be constructed using lunar materials, would spin to simulate gravity and would be solar powered.

These self-sustaining colonies would, supposedly, allow for the unlimited expansion of mankind: Mr. Bezos anticipates that "we can have a trillion humans in the Solar System, which means we'd have a thousand Mozarts and a thousand Einsteins." Indeed! Meanwhile, it should be noted, positively, that he has created the *Bezos Earth Fund* to address the climate crisis with a commitment of $10B to fund scientists, activists and nongovernmental organizations.

Commercial space activities would inevitably loom large in this future including the mining of Earth-approaching asteroids. Mr. Bezos has clearly been influenced by John Lewis, professor emeritus of the University of Arizona who is a meteorite specialist. John and Ruth Lewis argued the case for asteroid mining in their 1987 book *Space Resources: Breaking the Bonds of Earth* and John, in more detail, in his 1997 book *Mining the Sky: Untold Riches from the Asteroids, Comets, and Planets.* He included an estimate of the economic value of a small, 2 km metallic near-Earth asteroid: 3554 Amun. He calculated a mass of 30 billion tons and a 1996 market value of $8 trillion for its iron and nickel, another $6 trillion for its cobalt, and $6 trillion more for its platinum group metals. John proposed that solar powered colonies built with the resources of the asteroid belt could eventually support a vast civilization and that this could be a very good thing: "Intelligent life, once liberated by the resources of space, is the greatest resource in the Solar System ... the highest fulfillment of life is unbounded intelligence and compassion".

More than one commercial company has been formed to pursue the mining of asteroids so it will be of great interest to follow their fortunes. In 2000 Mr. Bezos founded a company, *Blue Origin,* that has made impressive progress in developing reusable rocket launch vehicles. According to *The Atlantic*, the funding of Blue Origin comes from the sale of $1 billion of Amazon stock each year. That can buy a lot of talent and experience. Nevertheless, the author's mind boggles at the vision in question. Perhaps a first small step for Blue Origin will be to purchase the International Space Station when the international Agency partners are ready to move on.

Mars Colonization!

The vision of mankind's expansive activity in space by billionaire Elon Musk is somewhat similar and is at least equally wildly ambitious – namely to colonize Mars. To that end, in 2002 he founded a very successful company – SpaceX – that builds a revolutionary line of powerful Falcon launch vehicles capable of vertical landing the first stage back at the launch site and subsequent re-use.

Elon Musk shares Jeff Bezos' concern about the climate crisis that, in spite of ample warning, has overtaken our planet and, in particular, the way in which we extravagantly use fossil fuels – something he regards with good reason as "the dumbest experiment in human history." Note, however, a certain irony in the hundreds of tons of kerosene burned in each launch. His own important visionary contribution to solving this global problem consists of various ventures – notably sustainable energy sources in the form of battery-electric cars, solar energy capture and improved energy storage. His *space* ambitions are best captured here by direct quotes from the SpaceX website and, to say the least, it will be fascinating to follow the progress of the company and its founder:

- SpaceX's Starship and Super Heavy Rocket represent a fully reusable transportation system designed to service all Earth orbit needs as well as the Moon and Mars. This two-stage vehicle – composed of the Super Heavy rocket (booster) and Starship (ship) – will eventually replace Falcon 9, Falcon Heavy and Dragon. By creating a single system that can service a variety of markets, SpaceX can direct resources from Falcon 9, Falcon Heavy and Dragon to Starship – which is fundamental in making the system affordable.
- Building Moon bases and Mars cities will require affordable delivery of significant quantities of cargo and people. The fully reusable Starship system uses in-space propellant transfer to enable the delivery of over 100t of useful mass to the surface of the Moon or Mars. This system is designed to ultimately carry as many as 100 people on long duration, interplanetary flights.

- Our aspirational goal is to send our first cargo mission to Mars in 2022. The objectives for the first mission will be to confirm water resources, identify hazards, and put in place initial power, mining, and life support infrastructure. A second mission, with both cargo and crew, is targeted for 2024, with primary objectives of building a propellant depot and preparing for future crew flights. The ships from these initial missions will also serve as the beginnings of the first Mars base, from which we can build a thriving city and eventually a self-sustaining civilization on Mars.[210]

Would that even the first steps could be achieved so easily! SpaceX has surely made a close reading of NASA's Mars Design Reference Mission and, among many things, should have noted the need for nuclear reactors (and White House permission to launch!) along with tanks of liquid hydrogen to support the accumulation of methane and liquid oxygen propellants. Regarding access to Martian water, ice is both widespread and abundant on the surface. Below 60 degrees of latitude, ice is concentrated in several regional patches, particularly around the Elysium volcanoes, Terra Sabaea, and northwest of Terra Sirenum, and exists in concentrations up to 18% ice in the subsurface. Above 60 degrees latitude, ice is highly abundant. Pole-ward of 70 degrees of latitude, ice concentrations exceed 25% almost everywhere, and approach 100% at the poles. The crew's required water resources could, most directly, be found in Korolev Crater at 71°N as discovered by ESA's Mars Express and also locked in the permanent residual ice caps at the two poles – but neither are practical locations to site a crewed base let alone a thriving city. ESA's Mars Express mission has found radar echo evidence of a 20 km wide liquid water lake 1.5 km deep at 81°S – a location also not well suited for access and exploitation.

Access to frozen water will not be a problem for Mars astronauts but it is the *liquid* phase that will be needed and here the problem is providing the source of heat. Unlike the polar regions of the Moon where water ice in the regolith can be heated using solar power, on Mars the Sun is too distant so a nuclear reactor would surely be needed.

More generally, the fantasy of building cities on Mars simply boggles the mind. Who would be crazy enough to choose to spend their lives under a small dome on an almost airless desert planet? Some time spent at Mars VR sites on the web

---

[210] www.globalsecurity.org>space>systems Big Falcon Rocket

that include an immersive tour of the Curiosity rover's traverse will convince anyone of the bleak nature of the Martian surface. Mars is at its best appreciated by most only when viewed from orbit.

In spite of the author's profound doubt about any dreams of colonizing Mars, it is likely that SpaceX can nevertheless make significant headlines in the coming years by carrying out crewed lunar missions much like those of the Apollo 8 and 10 missions. Similar flyby-return missions to Mars of a small crew would doubtless be possible and attract great attention. In an expensive advertisement, the maiden flight of his Falcon Heavy launched a red Tesla sports car with dummy driver on a successful November 2019 flyby mission to Mars. Making headlines is, likely, Mr. Musk's main purpose.

The Private Enterprise Puzzle

The author is not the only one to puzzle over the obsessions of Messrs Bezos and Musk. Dr Zeynep Tufekci who is an associate professor at the University of North Carolina School of Information and Library Science has written an article in the December 2019 edition of *Scientific American* (p71) titled *What's the Deal with Rich Men and Space?* she proposes that their obsession harks back to "classic" science fiction. Evidently, both men "have talked about their love of science fiction as part of their inspiration for investing in space." Professor Tufekci notes that, "At its best, science fiction is a brilliant vehicle for exploring not the far future or the scientifically implausible but the interactions among science, technology and society'. She also comments "one thing about being a billionaire is that it's probably not hard to find people who will encourage you to spend money chasing space operas that either will not happen because of scientific constraints or will end up in disaster." It is the present author's hope that disasters will be avoided.

Challenges to Security in Space

The US Defense Intelligence Agency has assessed the issues associated with challenges to security in space and provides the following Executive Summary:

Space-based capabilities provide integral support to military, commercial, and civilian applications. Longstanding technological and cost barriers to space are falling, enabling more countries and commercial firms to participate in satellite construction, space launch, space exploration, and human spaceflight. Although these advancements are creating new opportunities, new risks for space-enabled services have emerged. Having seen the benefits of space-enabled operations, some foreign governments are developing capabilities that threaten others' ability to use space. China and Russia, in particular, have taken steps to challenge the United States:

- Chinese and Russian military doctrines indicate that they view space as important to modern warfare and view counter-space capabilities as a means to reduce U.S. and allied military effectiveness. Both reorganized their militaries in 2015, emphasizing the importance of space operations.
- Both have developed robust and capable space services, including space-based intelligence, surveillance, and reconnaissance. Moreover, they are making improvements to existing systems, including space launch vehicles and satellite navigation constellations. These capabilities provide their militaries with the ability to command and control their forces worldwide and also with enhanced situational awareness, enabling them to monitor, track, and target U.S. and allied forces.
- Chinese and Russian space surveillance networks are capable of searching, tracking, and characterizing satellites in all earth orbits. This capability supports both space operations and counter-space systems.
- Both states are developing jamming and cyberspace capabilities, directed energy weapons, on-orbit capabilities, and ground-based anti-satellite missiles that can achieve a range of reversible to non-reversible effects.

Iran and North Korea also pose a challenge to militaries using space-enabled services, as each has demonstrated jamming capabilities. Iran and North Korea maintain independent space launch capabilities, which can serve as avenues for testing ballistic missile technologies.

The advantage the United States holds in space—and its perceived dependence on it—will drive actors to improve their abilities to access and operate in and through space. These improvements can pose a threat to space-based services across the military, commercial, and civil space sectors.[211]

Security issues like these would likely have been the main justification for the Trump Administration's establishment of a new arm of the US military in the form of the *Space Force* – ignoring President Eisenhower's specific decision to avoid militarization of the US Space Program. The coming decades promise to be very consequential for space exploration and for mankind's future.

A Mars Space Race with China?

The ongoing Cool War between the US and China remains a global problem. The international waters of the South China Sea are the most obvious place where, by accident, a dangerous confrontation may, one day, occur as a result of a US Navy vessel sailing close to islands patrolled by Chinese ships – with a chance of a collision. In the air there is a similar danger. In 2001 a Chinese fighter jet collided with a US Navy reconnaissance plane that led to the death of the Chinese pilot and the detention of the US crew. Since 2009 through 2018 there have been further close encounters as noted in an editorial 'opinion piece' in the New York Times of 4 February 2020 by Zhou Bo, a senior colonel in the People's Liberation Army of China. He comments, "As China's military strength continues to grow and it closes the gap with the United States, both sides will almost certainly need to put more rules in place, not only in areas like antipiracy or disaster relief – where the two countries already have been cooperating – but also regarding space exploration, cyberspace and artificial intelligence". The *New York Times* notes that "This essay reflects his opinion alone" but clearly a PLA colonel makes plain the Chinese government's position and is a call for the two nations to find common ground in various areas including space exploration. The Moon, where both mobile robots and, soon, landed astronauts will be active, and Mars where mobile robots from China will soon be operating are the two arenas of exploration where mutually beneficial rules could be established. The *Artemis Accords* proposed by NASA and discussed above make an excellent starting point in the case of the Moon.

The internationally competitive aspects of space exploration in 2020 are beginning to resemble a mirror image to the 1960s Space Race in that, presently, the US is dominant in space while its rival – China – is eager to catch up and surpass the US. China has already successfully undertaken landed lunar science missions and, at the time of writing, has a mobile lander on the way to Mars. While perhaps not a near term consideration, the possibility of a new space race like the 1960s race to land a man on the Moon has arisen. This time such a race, if it took place, would involve being the first to land a crew on Mars and, the big challenge: *return them safely to Earth.*

Given China's aggressive foreign policy, President Xi-Jinping has included competition with the US in space exploration and exploitation. This perhaps may include the prospect of winning a race to land the first man or woman on Mars. With no Congressional control over purse strings to constrain him, President Xi could certainly direct the resources to do so. With a minimal taikonaut crew, China could follow the plan described in some detail by NASA's Mars Design Reference Mission or by JPL's proposed approach to begin with a landing on Phobos.

---

[211] https://www.dia.mil/Portals/27/Documents/News/Military%20Power%20Publications/Space_Threat_V14_020119_sm.pdf

Revolutionary Developments since Apollo

There have been a number of revolutionary developments since the first journeys were undertaken into space beyond low Earth orbit – with important consequences for the next half-century of deep space missions by both astronauts and robotic spacecraft.

1. Today's digital information technology has advanced beyond all imagining from the 1960s and has enormously increased the capability of a spacecraft's onboard control and its scientific instrumentation while, at the same time, greatly diminishing the mass of the spacecraft and increasing its reliability. The best example of this is the New Horizons spacecraft that in 2015 encountered Pluto, its moon Charon and, subsequently, trans-Neptunian bodies yet more distant. A single launch, a 10+year journey and the spacecraft has not even needed to activate its backup subsystems! The great success of NASA's Curiosity rover in climbing and characterizing Mount Sharp on Mars is a prime example of progress in robotic spacecraft operations.

2. Artificial intelligence is increasingly and rapidly becoming a reality rather than just a science fiction trope. Its application to the scientific exploration of our Solar System is beginning and, inevitably, it will diminish any science benefit that an astronaut can bring to an exploration site. AI's principal application will be to Mars exploration where we now have vast stores of images and information collected by orbiters and mobile landers that are needed to "teach" a mobile surface laboratory how to go about its exploration in ways that would resemble that of an astrobiologist or geologist. This development is, in fact, well underway. Devon Island in the Canadian Arctic where Pascal Lee of the Mars Institute carries out simulations of astronaut field explorations would be an excellent site for such Mars AI training.

This revolutionary step is in the process of taking place: Masahiro (Hiro) Ono, group lead of the Robotic Surface Mobility Group at JPL, comments: "Our team is working on Mars robot autonomy to make future rovers more intelligent, to enhance safety, to improve productivity, and in particular to drive faster and farther".[212]

"Future missions would potentially use new high-performance, multi-core radiation hardened processors – chips that will provide about one hundred times the computational capacity of current flight processors using the same amount of power."

The team has been developing two novel capabilities for future Mars rovers, which they call Drive-By Science and Energy-Optimal Autonomous Navigation.

"We'd like future rovers to have a human-like ability to see and understand terrain," Ono said. "For rovers, energy is very important. There's no paved highway on Mars. The drivability varies substantially based on the terrain – for instance beach versus. bedrock. That is not currently considered. Coming up with a path with all of these constraints is complicated, but that's the level of computation that we can handle with the HPSC or Snapdragon chips. But to do so we're going to need to change the paradigm a little bit." Ono explains that new paradigm as commanding by policy, a middle ground between the human-dictated: "Go from A to B and do C," and the purely autonomous: "Go do science". Commanding by policy involves pre-planning for a range of scenarios, and then allowing the rover to determine what conditions it is encountering and what it should do. "The rover has the flexibility of changing the plan on board instead of just sticking to a sequence of pre-planned options," Ono said. "This is important in case something bad happens or it finds something interesting."[213]

---

[212] https://www.sciencedaily.com/releases/2020/08/200819120700.htm
[213] https://phys.org/news/2020-08-deep-future-mars-rovers-faster.html

Related research is also underway at NASA's Goddard Center:

NASA has stepped closer to allowing remote onboard computers to direct the search for life on other planets. Scientists from the NASA Goddard Space Flight Center have announced first results from new intelligent systems, to be installed in space probes, capable of identifying geochemical signatures of life from rock samples. Allowing these intelligent systems to choose both what to analyse and what to tell us back on Earth will overcome severe limits on how information is transmitted over huge distances in the search for life from distant planets. The systems will debut on the 2022/23 ExoMars mission, before fuller implementation on more distant bodies in the Solar System.[214]

3. Launch vehicles are now capable of recovery at their launch site for refueling and reuse – with obvious major savings in cost. A Saturn V-class launcher that can be recovered and refueled has not yet been developed and, no doubt, would be valued for astronaut missions to the Moon. NASA's giant Space Launch System (SLS) is almost ready after an embarrassingly slow development that would have made Von Braun blush. For the robotic missions that the author expects will make up all of the science-driven deep space missions, re-useable launchers that already exist should more than suffice.

4. The exposure of astronauts in deep space to solar and cosmic radiation is much better understood than when the Apollo 11 crew was launched half a century ago. It is very hostile when accumulated over deep space missions lasting many hundreds of days. The travel of astronauts to destinations beyond the Moon carries a significant degree of risk to health as well as along with the many other risks to life that such journeys impose (see Apollo 13). Recalling John Kennedy's announcement of the Apollo ambition – it included *returning the crew safely back to Earth*.

5. We have characterized all of the planets and moons in our Solar System (as well as representative comets and asteroids) in sufficient detail to understand the most promising sites where we may discover evidence of life – the outstanding science (and philosophical) question of our time. These key sites include Mars especially, together with several outer planet moons whose make-up includes a subsurface ocean of water. The return of samples from Mars by robotic spacecraft is planned to take place within a dozen years. The issue of planetary protection from extra-terrestrial microbial contamination is a major responsibility of the implementing Agencies: NASA and ESA. Protection of Mars from terrestrial contamination is also essential and argues against landing astronauts on the surface.

6. The recent successful return of surface samples from the Moon by the Chinese Space Agency has been followed – as described in chapter 11 – by the welcome offering of an olive branch by Xu Yansong, who Director-General of the Asia-Pacific Space Cooperation Organization and Director for International Cooperation at China's National Space Administration.

Xu Yansong "underlines the importance of nations working together as humans start to look beyond their planet again: Not so much a space race as a joint quest".

"Global cooperation is essential for exploration missions," Xu insists. "The moon is not far away from Earth, we can have independent missions – but to go further, we need to collect all the strength that we have on Earth."

Mentioning previous cooperation with the European Space Agency – Sweden, Germany, France and Belgium had instruments and payload on board the earlier Chang'e-4 mission – Xu says China is "looking forward to more extensive international cooperation."

That includes the U.S. now that the divisive era of President Donald Trump is over. "With the new administration, I believe that the bilateral cooperation between China and the U.S. will continue," Xu says. "Globally speaking, China is fully open for international cooperation."

---

[214] https://phys.org/news/2020-06-nasa-life-mars.html

A Summary Perspective

For the last several decades NASA has sought the advice of committees of blue ribbon experts concerning the goals and timetable of future manned spaceflight – and invariably has been told that Mars remains the ultimate goal – but *not* an immediate goal. Presently NASA is looking to satisfy that ambition in the 2030s. Meanwhile Project Artemis is getting underway in much the same way that NASA and international partners built and operated the International Space Station beginning in the 1980s. The construction and operation of an international lunar base could well be a project that continues into the 2050s. Whereas the ISS has not led to any commercial enterprises that would take advantage of a microgravity manufacturing environment, a lunar base may lead to a commercial enterprise – the processing of lunar water ice into liquid form and into LOX-Hydrogen rocket fuel. If so, the ongoing evolution of the manned space program could be regarded as a long-term investment. Presumably China's planning for its taikonaut program may have similar motivation. The coming decades hold promise of much interesting competition focused on the Moon.

An International Lunar Base and supporting Gateway Orbiter will also provide the means by which the feasibility (purpose, crew safety, risk, cost) of a manned mission to Mars can be evaluated. Successful robotic Mars missions like Curiosity, the 2020 Perseverance rover and the first return of Mars samples in 2031 will provide – or not – the scientific justification for launching one or more crews to further explore Mars. As discussed above, manned Mars exploration, if it ever takes place, would be more effective, much less risky and very much less expensive if carried out from the inner moon, Phobos. Here Daniel Goldin's *faster, cheaper, better* mantra might find new and more appropriate application.

A question arises – where do Elon Musk's plans for cities on Mars fit in? Answer: Nowhere.

INDEX